# Lecture Notes in Control and Information Sciences 266

Editors: M. Thoma • M. Morari

Springer
*London*
*Berlin*
*Heidelberg*
*New York*
*Barcelona*
*Hong Kong*
*Milan*
*Paris*
*Singapore*
*Tokyo*

Stefano Stramigioli

# Modeling and IPC Control of Interactive Mechanical Systems – A Coordinate-Free Approach

With 98 Figures

Springer

**Author**
Stefano Stramigioli, MSc,PhD
Mekelweg 4, 2628CD Delft, The Netherlands

ISBN 1-85233-395-2 Springer-Verlag London Berlin Heidelberg

British Library Cataloguing in Publication Data
Stramigioli, Stefano
Modeling and IPC control of interactive mechanical systems
a co-ordinate free approach. - (Lecture notes in control
And information sciences ; 266)
1.Production control
658.5
ISBN 1852333952

Library of Congress Cataloging-in-Publication Data
A catalog record for this book is available from the Library of Congress



Typesetting: Electronic text files prepared by author
Printed and bound at the Athenæum Press Ltd., Gateshead, Tyne & Wear
69/3830-543210 Printed on acid-free paper SPIN 10776904

*Alle mie 'prinsesjes':*

*Emily & Fiona*

# Preface

This work is based on my Ph.D. doctoral thesis, which I defended in Delft, The Netherlands in 1998 [1]. The thesis was about the modeling and control of robots interacting with the environment. The dissertation presented a new paradigm called Intrinsically Passive Control (IPC) based on the ideas initially introduced by Neville Hogan in *impedance control* [2, 3, 4] with whom I had the pleasure and honor to work with during a visit of a few months at MIT. This paradigm is about approaching robot control as a power consistent port interconnection between the robot to be controlled and the controller, and splits the complete controller in two parts called the IPC and the Supervisor [5]. This structure is a must for robots interacting with the environment for stability reasons. For such systems, it does not make much sense to only control their position or their force as it is explained in Sect. 4.1 and mathematically treated in [6].

After having sent my dissertation to a colleague and friend, Prof. Bruno Siciliano, during the International Conference of Robotics and Automation 1998 in Detroit, to my great pleasure, he expressed his positive opinion about the work and suggested that I publish it as a book. I am very thankful to him for his advice and support during this process. It took me a while to find the time to work on it and get to its current state. Since my dissertation, some new results have been included and some old ones have been left out. New results are included in Chapter 1, which are the outcome of an extremely pleasant and fruitful collaboration with Bernhard Maschke and Catherine Bidard [7, 8]. Furthermore new results are also presented in Chapter 3 where Dirac structures have been used to model rigid mechanisms. In Chapter 5 a new compensation has been introduced for the control of multi-limbed mechanisms, which has been also tested with success and was carried out during the visit of Cristian Secchi from the University of Bologna to my lab. Eventually, Chapter 6 presents results concerning the tele-manipulation of Hamiltonian systems. These results have been presented in [9] and are the outcome of a collaboration with Arjan van der Schaft, Bernhard Maschke, Stefano Andreotti and Claudio Melchiorri. In this chapter a new system theoretic condition is presented to analyse the matching of a generalized Hamiltonian system to a communication line.

Writing a book is a big task and even if I am the only author, a lot of people have contributed to the results presented in this work and among others there are a few people that I would like to thank for fruitful discussions. Arjan van der Schaft and Bernhard Maschke besides being excellent scientists, are two extremely pleasant people to work with, but not only. They always make time for discussions and have stimulated my work continuously. I also have to thank Herman Bruyninckx with whom I give a robotics course at the Dutch Institute of Systems and Control, for the extremely pleasant and useful discussions which took place mostly by e-mail. Our collaboration has started recently and it is, to my great pleasure, increasing constantly. Other colleges that have inspired and supported my work, and deserve a mentiom, are Romeo Ortega, Neville Hogan and Ernie Fasse. Furthermore I would like to thank Johan Hamberg of the Swedish Defence Research Agency for pointing out a problem which has been solved in Chapter 6. Writing a book is nevertheless not only about scientific discussions but it inevitably has a price to pay in time that I had to withdraw from my family. My wife Fiona gave me the time and the love which are also essential to complete such a big task and I really thank her a lot for that. I will be now able to spend some more time with her and my beautiful daughter Emily to whom this book is dedicated.

I hope that this work will inspire or help the research of other people as a lot of other books have done for me during the past years.

Delft, December 2000 *Stefano Stramigioli*

# Contents

# 1. Rigid Bodies and Motions

This chapter presents a complete framework for defining motion between objects and it explicitly states assumptions often forgotten or not considered in the use of Lie groups for rigid body motions [10, 11]. The presentation will start with a formal definition of projective, Euclidean spaces and observers to define relative motion and twists as intrinsic coordinate-free entities. The chapter contributes by clarifying the intrinsic assumptions made when rigid body motion theory is used and by formally relating screw theory to the theory of Lie groups [7].

A lot of objects we have to deal with every day can be considered *rigid bodies*. Intuitively, we say that an object is rigid if the distances between its points do not change perceptibly. This means that the concept of rigidity is a model hypothesis and an abstraction: in reality nothing is completely rigid as each object always has, often imperceptible, deflections. Nevertheless, the abstraction of considering an object rigid is very useful and practical. In this chapter we will show how we can formally define rigid bodies and how we can describe their relative motion.

There are two main ways to describe rigid body motion: screw theory, which is based on the original work of Ball [12] and was then used by Lipkin [13], [14], [15], and Lie groups theory [16], [17], which is used for kinematics in [10] (see Sect. A.8). Screw theory is based on three elements: 1. on the geometry of lines of the projective extension of the Euclidean three-dimensional space, 2. on Chasles's fundamental theorem on motions of rigid bodies and 3. on Poinsot's theorem, which is concerned with a system of forces applied to a rigid body [10]. A projective extension of the Euclidean three-dimensional space is needed in order to describe pure rigid body translations as will be shown in the sequel. A pure translation can be associated with a line at infinity called the screw axis. In projective terms, motions are described as metric-invariant collineations [13, 14, 15]. We can then associate either a twist or a wrench with a screw. A twist describes the motion of a body, whereas a wrench describes the system of forces applied to a body.

In the Lie group approach, the basic starting point is the isometry group $SE(3)$ of the Euclidean three-dimensional spaces which can be used to describe motions. Here twists are seen as elements of the Lie algebra $se(3)$ which correspond to the Lie group $SE(3)$ (see Sect. A.6). Wrenches are defined as

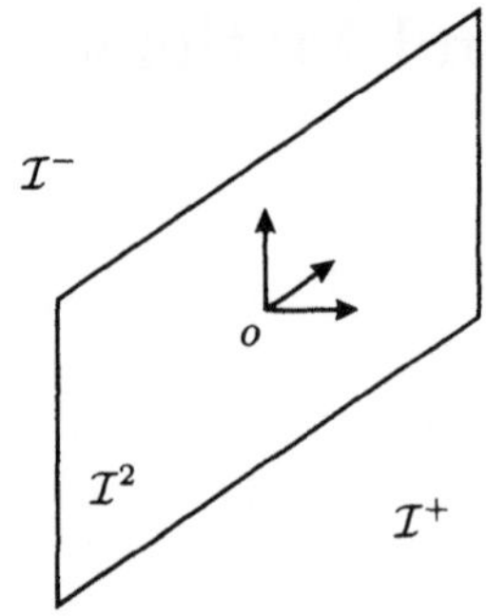


**Fig. 1.1.** A two-dimensional projective space.

elements of the dual (see Def. A.5.3) of $se(3)$, which is indicated as $se^*(3)$. Unfortunately, in this approach, the position of a rigid body is identified by an element of $SE(3)$, and a reference frame is chosen the choice of which is not intrinsic. Furthermore, almost everywhere, a framework with coordinates is used and thus certain important hypotheses are hidden. The problem is caused by the fact that from the start, the group of motions in the Euclidean space is identified with the set of homogeneous matrices $H$ (see Sect. 1.2.1) of the form

$$H = \begin{pmatrix} R & p \\ 0 & 1 \end{pmatrix}$$

where $R$ is an orthonormal matrix representing a rotation and $p$ a translation vector. This also happens in [10], and here simple but important hypotheses are hidden which are clear in a coordinate-free framework.

Only in [18, 19, 1], the importance and necessity of a NOT intrinsic reference is presented.

In this chapter we will try to make all the assumptions explicit in order to build a clear and formally correct framework. After a coordinate-free treatment, coordinates will be added and explicit assumptions will be made.

In this framework we will consider a classical mechanical framework with the Galilean separation of space-time[1].

[1] Thanks to the special theory of relativity, it is well accepted that space is better modeled as a pseudo-Riemannian four-dimensional manifold endowed with the Minkowsky pseudo-metric [20]. The advantages of a space-time-coupled four-dimensional space is nevertheless only useful when the speed of motions is so-called "relativistic" i.e. when the motions approach the speed of light. For our context, such a detailed consideration of space would be cumbersome and useless so the older Galilean three-dimensional approach will be used.

## 1.1 Projective geometry and Euclidean spaces

Screw theory uses the projective extension of the Euclidean three-dimensional space. Since in this chapter we relate the two approaches, we first start with a formal explanation relating the usual conception of an Euclidean space to its projective extension.

In what follows, it is easy to grasp the idea of the lower two-dimensional case when one considers it as a cinema with an infinitely extended screen (the two-dimensional Euclidean space). A point on the screen is actually a beam passing through the projector (the origin of the vector space of dimension three which embeds the screen). The plane that is parallel to the screen and passes through the projector is called the improper hyperplane.

This concept is formalized using projective geometry[2]. To talk about the 3D Euclidean world in a projective setting, we need three ingredients:

- A real vector space $\mathcal{V}^4$ called the *supporting vector space* of dimension 4 from which we exclude the origin.
- An equivalence relation on $\mathcal{V}^4 - \{0\}$: $v_1 \sim v_2 \Leftrightarrow \exists \alpha \in \mathbb{R} \neq 0$ s.t. $v_1 = \alpha v_2$.
- A *polarity* $P$, which is a 2 covariant, symmetric tensor defined on $\mathcal{V}^4$, which in the sequel will be taken to be semi-positive defined and of rank 1.

The basic transformations between points of projective spaces are defined as injective linear transformations between the supporting vector spaces. These transformations must be injective to prevent that the subspace corresponding to the kernel of the transformation is mapped to the 0 element of the codomain which is NOT a valid element of the projective space[3]. These kinds of transformations are called *homographies* or *collineations* and in our case they are mappings from $\mathcal{V}^4$ to $\mathcal{V}^4$. Maps from $\mathcal{V}^4$ to the dual $\mathcal{V}^{4*}$ are called *correlations*. A symmetric correlation as $P$ is called *polarity*.

**Improper hyperplane.** Using the polarity $P$, it is possible to consider the vectors $p^i$ belonging to the quadric defined by the polarity $P$ which is called *the absolute*:

$$P_{ij}p^ip^j = 0 \qquad p^i \in \mathcal{V}^4.$$

The absolute is a three-dimensional subspace of $\mathcal{V}^4$ which is called the *improper hyperplane* and it is indicated with $\mathcal{I}^3 \subset \mathcal{V}^4$. This hyperplane represents the "points at infinity".

The improper hyperplane splits $\mathcal{V}^4$ in two disjointed semi-spaces, which we will call *positive* semi-space, indicated with $\mathcal{I}^+$, and the *negative* semi-space, indicated with $\mathcal{I}^-$.

[2] In 1859 Caley, by introducing the concept of an "absolute", showed that projective geometry is the most general geometry.

[3] It will be shown later that to obtain a proper definition of the projective space, the 0 element must be excluded.

The three-dimensional projective space is defined as the quotient space of $\mathcal{V}^4$ excluding the origin with respect to the defined equivalent relation:

$$\mathcal{P}^3 := \frac{\mathcal{V}^4 - \{0\}}{\sim}$$

A representation of the two-dimensional case is reported in Figure 1.1. In a purely projective setting where the polarity is not considered, all points are of the same type. Considering the polarity, we can make a distinction between finite points and infinite points. Infinite points are those whose representative in $\mathcal{V}^4$ belong to $\mathcal{I}^3$ and finite points are the others. We indicate with $\mathcal{P}_F$ the finite points and with $\mathcal{P}_\infty$ the points at infinity. Clearly it is true that $\mathcal{P} = \mathcal{P}_F \cup \mathcal{P}_\infty$. Furthermore, for each finite point $p \in \mathcal{P}_F$ $(P_{ij}p^ip^j \neq 0)$, there are representatives $v \in \mathcal{V}^4$ belonging either to $\mathcal{I}^+$ or to $\mathcal{I}^-$. We can define the sign function $\sigma$ for elements $v \in \mathcal{V}^4$:

$$\sigma(v) := \begin{cases} +1 & v \in \mathcal{I}^+ \\ 0 & v \in \mathcal{I}^3 \\ -1 & v \in \mathcal{I}^- \end{cases} \tag{1.1}$$

**Adjoint polarity.** Together with the polarity $P$, one defines its adjoint[4] $Q$, which is a 2 contravariant, symmetric, semi-positive tensor of rank 3. Once a proper base $\{e_x, e_y, e_z, e_0\}$ for $\mathcal{V}^4$ is chosen, the representations of $P$ and $Q$ become:

$$P = \begin{pmatrix} 0&0&0&0 \\ 0&0&0&0 \\ 0&0&0&0 \\ 0&0&0&1 \end{pmatrix} \qquad Q = \begin{pmatrix} 1&0&0&0 \\ 0&1&0&0 \\ 0&0&1&0 \\ 0&0&0&0 \end{pmatrix} \tag{1.2}$$

In the same coordinates, a vector of $\mathcal{V}^4$ has the form $(x, y, z, \alpha)^T$ and if $\alpha = 0$ the vector belongs to $\mathcal{I}^3$.

**Points and Free vectors.** It is possible to associate with each pair of finite vectors in $\mathcal{P}_F$ a unique element of $\mathcal{I}$:

$$f : \mathcal{P}_F \times \mathcal{P}_F \to \mathcal{I} \; ; \; (p, q) \mapsto \frac{p}{||p||\sigma(p)} - \frac{q}{||q||\sigma(q)} \tag{1.3}$$

where $p, q$ are any representatives and $|| \cdot ||$ represents the $P$-norm. It is possible to see that the previous operation is indeed independent of the representatives of the points and therefore well defined. Note that the difference of two points can be calculated using the vector structure of $\mathcal{V}^4$. Furthermore, the improper hyperplane without the equivalence relation and with the origin of $\mathcal{V}^4$ gets the meaning of the vector space of free vectors. In the usual coordinates, this means that if

[4] If instead of 0 on the diagonal elements of $P$ and $Q$ we use $\epsilon$, we see that $QP = PQ = \epsilon I$, where $I$ is the identity matrix.

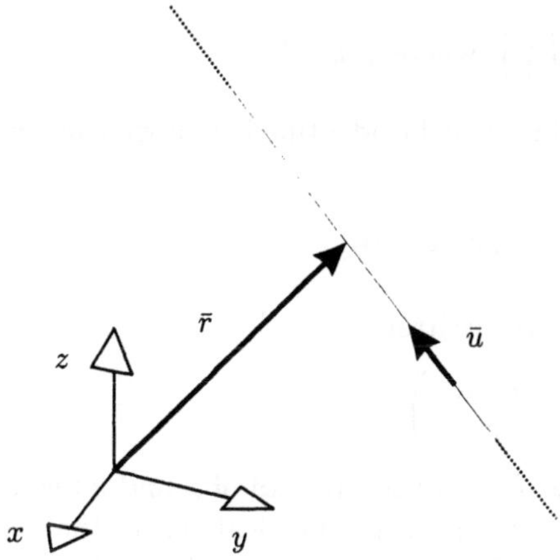


**Fig. 1.2.** Coordinates of a line

$$p = \begin{pmatrix} \alpha_p x_p \\ \alpha_p y_p \\ \alpha_p z_p \\ \alpha_p \end{pmatrix} \text{ and } q = \begin{pmatrix} \alpha_q x_q \\ \alpha_q y_q \\ \alpha_q z_q \\ \alpha_q \end{pmatrix}$$

we have $||p||\sigma(p) = \alpha_p$ and $||q||\sigma(q) = \alpha_q$ and therefore

$$f(p,q) = \begin{pmatrix} x_p - x_q \\ y_p - y_q \\ z_p - z_q \\ 0 \end{pmatrix}$$

Note that indeed the last component is equal to zero, which confirms the fact that the element belongs to $\mathcal{I}$. It is common to use the notation

$$p - q := f(p,q)$$

for obvious reasons, but the normalization of Eq. (1.3) before the subtraction is essential to make the operation intrinsically defined.

**Lines.** A line $l \subset \mathcal{P}^3$ in a projective context is nothing else than a one-dimensional subspace which corresponds to a two-dimensional subspace $L \subset \mathcal{V}^4$ (without the origin) of the supporting vector space $\mathcal{V}^4$. In a projective context it is also possible to talk about lines at infinity when $L \subset \mathcal{I}^3$. Furthermore, a line can be described both as the subspace spanned by two points of $\mathcal{V}^4$ and as the intersection of two hyperplanes of $\mathcal{V}^4$. As in every $2n$-dimensional vector space an $n$-dimensional subspace is self-dual, so is a line in $\mathcal{P}^3$ (defined using $\mathcal{V}^4$) a self dual entity.

It is possible to show that using the previous coordinates, given two distinct points $x, y$ belonging to a line where

$$x = \begin{pmatrix} \bar{x} \\ 1 \end{pmatrix} \text{ and } \begin{pmatrix} \bar{y} \\ 1 \end{pmatrix} \text{ where } \bar{x}, \bar{y} \in \mathbb{R}^3$$

the corresponding line can be identified homogeneously to a vector of the form

$$\alpha \begin{pmatrix} \bar{y} - \bar{x} \\ \bar{x} \wedge \bar{y} \end{pmatrix} \text{ where } \alpha \in \mathbb{R} - \{0\}. \tag{1.4}$$

Eq. (1.4) can be clearly written as:

$$\begin{pmatrix} \bar{y} - \bar{x} \\ \bar{x} \wedge (\bar{y} - \bar{x}) \end{pmatrix} = \begin{pmatrix} \bar{u} \\ \bar{r} \wedge \bar{u} \end{pmatrix}. \tag{1.5}$$

It is easy to see that if we choose instead of $\bar{x}$ and $\bar{y}$ two other distinct points on the same line in the usual geometrical sence, the vector $l$ as defined in Eq. (1.4) is the same if a proportional multiplication constant is used. This implies that the space of lines has somehow a projective nature by itself, but it should be noticed that the linear combination of two lines as expressed in Eq. (1.4) is in general NOT a line. Using the usual interpretation of $\mathcal{E}^3$, we can interpret $\bar{u}$ and $\bar{r}$ of Eq. (1.5) using Figure 1.2. Note the projective nature of a line: if instead of $\bar{u}$ we would use $\alpha\bar{u}$, we would effectively describe:

$$\alpha \begin{pmatrix} \bar{u} \\ \bar{r} \wedge \bar{u} \end{pmatrix}, \tag{1.6}$$

but this represents the same physical line using the representation of Figure 1.2.

If we add to a physical line a direction and a module, we get rid of its projective nature and have what is called a *rotor*. This can be done by considering the direction and module of $\bar{u}$.

**Euclidean product on the improper hyperplane.** It is now possible to define a proper non-singular, internal product for $\mathcal{I}^3$, which gives rise to the scalar product which characterizes a proper Euclidean Space. In the previous canonical coordinates, an element $v_i \in \mathcal{I}^3$ is characterized by the last component equal to zero. We can associate with $v_i \in \mathcal{I}^3$ the subspace of hyperplanes $\mathfrak{H}_{v_i}$ for which $v_i$ is what is called the *polar* with respect to $Q$:

$$h \in \mathfrak{H}_{v_i} \Leftrightarrow v_i = Qh$$

where $Q$ is the adjoint of $P$. We can now define the scalar product of two vectors $v_1, v_2$ in $\mathcal{I}^3$ as:

$$\langle v_1, v_2 \rangle := \sqrt{h_1^T Q h_2} \qquad h_1 \in \mathfrak{H}_{v_1}, h_2 \in \mathfrak{H}_{v_2} \tag{1.7}$$

It is easy to see that the previous definition is well posed since it is independent of the elements $h_1$ and $h_2$ due to the structure of $Q$.

It is now possible to define the distance between the finite points using the norm which can be defined based on the scalar product:

$$d(p_1, p_2) := \|p_1 - p_2\| := \sqrt{\langle p_1 - p_2, p_1 - p_2 \rangle} \tag{1.8}$$

**The Euclidean space .** It is now possible to define the following Euclidean three-dimensional space as the pair $(\mathcal{E}^3, \langle\rangle)$ such that $\mathcal{E}^3 := \mathcal{P}^3$, and $\langle,\rangle$ is the scalar product just defined on $\mathcal{E}^3_* := \mathcal{I}^3$ which is treated as a vector space. As with $\mathcal{P}$, we indicate with $\mathcal{E}_F$ finite points and with $\mathcal{E}_\infty$ infinite points and again:

$$\mathcal{E} = \mathcal{E}_F \cup \mathcal{E}_\infty .$$

Proper collineations of the Euclidean space are defined as the collineations which keep the polarity $P$, and therefore the improper hyperplane $\mathcal{I}^3$ invariant. It can be seen that in the previous coordinates a proper Euclidean collineation has the form:

$$c : \mathbb{R}^4 \to \mathbb{R}^4; q \mapsto \begin{pmatrix} \sigma R & p \\ 0 & 1 \end{pmatrix} q \tag{1.9}$$

where $R \in SO(3)$, the set of positively oriented orthonormal matrices, $\sigma$, can be either equal to $+1$ for an orientation-preserving transformation or $-1$ for a transformation-changing orientation and $p \in \mathbb{R}^3$. Using an abuse of notation, we will often indicate in the sequel with $\mathcal{E}$ only the finite points, even if they where previously denoted as $\mathcal{E}_F$.

Looking at the expression of Eq. (1.4) for a line, we can see how these coordinates transform. Using Eq. (1.9) and Eq. (1.4), it is possible to see that the mapping of a line that corresponds to the mapping of points as reported in Eq. (1.9) becomes:

$$\begin{pmatrix} \bar{y}_2 - \bar{x}_2 \\ \bar{x}_2 \wedge \bar{y}_2 \end{pmatrix} = \begin{pmatrix} \sigma R & 0 \\ \sigma \tilde{p} R & R \end{pmatrix} \begin{pmatrix} \bar{y}_1 - \bar{x}_1 \\ \bar{x}_1 \wedge \bar{y}_1 \end{pmatrix} \tag{1.10}$$

where the operator tilde is such that $\tilde{p}$ is a skew-symmetric matrix and it is such that $\tilde{p}a = p \wedge a$ for each $a \in \mathbb{R}^3$.

Using the matrix rotation group $SO(3)$ and the intrinsically defined operation of Lie brackets [18], it is possible to orient $\mathcal{E}$ by defining a 3−form $\Omega$ on $\mathcal{I}^3$. We suppose that the coordinates $(e_x, e_y, e_z, e_0)$ are chosen such that $\Omega(e_x, e_y, e_z) = +1$ and $e_z = [e_x, e_y]$, where $[,]$ represents the Lie algebra commutator of $so(3)$. In these coordinates, the operation of vector product "$\wedge$" in $\mathbb{R}^3$ results in the usual product. More details will be given later on this issue.

**Coordinate functions.** A coordinate $\Psi$ for the space $\mathcal{E}$ is a 4-tuple of the form

$$\Psi = (o, \hat{x}, \hat{y}, \hat{z}) \qquad o \in \mathcal{E}, (o \notin \mathcal{I}) \qquad \hat{x}, \hat{y}, \hat{z} \in \mathcal{E}_*$$

where $\hat{x}, \hat{y}, \hat{z}$ should be linearly independent[5]. If, furthermore $\hat{x}, \hat{y}, \hat{z}$ are orthonormal, we call the coordinates *orthonormal.* Eventually we call the coordinates positively oriented iff $\Omega(\hat{x}, \hat{y}, \hat{z}) = +1$, where $\Omega$ is the form orientating

[5] Properly speaking, in projective geometry, the concept of an homogeneous Cartesian reference can be introduced, which is composed on $n + 2$ projective points one of which is called the identity. For our needs this is not necessary

the Euclidean space. If not otherwise specified, we will always consider positively oriented, orthonormal coordinates systems. We can now define a base for $\mathcal{V}^4$ that is dependent on $\Psi$ using the four linearly independent vectors:

$$\hat{x}, \hat{y}, \hat{z}, \frac{o}{||o||\sigma(o)} \in \mathcal{V}^4$$

of $\mathcal{V}^4$.

The map which associates a representative of a point of $\mathcal{E}$ with its numerical representation using this base is indicated with $\psi_i$:

$$\psi(p) = (p_x, p_y, p_z, \alpha_p) \Rightarrow p = p_x\hat{x} + p_y\hat{y} + p_z\hat{z} + \alpha\frac{o}{||o||\sigma(o)}$$

Clearly we have that:

$$\psi_i^{-1}(\cdot,\cdot,\cdot,0) = \mathcal{I}^3 \qquad \forall i = 1,\ldots,m \tag{1.11}$$

and therefore these coordinates are characterized by the fact that the last component that is equal to zero represents a point at infinity.

## 1.2 Rigid bodies and Observers

The concept of an observer is very important and is coupled with the concept of space: an observer is identified with a Euclidean space to which he is rigidly connected together with all those objects which are always stationary in relation to him. A set of objects which never change position in relation to one another can be considered as a single entity from a kinematic point of view.

For these reasons and for more formal ones that should become clear in the sequel, we consider as many Euclidean spaces as there are bodies moving relative to one another. In the projective setting, these Euclidean spaces can be seen as a set of four-dimensional vector spaces with a common polarity $P$ and a common improper hyperplane.

An object will be a subset of a Euclidean space together with a mass density function which associates a value with each of the body's points. The mass density function is relevant for dynamic considerations. For kinematic considerations, it is sufficient to define a proper body $B_i$ as a subset of a Euclidean space $\mathcal{E}_i$.

Once we have defined the geometry of bodies and spaces, we can introduce time as an absolute[6] scalar variable. We can now define what we will call the Euclidean system.

---

[6] The fact that we will consider time as an absolute is a consequence of the classical separation in mechanics of space and time, which is due to the Galilean hypothesis.

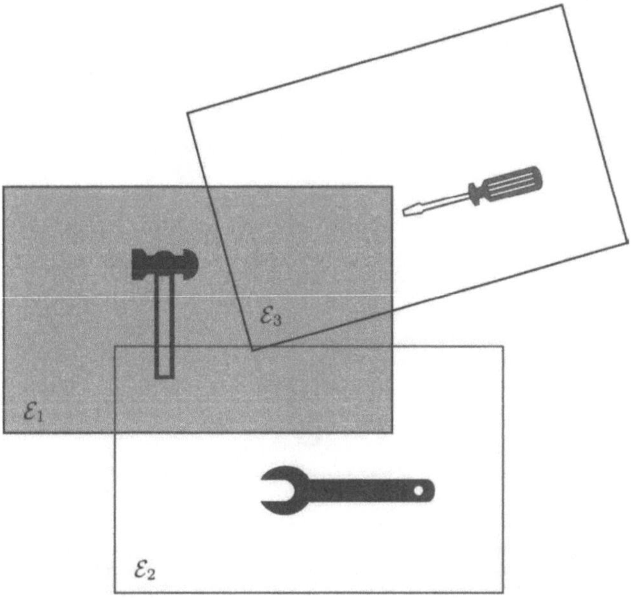


**Fig. 1.3.** An example of a two-dimensional Euclidean system.

**Definition 1.2.1 (Euclidean system).** *The ordered set of m, n-dimensional Euclidean oriented spaces:*

$$S^m(n) := \{\mathcal{E}_1, \mathcal{E}_2, \ldots, \mathcal{E}_m\}$$

*is called a* Euclidean system of dimension $n$ and order $m$

We use the Euclidean system as the basic structure to consider relative motions. All the material which will be treated will be general and applicable to $n$ dimensions. Clearly in 3D mechanics, we will have $n = 3$. The number $m$ corresponds to the number of objects and/or observers we consider in our study of motion.

To understand the concepts intuitively, we consider the two-dimensional Euclidean system $S^3(2)$ of Figure 1.3. We use a two-dimensional example for simplicity and to illustrate some concepts which will be introduced later on. We can consider the Euclidean spaces as sheets of paper having an infinite extension which can be placed on one another. In the figures, the sheets are drawn as finite rectangles for the sake of clarity, but they should be thought of as extending infinitely.

The following definition is necessary to understand relative positions:

**Definition 1.2.2 (Isometry).** *Let $\mathcal{E}_i, \mathcal{E}_j \in S^m(n)$ be 2 Euclidean spaces with distance functions $d_l : \mathcal{E}_l \times \mathcal{E}_l \to \mathbb{R}^+$ with $l = i, j$. An isometry between $\mathcal{E}_i$ and $\mathcal{E}_j$ is defined as a bijective application $h_i^j : \mathcal{E}_i \to \mathcal{E}_j$ ; $p \mapsto h_i^j(p)$ such*

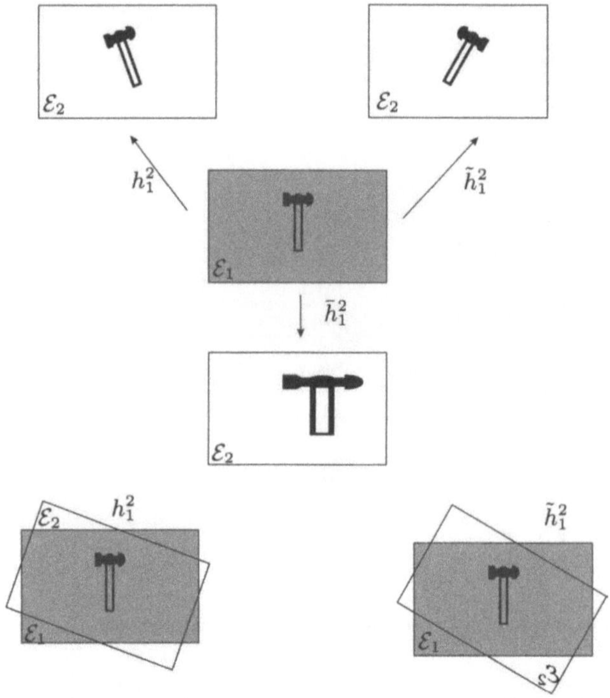


**Fig. 1.4.** Maps between Euclidean spaces.

*that* $\forall p, q \in \mathcal{E}_i$, $d_j(h_i^j(p), h_i^j(q)) = d_i(p,q)$. *If* $i = j$, *we will call the isometry auto-isometry or motion.*

It is well-known [20] that an isometry does not imply anything about orientational issues. The preservation of orientation can only be defined if "reference orientations" are defined on the spaces under consideration by means of an $n-$form, which is in our case $\Omega_i$. A reference orientation is only necessary for isometries defined between different spaces: a motion can be defined to be orientation preserving even if no reference orientation is chosen for it.

**Definition 1.2.3 (Positive isometry).** *A positive isometry from* $\mathcal{E}_i$ *with associated orientation* $\Omega_i$ *to* $\mathcal{E}_j$ *with associated orientation* $\Omega_j$ *we call an isometry* $h_i^j : \mathcal{E}_i \to \mathcal{E}_j$ *such that there exists* $\lambda \in C^\infty(\mathcal{E}_i)$ *s.t.* $\lambda(\cdot) > 0$ *and* $\lambda \Omega_i = (h_i^j)^* \Omega_j$ *where with* $(h_i^j)^*$ *we indicate the pullback operator (see Def. A.7.20).* $SE_i^j(n)$ *denotes the set of positive isometries from* $\mathcal{E}_i$ *to* $\mathcal{E}_j$. *The subscript* $i$ *of* $h_i^j$ *corresponds to the domain and the superscript* $j$ *corresponds to the range space. The set of motions* $h_i^i$ *of space* $\mathcal{E}_i$ *is indicated with* $SE_i(n)$.

To improve our intuition, we could consider the orientation by writing the name of the Euclidean space on each of the sheets (see Figure 1.4). It is then possible to see that in Figure 1.4, $h_1^2$ is a positive isometry and this can be seen by shifting the two sheets of paper such that the two hammers coincide, as reported bottom left in the figure. Furthermore, $\tilde{h}_1^2$ is still an isometry, but it is not positive since we must first turn over the sheet before all the points of one sheet correspond to those on the other sheet. This is reported bottom right of the figure. The last mapping, $\bar{h}_1^2$, is not even an isometry since the distances between points are changed and there is no way that the two sheets can be placed such that the points on one correspond to the points on the other.

**Definition 1.2.4 (Relative Position).** *An element $h_i^j \in SE_i^j(n)$ is called a relative position of $\mathcal{E}_i$ with respect to $\mathcal{E}_j$.*

It is important to reflect a moment on the previous definitions. Observers and/or objects are associated with a Euclidean space to which they are rigidly connected. We can think of all these spaces as interpenetrating. In the example, this "interpenetration" can be understood by considering how the sheets overlap. In each instant, all the points of one of the spaces will have associated points on each of the other spaces. In the example "the association" is the one which assigns to all the points of one sheet the corresponding points underneath on another sheet. This map is represented by a positively oriented isometry. It is important at this point to reason without any use of coordinates because it enables us to define entities which are intrinsic. We can now define a *kinematic state*.

**Definition 1.2.5 (Kinematic State).** *We call a kinematic state for a Euclidean system $S^m(n)$ a point:*

$$h := (h_2^1, h_3^2, \ldots, h_m^{m-1}) \in SE_2^1(n) \times SE_3^2(n) \times \ldots \times SE_m^{m-1}.$$

Most of the Euclidean spaces belonging to $S^m(n)$ will represent physical bodies. These spaces have a subset of points $B_i \subset \mathcal{E}_i$ where matter is present. This set $B_i$ is called the *matter set* and it corresponds to the Euclidean space $\mathcal{E}_i$. We say that a Euclidean space in $S^m(n)$ is a *pure observer* iff its matter set is the empty set. In the working example, the set $B_1$ will be the set of points where the hammer is, $B_2$ where the screw driver is and $B_3$ where the spanner is. These are sets in two-dimensional Euclidean spaces. In the example there are no pure observers which could be thought of as additional sheets with no objects on them. We can now talk about a compatible kinematic state:

**Definition 1.2.6 (Compatible Kinematic State).** *We will say that a kinematic state $h$ as defined in Def. 1.2.5 is compatible iff $h_i^j(B_i) \cap B_j = \emptyset \ \forall i \neq j \ i, j = 1, \ldots, m$.*

The previous definition can be used to describe object impenetrability. This is the simplest and most basic constraint which has to be satisfied at this stage: rigid objects will not penetrate one another[7].

[7] In reality, when bodies are in contact, they will slightly deflect and generate an elastic force which will oppose the compenetration.

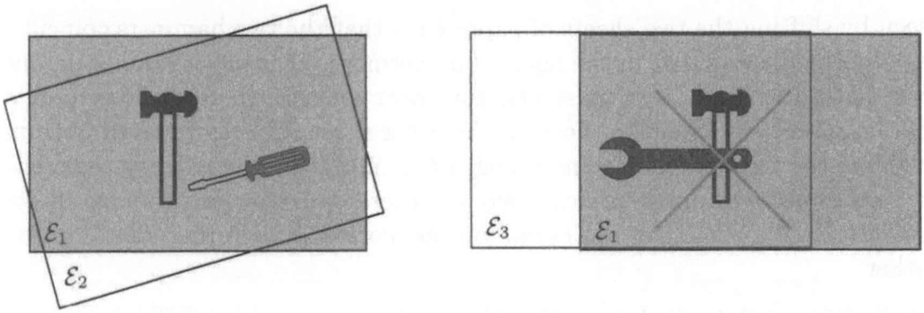


**Fig. 1.5.** Compatibility of the kinematic state.

The idea of compatibility of state is illustrated in Figure 1.5. The relative position of $\mathcal{E}_1$ and $\mathcal{E}_2$ does not create any problem for the kinematic state compatibility, but the relative position of $\mathcal{E}_1$ and $\mathcal{E}_3$ is not compatible since it maps points of an object belonging to one space to points of another object in another space. Remember that in the example, the previously considered map, corresponds to the one which maps points of one sheet (Euclidean space) to the points underneath on another sheet.

Here a remark is in order: at first glance the presented framework may seem unnecessarily theoretical and formal. Nevertheless, such an approach is not just fundamentally meaningful; it could be used as the basis for implementing computer tools for the analysis of mechanisms. To be able to describe a mechanism, we need to have formal and exact ways to represent information, and the definitions proposed here help to give such a structure.

We can now define the corresponding left, right and conjugation maps between two spaces. Note that if the spaces are different, these operations are neither endomorphisms nor group operations. This is why we use the adjective "hybrid" for the following definitions.

**Definition 1.2.7 (Hybrid Right and Left Maps).** *We call*

$$R_{h_i^j} : SE_j(n) \to SE_i^j(n) \; ; h_j \mapsto h_j \, o \, h_i^j$$

*the right hybrid map for the pair of spaces* $(\mathcal{E}_i, \mathcal{E}_j)$ *at* $h_i^j \in SE_i^j(n)$ *and*

$$L_{h_i^j} : SE_i(n) \to SE_i^j(n) \; ; h_i \mapsto h_i^j \, o \, h_i$$

*the left hybrid map for the pair of spaces* $(\mathcal{E}_i, \mathcal{E}_j)$ *at* $h_i^j \in SE_i^j(n)$.

The " $o$ " indicates the composition of maps. The left hybrid map can be used to create a bijection between $SE_i^j(n)$ and $SE_i(n)$ and the right hybrid map can be used to create a bijection between $SE_i^j(n)$ and $SE_j(n)$. Remember that $SE_i^j(n)$ represents the space of absolute relative positions of $i$ and $j$ independent of any relative reference position. Then it is possible to consider how the corresponding representations in $SE_i(n)$ and $SE_j(n)$ are related given a reference relative position $h_i^j$. This is done using the hybrid conjugation map.

**Definition 1.2.8 (Hybrid Conjugation map).** *We will call the map* $K_{h_i^j} := R^{-1}_{h_i^j} L_{h_i^j}$ *hybrid conjugation map for the pair of spaces* $(\mathcal{E}_i, \mathcal{E}_j)$ *in* $h_i^j \in SE_i^j(n)$. *This map can be expressed as*

$$K_{h_i^j} : SE_i(n) \to SE_j(n) \; ; \; h_i \mapsto h_i^j \; o \; h_i \; o \; (h_i^j)^{-1}.$$

With respect to the previous definitions of hybrid right, left and conjugation maps, it is important to note that if the domain and range spaces coincide (if they are the same Euclidean space), these maps become endomorphisms and correspond to the standard Lie group maps on a special isometry group (see Sect. A.8.1).

*Remark 1.2.1 (Swap domain range).* If in the previous definition we were to swap the domain space with the range ($i \leftrightarrow j$), the result would be that $r \to r^{-1}$, $h \to h^{-1}$ and the right maps would correspond to the previous left maps and so on. This can be formally stated as follows:

$$(L_{\bar{h}}^{-1}(r))^{-1} = (h^{-1} \; o \; r)^{-1} = (r^{-1} \; o \; h) = R^{-1}_{\overline{(h^{-1})}}(r^{-1})$$

where $r, h \in SE_i^j(n)$ and the previous elements all belong to $SE_i(n)$, or dually:

$$(L^{-1}_{\overline{(h^{-1})}}(r^{-1}))^{-1} = (h \; o \; r^{-1})^{-1} = (r \; o \; h^{-1}) = R_{\bar{h}}^{-1}(r)$$

where $r, h \in SE_i^j(n)$ and the previous elements all belong to $SE_j(n)$.

This remark has important implications when twists are introduced.

It can be easily verified that $K_{h_i^j}(e_i) = e_j \;\; \forall h_i^j \in SE_i^j(n)$, where $e_k \in SE_k(n)$ is the identity element of the group $SE_k(n)$. We can consider the tangent map of $K_{h_i^j}$:

$$\left(K_{h_i^j}\right)_* : TSE_i(n) \to TSE_j(n) \; ; \; (h_i, v_i) \mapsto \left(h_i^j \; o \; h_i \; o \; (h_i^j)^{-1}, \frac{d}{dt}\left((h_i^j \; o \; e^{v_i t} h_i \; o \; (h_i^j)^{-1})\right)\bigg|_{t=0}\right) \quad (1.12)$$

where with $e^{v_i t} h_i$ we used the usual exponential map (Sect. A.8.6) to consider a curve in $SE_i(n)$ parameterized by $t$ and passing through $h_i$ at time $t = 0$ with velocity $v_i$. It is now possible to define the hybrid adjoint.

**Definition 1.2.9 (Hybrid adjoint).** *We call hybrid adjoint from space* $se_i(n)$ *to space* $se_j(n)$ *the following map:*

$$Ad_{h_i^j} : se_i(n) \to se_j(n) \; ; \; t_i \mapsto \Pi\left(\left(K_{h_i^j}\right)_* (e_i, t_i)\right)$$

*where* $\Pi(\cdot)$ *indicates the fiber projection (see Def. A.7.6), which in this case corresponds to the tangent space in* $e_j$, *the identity element of* $SE_j(n)$.

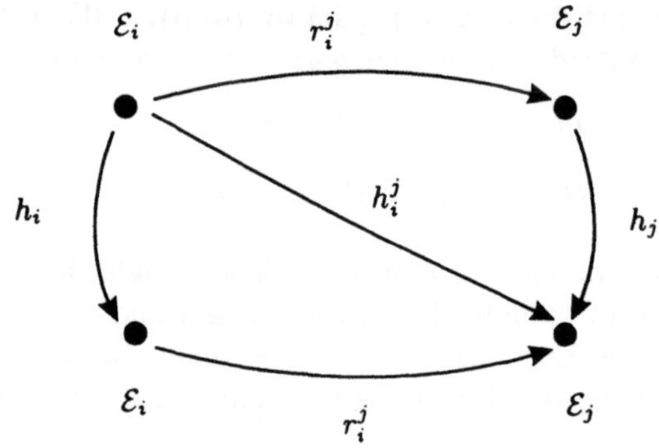


**Fig. 1.6.** The commutation diagram of the relation among elements of the Lie groups and the reference.

It is important to note that $se_i(n)$ and $se_j(n)$ are NOT the same and that there is no natural identification for these algebras[8]. Furthermore, if the domain and range space of $Ad_{h_i^j}$ coincide, the hybrid adjoint corresponds to the usual adjoint group operation (see Def. A.8.5). With reference to Figure 1.6, we can associate an element of $SE_i(n)$ or $SE_j(n)$ with a relative position of $i$ with respect to $j$ that corresponds to a $h_i^j \in SE_i^j(n)$ by using a reference $r_i^j \in SE_i^j(n)$:

$$h_i = L_{r_i^j}^{-1}(h_i^j) = (r_i^j)^{-1} \circ h_i^j \tag{1.13}$$

$$h_j = R_{r_i^j}^{-1}(h_i^j) = h_i^j \circ (r_i^j)^{-1}. \tag{1.14}$$

*Remark 1.2.2.* Note that the identification of a relative position by an element of $SE_i(n)$ or an element of $SE_j(n)$ is **NOT** intrinsic and depends on the choice of a reference-relative position $r_i^j$. This is often not stated in the literature, even though the use of Lie group theory to describe rigid body motions depends on it !

### 1.2.1 Numerical representation of configurations

Up to this moment, we have considered relative positions without using coordinates. For each of the Euclidean spaces $\mathcal{E}_i$ belonging to the Euclidean system, we can choose a right-handed orthonormal coordinate system $\Psi_i$ as explained in Sect. 1.1. Doing so, we can identify certain operations to operations within a matrix algebra. We will see that an intelligent choice of the

[8] The vector spaces $se_i(n)$ and $se_j(n)$ are algebras since they have a commutator operator derived from the Lie bracket of vector fields on $SE_i(n)$ and $SE_j(n)$. For details see Def. A.7.12 and Def. A.6.1.

reference $r_i^j$, which is dependent on the choice of the coordinates, will yield standard Lie group operations of the matrix group $SE(n)$ as presented in [10].

**Isometries in $\mathbb{R}^n$.** The positively oriented motions in $\mathbb{R}^n$ will be essential to what follows. It is well known that this set is homeomorphic to the special isometries matrix group $SE(n)$, which is defined in the following way:

$$SE(n) := \left\{ \begin{pmatrix} R & p \\ 0 & 1 \end{pmatrix} \ s.t. \ R \in SO(n), p \in \mathbb{R}^n \right\} \tag{1.15}$$

where with $SO(n)$ we indicate the special orthonormal group of orthonormal matrices defined as:

$$SO(n) := \left\{ R \in \mathbb{R}^{n \times n} \ s.t. \ \det(R) = 1, R^{-1} = R^T \right\}. \tag{1.16}$$

At page 7 we have seen that collineations are defined in such a way that they keep the polarity invariant. This implies that such mappings do not change the distances between points since the polarity is the means that is used to define the Euclidean scalar product, and therefore the norm and distance. This also implies that a collineation corresponds to an isometry as defined in Def. 1.2.2 within the same space; therefore the collineation has a representation as given in Eq. (1.9). If, furthermore we require that this isometry is positive, we have $\sigma = 1$ in Eq. (1.9) and therefore the operation $y = h(x)$ for a positive isometry $h$ from $\mathbb{R}^n$ to itself corresponds to a matrix multiplication of the form:

$$\begin{pmatrix} y \\ 1 \end{pmatrix} = \begin{pmatrix} R & p \\ 0 & 1 \end{pmatrix} \begin{pmatrix} x \\ 1 \end{pmatrix}$$

where the matrix belongs to $SE(n)$.

**Representation of relative positions with coordinates.** Suppose we associate with each $\mathcal{E}_i \in S^m(n)$ a right-handed coordinate frame $\Psi_i$. We can, therefore, assign an element of $H_i^j \in SE(n)$ to each element $h_i^j \in SE_i^j(n)$ which corresponds to the following isometry of $\mathbb{R}^n$:

$$H_i^j = \Psi_j \, o \, h_i^j \, o \, \Psi_i^{-1}. \tag{1.17}$$

The element $H_i^j$ is an element of $SE(n)$ because it is defined as a composition of positively oriented isometries and it is therefore a positively oriented isometry of $\mathbb{R}^n$.

## 1.3 Twists

So far we have considered relative positions without introducing motion. The latter is clearly a change of relative position. To be able to talk about change, we need to consider a time set $I$ which will be an open interval of $\mathbb{R}$. We can then consider curves in $SE_i^j(n)$ parameterized by $t \in I$.

**Definition 1.3.1 (Relative Motion).** *We call a differentiable function of the following form*[9]*:*

$$h_i^j : I \to SE_i^j(n)$$

*a relative motion of space i with respect to j*

Since relative motions are differentiable, we can consider the velocity vector of the curve at $h_i^j$, which we indicate as

$$\dot{h}_i^j \in T_{h_i^j} SE_i^j(n).$$

Considering Eq. (1.13) (respectively Eq. (1.14)) we can map these relative motion velocities to elements of the tangent bundle $TSE_i(n)$ (respectively $TSE_j(n)$):

$$\left(L_{r_i^j}^{-1}\right)_* : TSE_i^j(n) \to TSE_i(n) :$$

$$(h_i^j, v_i^j) \mapsto \left( L_{r_i^j}^{-1}(h_i^j), \frac{d}{dt}\left(\left(L_{r_i^j}^{-1}(e^{v_i^j t} h_i^j)\right)\right)\Big|_{t=0} \right) \qquad (1.18)$$

where with $t \mapsto e^{v_i^j t} h_i^j$ we indicate any curve in $SE_i^j(n)$ that passes through $h_i^j$ at time $t = 0$ with velocity $v_i^j$. An analogous map can be used to map to $TSE_j(n)$ using the hybrid right translation.

*Remark 1.3.1 (Inverse mapping).* It is important to realize that each $h_i^j : I \to SE_i^j(n)$ is bijective to a $h_j^i : I \to SE_j^i(n) \; ; \; t \mapsto (h_i^j(t))^{-1}$. This implies also that to each $\dot{h}_i^j \in T_{h_i^j} SE_i^j(n)$ an element $\dot{h}_j^i \in T_{h_j^i} SE_j^i(n)$ will uniquely correspond. This means that whenever we will talk about vector fields or distributions on $SE_i^j(n)$, we will also have uniquely corresponding vector fields or distributions defined on $SE_j^i(n)$.

So far, we have shown a way to map relative velocities to elements of $TSE_i(n)$ or $TSE_j(n)$ (see Figure 1.9). Both of these maps are not intrinsic since they depend on a choice of a reference-relative position $r_i^j$. These elements belong to tangent bundles of Lie groups and therefore it is meaningful to transport these tangent vectors to the group identities by means of either left or right translations (see Sect. A.8.1).

### 1.3.1 Transports to the identity of the domain Lie group

Consider the element $\left(L_{r_i^j}^{-1}\right)_* (h_i^j, v_i^j)$ which is the representative in $TSE_i(n)$ of the relative motion of $\mathcal{E}_i$ with respect to $\mathcal{E}_j$ in the relative position $h_i^j$ with

[9] Note the abuse of notation here: $h_i^j$ has been used both as an element of $SE_i^j(n)$ and as a function from $I$ to $SE_i^j(n)$.

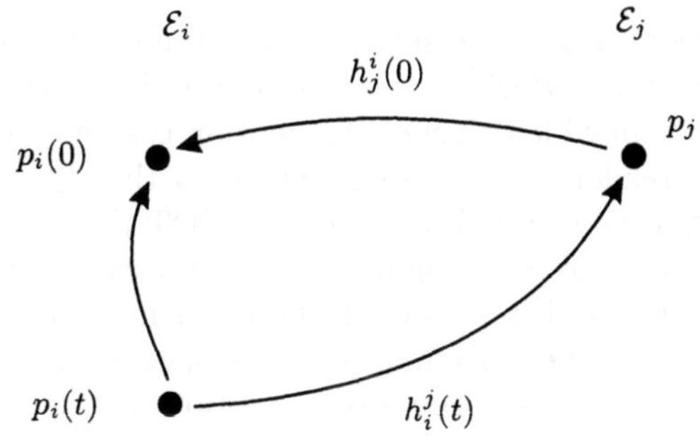


**Fig. 1.7.** Commutation diagram of the intrinsic left translations.

a relative velocity $v_i^j$. We can transport this tangent vector to $e_i$, the identity element of $SE_i(n)$ either by left or right translation. An important result is the following:

**Theorem 1.3.1.** *The left translation to the identity of the left hybrid translated tangent vector representative of a relative motion is defined intrinsically and independent of any choice of a fixed reference* $r_i^j$.

**Proof**. Consider an element $(h_i^j, v_i^j) \in T_* SE_i^j(n)$. Consider a relative motion

$$\tilde{h}_i^j : [-a, a] \to SE_i^j(n)$$

with $a > 0$ and such that $\tilde{h}_i^j(0) = h_i^j$ and $\frac{d}{dt}\left(\tilde{h}_i^j(t)\right)\Big|_{t=0} = v_i^j$. Defining

$$(h_i, v_i) := \left(L_{r_i^j}^{-1}\right)_* (h_i^j, v_i^j)$$

it is possible to see that $h_i = \tilde{h}_i(0)$ and $v_i = \frac{d}{dt}\left(\tilde{h}_i(t)\right)\Big|_{t=0}$ where $\tilde{h}_i(t) = L_{r_i^j}^{-1}(\tilde{h}_i^j(t))$.

To calculate the left translation at the identity of $(h_i, v_i)$, we have, therefore:

$$t_i^{i,j} := \Pi\left(\left(L_{h_i^{-1}}\right)_* (h_i, v_i)\right) = \frac{d}{dt}\left((\tilde{h}_i^{-1}(0) \; o \; \tilde{h}_i(t))\right)\Big|_{t=0} =$$
$$\frac{d}{dt}\left((\tilde{h}_j^i(0) \; o \; r_i^j \; o \; r_j^i \; o \; \tilde{h}_i^j(t))\right)\Big|_{t=0} = \frac{d}{dt}\left((\tilde{h}_j^i(0) \; o \; \tilde{h}_i^j(t))\right)\Big|_{t=0}. \quad (1.19)$$

■

It is possible to see that the last expression is independent of $r_i^j$ and therefore defined intrinsically. We will call the last quantity $t_i^{i,j} \in se_i(n)$ the intrinsic twist of the motion of $\mathcal{E}_i$ with respect to $\mathcal{E}_j$ expressed in the space $se_i(n)$.

Note the notation: in $t_i^{i,j}$, the subscript and the second superscript indicate the relative motion, which in this case is $i$ with respect to $j$. The first superscript indicates in which algebra this motion is expressed, which in this case is $se_i(n)$. The reader could be tempted to say that $t_i^{i,j}$ is the twist that an observer sitting in $\mathcal{E}_i$ would observe. This is NOT the case, as can be seen in the commutation diagram of Figure 1.7: an observer sitting in $\mathcal{E}_i$ who is looking at a point $p_j$ fixed in $\mathcal{E}_j$ would observe a motion that is the exact opposite of the one described. To understand this, consider that the observer $\mathcal{E}_i$ for whom the time increases would see a motion corresponding to a mapping from $p_i(0)$ to $p_i(t)$ with $t > 0$, which has a direction that is opposite to the one reported in Figure 1.7 and corresponds to $(h_j^i(0) \; o \; \tilde{h}_i^j(t))$.
If instead of left translating to the identity we right translate, we do NOT get a quantity that is intrinsically defined. This is formally stated below:

**Theorem 1.3.2.** *The right translation to the identity of the left hybrid translated tangent vector representative of a relative motion is NOT intrinsically defined and is equal to*

$$\bar{t}_i^{i,j} = Ad_{r_j^i} t_i^{j,j}$$

*where*

$$t_i^{j,j} := \frac{d}{dt}\left((\tilde{h}_i^j(t) \; o \; \tilde{h}_j^i(0))\right)\Big|_{t=0}. \tag{1.20}$$

We will indicate with the simplified notation $t_i^j = t_i^{j,j}$ the case in which the two superscripts coincide. **Proof.** With similar definitions as in the previous theorem, we show that:

$$\bar{t}_i^{i,j} := \Pi\left(\left(R_{h_i^{-1}}\right)_* (h_i, v_i)\right) =$$
$$\frac{d}{dt}\left((\tilde{h}_i(t) \; o \; \tilde{h}_i^{-1}(0))\right)\Big|_{t=0} = \frac{d}{dt}\left((r_j^i \; o \; \tilde{h}_i^j(t) \; o \; \tilde{h}_j^i(0) \; o \; r_i^j)\right)\Big|_{t=0} =$$
$$Ad_{r_j^i} t_i^j. \tag{1.21}$$

■

### 1.3.2 Transports to the identity of the range Lie group

Like we did in the previous subsection, we can first use the hybrid right translation to map relative motions to $SE_j(n)$, and then Lie group right or left translations within $SE_j(n)$ to map to the identity. The following two theorems then correspond to the previous ones.

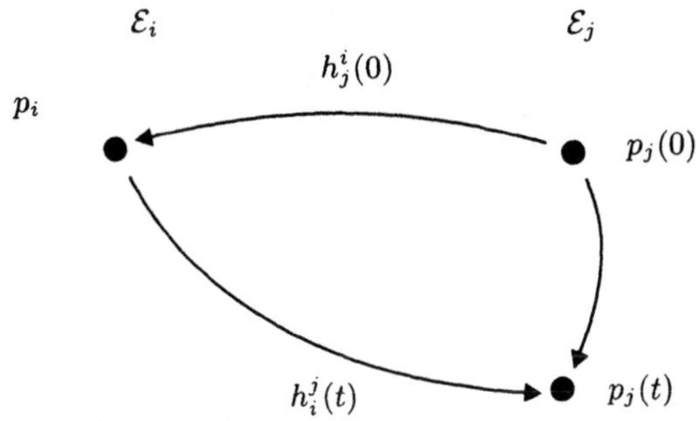


**Fig. 1.8.** Commutation diagram of the intrinsic right translations.

**Theorem 1.3.3.** *The right translation to the identity of the right hybrid translated tangent vector representative of a relative motion is defined intrinsically, independent of any choice of reference $r^j_i$ and equal to $t^j_i$ as defined in Eq. (1.20).*

**Proof.** See proof of Theorem 1.3.1. ∎

Observing the commutation diagram of Figure 1.8, we can see that the twist $t^j_i$ is indeed the twist that an observer sitting in $\mathcal{E}_j$ would see. The diagram shows that a fixed point $p_i \in \mathcal{E}_i$ is indeed mapped from a point $p_j(0) \in \mathcal{E}_j$ to future points $p_i(t)$. This means that the right translation is somehow more representative. On the other hand, as a consequence of what was said in Remark. 1.2.1, p.13, we could consider that the isometries going from $j$ to $i$ correspond to the inverse of what done so far. This would result in a right translation which would give the intrinsic twist $t^i_j$. This will be discussed in detail later.
If instead of right translating to the identity we left translate, we do NOT get an intrinsically defined quantity. This is formally stated in what follows:

**Theorem 1.3.4.** *The left translation to the identity of the right hybrid translated tangent vector representative of a relative motion is NOT defined intrinsically and is equal to*

$$\bar{t}^j_i = Ad_{r^j_i} t^{i,j}_i.$$

**Proof.** See proof of Theorem 1.3.2. ∎

In a similar way, it is also possible to find an intrinsic mapping which maps intrinsic twists of one space to the other. This is formally stated in the following theorem, which can be proven along the same lines as the previous ones.

**Theorem 1.3.5 (Map between intrinsic twists).** *For any relative position $h^j_i$, a bijection exists between the intrinsic twists in $se_i(n)$ and the ones in $se_j(n)$. This map is given by:*

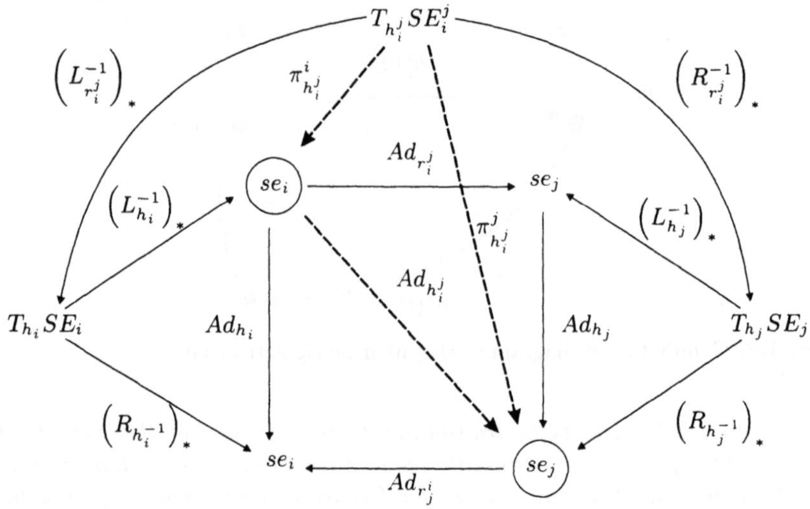

Fig. 1.9. The complete commutation diagram.

$$t_i^j = Ad_{h_i^j} t_i^{i,j}.$$

$Ad_{h_i^j}$ is to the hybrid adjoint of Def. 1.2.9. The presented results are summarized in Figure 1.9.

The thick cross-hatched lines show the intrinsic maps to and between the Lie algebras of the Lie groups of the motions of the two spaces under consideration. Only those twists are intrinsic and independent of the reference $r_i^j$ which are obtained by following the maps to the circled $se_i$ and $se_j$.

Furthermore, it has been shown that the standard adjoint maps within the Lie algebras of the two spaces are not maps between intrinsically defined quantities. This means that in a coordinate-free framework, it is not correct to say that the adjoint group operation maps twists from one reference to the other since one of the two elements is not intrinsic and dependent on the reference $r_i^j$. We denoted these nonintrinsic elements with $\bar{t}_i^j$ and $\bar{t}_i^{i,j}$.

### 1.3.3 Relations between twists

It is possible to see that the complete right translation is somehow more representative than the left one since it gives the motion of $\mathcal{E}_i$ that an observer sitting in $\mathcal{E}_j$ would directly observe. From the considerations of Figure 1.7 and Figure 1.8, it should be clear that

$$t_i^j = \frac{d}{dt}\left((\tilde{h}_i^j(t) \circ \tilde{h}_j^i(0))\right)\Big|_{t=0}$$

represents the motion of $\mathcal{E}_i$ for an observer sitting in $\mathcal{E}_j$. In exactly the same way, namely by inverting the domain and range space as explained in Remark. 1.2.1, p.13, we could arrive at the definition of

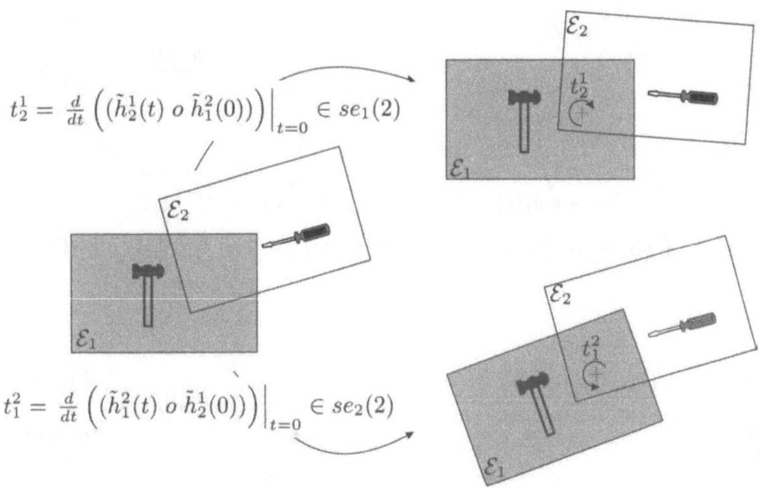


**Fig. 1.10.** Intrinsic twist maps.

$$t_j^i = \frac{d}{dt}\left((\tilde{h}_j^i(t)\ o\ \tilde{h}_i^j(0))\right)\Big|_{t=0}$$

which would again correspond to the right translation to the identity of the right hybrid translated tangent vector of a motion, but now in the opposite direction: for the isometries from $\mathcal{E}_j$ to $\mathcal{E}_i$.

A representation of the two intrinsic right translated twists $t_i^j$ and $t_j^i$ is shown in Figure 1.10. It is possible to see that the direction of $t_1^2$ is opposite to the one of $t_2^1$. This relation can be formally stated in the following theorem.

**Theorem 1.3.6 (Twist relations).** *With the notation used previously, the following identities hold:*

$$-t_j^{j,i} = t_i^j = Ad_{h_i^j} t_i^{i,j} = -Ad_{h_i^j} t_j^i \tag{1.22}$$

*and*

$$t_j^i = -t_i^{i,j} = -Ad_{h_j^i} t_i^j = Ad_{h_j^i} t_j^{j,i}. \tag{1.23}$$

Instead of giving a formal proof of the previous theorem, we use Figure 1.11, Figure 1.12, Figure 1.13, Figure 1.14 to understand the result. The top figures show the right translated relative motion in the two directions and the bottom ones show the left translated. The left figures correspond to the intrinsic map $\pi^j_{h_i^j}$ which is also denoted with $\pi_{h_i^j}$(Figure 1.11) and $\pi^i_{h_i^j}$ (Figure 1.13) of Figure 1.9 and the right figures correspond to the intrinsic maps $\pi^i_{h_j^i}$ (Figure 1.12) and $\pi^j_{h_j^i}$ (Figure 1.14) of a diagram similar to Figure 1.9, but now the considered isometries would belong to $SE_j^i(n)$ instead of $SE_i^j(n)$.

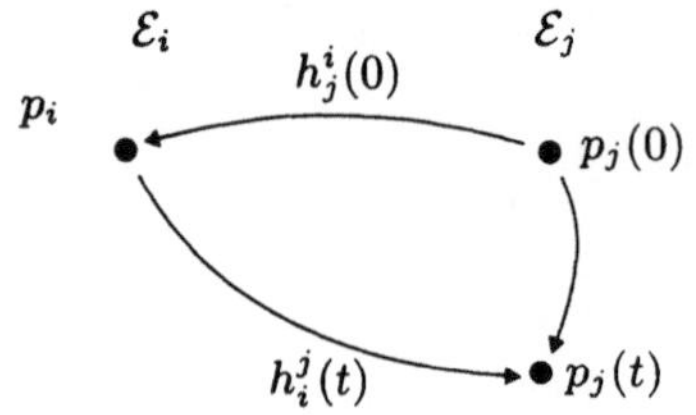


**Fig. 1.11.** $t_i^j = \frac{d}{dt}\left(h_i^j(t) \; o \; h_j^i(0)\right)\big|_0$.

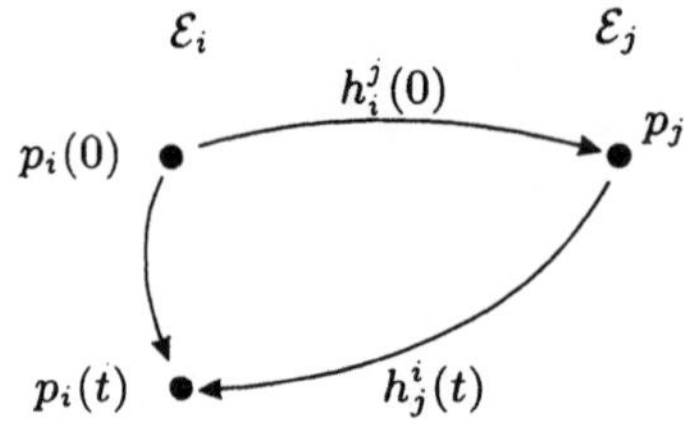


**Fig. 1.12.** $t_j^i = \frac{d}{dt}\left(h_j^i(t) \; o \; h_i^j(0)\right)\big|_0$.

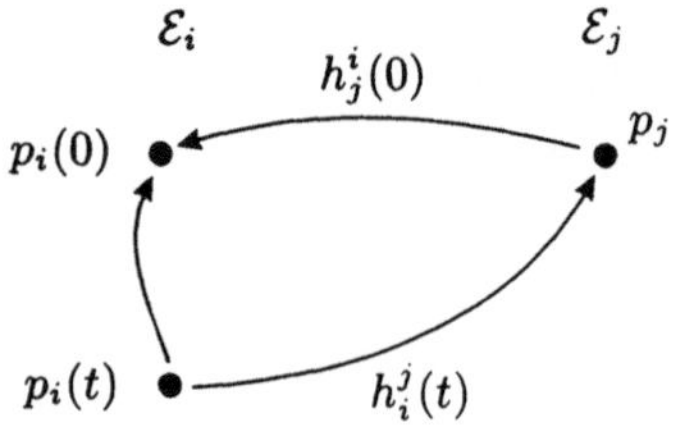


**Fig. 1.13.** $t_i^{i,j} = \frac{d}{dt}\left(h_j^i(0) \; o \; h_i^j(t)\right)\big|_0$.

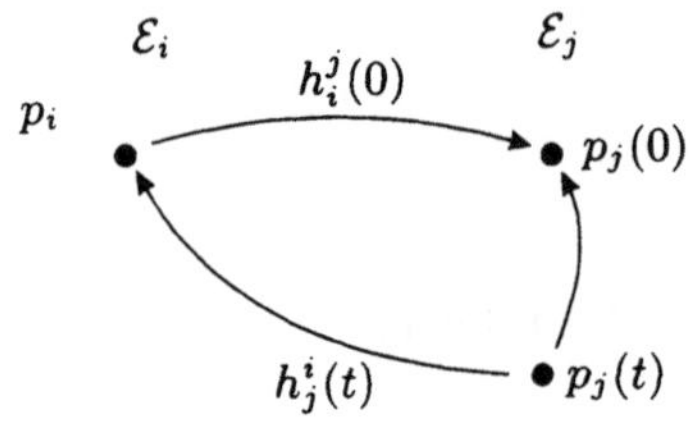


**Fig. 1.14.** $t_j^{j,i} = \frac{d}{dt}\left(h_i^j(0) \; o \; h_j^i(t)\right)\big|_0$.

### 1.3.4 Numerical representation of twists

We have seen in Sect. 1.2.1 that once we have chosen coordinates, we can relate a relative configuration $h_i^j \in SE_i^j(n)$ to a matrix belonging to $SE(3)$. We can associate the relative position $h_i^j \in SE_i^j(n)$ with the matrix $H_i^j$, which corresponds to the change of coordinates from $\Psi_i$ to $\Psi_j$.

Now consider a point $p \in \mathcal{E}_i$. If we take the numerical representation $P^i$ of this point $p$ using the coordinates $\Psi_i$, we clearly have that $\dot{P}^i = 0$ since $p$ is a fixed point of $\mathcal{E}_i$. We can then write the numerical representation of this point in the coordinate $\Psi_j$ fixed with $\mathcal{E}_j$. Using the coordinate-change matrix $H_i^j := \psi_j \; o \; h_i^j \; o \; \psi_i^{-1}$, we obtain:

$$P^j = H_i^j P^i$$

Differentiating over time, we have:

$$\dot{P}^j = \dot{H}_i^j P^i$$

where we used the fact that $\dot{P}^i = 0$ and consider the homogeneous coordinates introduced in Sect. 1.1 for points:

$$P^k := \begin{pmatrix} p^k \\ 1 \end{pmatrix} \qquad k = i, j$$

Before proceeding, we need a result which is well known and useful for what follows.

**Table 1.1.** The possible forms of twists

| $t$ | $\tilde{T}$ | $\dot{H}_i^j$ | | Notation |
|---|---|---|---|---|
| $t_i^{i,j} \Rightarrow$ | $\tilde{T} := H_j^i \dot{H}_i^j$ | $\dot{H}_i^j = H_i^j \tilde{T}$ | $\dot{P}^j = H_i^j(\tilde{T}P^i)$ | $\tilde{T}_i^{i,j}$ |
| $t_i^{j,j} \Rightarrow$ | $\tilde{T} := \dot{H}_i^j H_j^i$ | $\dot{H}_i^j = \tilde{T} H_i^j$ | $\dot{P}^j = \tilde{T}(H_i^j P^i)$ | $\tilde{T}_i^{j,j}$ |

**Theorem 1.3.7.** *Given a matrix $H \in SE(3)$ differentiabily depending on a scalar $t$, we have that:*

$$\dot{H}H^{-1}, H^{-1}\dot{H} \in se(3).$$

*where*

$$se(3) := \left\{ \begin{pmatrix} \tilde{\omega} & v \\ 0 & 0 \end{pmatrix} \in \mathbb{R}^{4\times 4} \; ; \; \omega, v \in \mathbb{R}^3 \right\}$$

*and $\tilde{\omega}$ is the skew-symmetric matrix such that for any vector $x$, $\tilde{\omega}x = \omega \wedge x$.*

The set of matrices $se(3)$ which result from the product of a matrix belonging to $SE(3)$ times the derivative of its inverse (or vice versa) is called the Lie algebra of $SE(3)$. The elements belonging to $se(3)$ are the numerical representative of the twists in $\imath$–frame or in $\jmath$–frame. We will use the following notation:

$$T := \begin{pmatrix} \omega \\ v \end{pmatrix} \Rightarrow \tilde{T} = \begin{pmatrix} \tilde{\omega} & v \\ 0 & 0 \end{pmatrix} \in se(3) \tag{1.24}$$

The vector $T$ is the vector of Plücker coordinates of the twist $\tilde{T}$, and we will abusively call this a twist as well. We therefore have a vector representation of a twist and a matrix representation: $T \in \mathbb{R}^6$ and $\tilde{T} \in \mathbb{R}^{4\times 4}$ respectively. For the twist of $\mathcal{E}_i$ with respect to $\mathcal{E}_j$ we consider $\dot{P}^j = \dot{H}_i^j P^i$ and we have the two possibilities reported in Table 1.1. Furthermore

$$\dot{P}^j = H_i^j(\tilde{T}P^i) \Rightarrow \begin{pmatrix} \dot{p}^j \\ 0 \end{pmatrix} = \begin{pmatrix} R_i^j & p_i^j \\ 0 & 1 \end{pmatrix} \begin{pmatrix} \tilde{\omega} & v \\ 0 & 0 \end{pmatrix} \begin{pmatrix} p^i \\ 1 \end{pmatrix}$$
$$\Rightarrow \dot{p}^j = R_i^j(\omega \wedge p^i) + R_i^j v$$

and

$$\dot{P}^j = \tilde{T}(H_i^j P^i) \Rightarrow \begin{pmatrix} \dot{p}^j \\ 0 \end{pmatrix} = \begin{pmatrix} \tilde{\omega} & v \\ 0 & 0 \end{pmatrix} \begin{pmatrix} R_i^j & p_i^j \\ 0 & 1 \end{pmatrix} \begin{pmatrix} p^i \\ 1 \end{pmatrix}$$
$$\Rightarrow \dot{p}^j = \omega \wedge (R_i^j p^i + p_i^j) + v$$

*Remark 1.3.2 (Interpretation twists).* The rotational part is the usual angular rotation. The linear velocity part of the twist $T_j^{i,i}$, $v = p_j^i \wedge \omega_j^{i,i} + \dot{p}_j^i$ is the velocity of a point $q$ that is rigidly connected to $\Psi_j$ and passing through the origin of $\Psi_i$, and NOT the velocity of the origin of $\Psi_j$ with respect to $\Psi_i$: this is NOT a geometric entity and it would be conceptually wrong !!

**Changes of coordinates for twists: the adjoint representation.** The twists in $i$–frame and in $j$–frame are related by a linear map called *hybrid adjoint mapping* which was defined in Def. 1.2.9. To find a numerical representation of this map, just look at Table 1.1 and consider the relation between $t_i^{i,j}$ and $t_i^{j,j}$ reported in Eq. (1.22):

$$t_i^{j,j} := t_i^j = Ad_{h_i^j} t_i^{i,j} \tag{1.25}$$

Numerical representations of $t_i^{i,j}$ and $t_i^{j,j}$ are $\tilde{T}_i^{i,j}$ and $\tilde{T}_i^{j,j}$ respectively, as defined in Table 1.1. This implies that the numerical representation of the Adjoint is such that:

$$Ad_{H_i^j} : se(3) \to se(3); \tilde{T} \mapsto H_i^j \tilde{T} H_j^i$$

Since this map is linear, the question arises how we can express the previous map as a matrix operator on twists as expressed as six-dimensional vectors $T$ as defined in Eq. (1.24). This can be done with a straightforward calculation. Starting from the previous definition

$$\dot{H}_i^j = \tilde{T}_i^{j,j} H_i^j = H_i^j \tilde{T}_i^{i,j} \Rightarrow \tilde{T}_i^{j,j} = H_i^j \tilde{T}_i^{i,j} H_j^i \tag{1.26}$$

which represents the mapping of twists we can write out the matrices using their form as seen in Theorem 1.3.7 and obtain:

$$\begin{pmatrix} \tilde{\omega}_i^{j,j} & v_i^{j,j} \\ 0 & 0 \end{pmatrix} = \begin{pmatrix} R_i^j & p_i^j \\ 0 & 1 \end{pmatrix} \begin{pmatrix} \tilde{\omega}_i^{i,j} & v_i^{i,j} \\ 0 & 0 \end{pmatrix} \begin{pmatrix} R_j^i & -R_j^i p_i^j \\ 0 & 1 \end{pmatrix}$$

which implies after few steps[10]

[10] Note that for any orthonormal matrix $R$, the following identity holds: $R\tilde{\omega}R^T = (R\omega)\tilde{}$ where the operator 'tilde' yields a skew symmetric matrix that corresponds to its argument.

$$
\begin{aligned}
&\begin{pmatrix} \tilde{\omega}_i^{j,j} & v_i^{j,j} \\ 0 & 0 \end{pmatrix} = \begin{pmatrix} R_i^j & p_i^j \\ 0 & 1 \end{pmatrix} \begin{pmatrix} \tilde{\omega}_i^{i,j} R_j^i & \tilde{\omega}_i^{i,j} R_j^i p_i^j + v_i^{i,j} \\ 0 & 0 \end{pmatrix} = \\
&\quad \begin{pmatrix} R_i^j \tilde{\omega}_i^{i,j} R_j^i & -(R_i^j \tilde{\omega}_i^{i,j} R_j^i) p_i^j + R_i^j v_i^{i,j} \\ 0 & 0 \end{pmatrix} = \\
&\quad \begin{pmatrix} (R_i^j \omega_i^{i,j})^{\sim} & -(R_i^j \omega_i^{i,j})^{\sim} \; p_i^j + R_i^j v_i^{i,j} \\ 0 & 0 \end{pmatrix} = \\
&\quad \begin{pmatrix} (R_i^j \omega_i^{i,j})^{\sim} & p_i^j \wedge (R_i^j \omega_i^{i,j}) + R_i^j v_i^{i,j} \\ 0 & 0 \end{pmatrix} = \\
&\qquad\qquad\qquad \begin{pmatrix} (R_i^j \omega_i^{i,j})^{\sim} & \tilde{p}_i^j R_i^j \omega_i^{i,j} + R_i^j v_i^{i,j} \\ 0 & 0 \end{pmatrix}
\end{aligned}
$$

Finally we have obtained

$$\tilde{T}_i^{j,j} = \begin{pmatrix} \tilde{\omega}_i^{j,j} & v_i^{j,j} \\ 0 & 0 \end{pmatrix} = \begin{pmatrix} (R_i^j \omega_i^{i,j})^{\sim} & \tilde{p}_i^j R_i^j \omega_i^{i,j} + R_i^j v_i^{i,j} \\ 0 & 0 \end{pmatrix} = H_i^j \tilde{T}_i^{i,j} H_j^i$$

which implies:

$$T_i^{j,j} = \begin{pmatrix} \omega_i^{j,j} \\ v_i^{j,j} \end{pmatrix} = \begin{pmatrix} R_i^j & 0 \\ \tilde{p}_i^j R_i^j & R_i^j \end{pmatrix} \begin{pmatrix} \omega_i^{i,j} \\ v_i^{i,j} \end{pmatrix}$$

We indicate this now as:

$$T_i^{j,j} = Ad_{H_i^j} T_i^{i,j}$$

where

$$Ad_{H_i^j} := \begin{pmatrix} R_i^j & 0 \\ \tilde{p}_i^j R_i^j & R_i^j \end{pmatrix}$$

is the matrix representation of the linear map we where looking for, and in the Lie group setting it is called the *adjoint representation of the Lie group* $SE(3)$ *on* $se(3)$ [18, 1].
The matrix $Ad_{H_i^j}$ is dependent on $H_i^j$ and this implies that if the relative position of $\mathcal{E}_i$ changes with respect to $\mathcal{E}_j$, $H_i^j$ will change and as a consequence, the matrix $Ad_{H_i^j}$ will be time varying. We will need the time derivative of $Ad_{H_i^j}$ for expressing the dynamics of a rigid body.

*Remark 1.3.3.* It is important to note that the adjoint map corresponds to a positive ($\sigma = +1$) collineation of lines, as shown in Eq. (1.10). This shows a first relation between Lie group elements of $se(3)$ and the projective geometry of lines.

We can generalize this result by looking at more Euclidean spaces belonging to a Euclidean system. The main result follows:

**Theorem 1.3.8 (Twist composition).** *Consider a Euclidean System $S^m(n)$ and two positive integers $k, i$ such that $k < i < m$. We find*

$$t_i^k = t_{k+1}^k + Ad_{h_{k+1}^k} t_{k+2}^{k+1} + Ad_{h_{k+2}^k} t_{k+3}^{k+2} + \ldots + Ad_{h_{i-1}^k} t_i^{i-1}$$

*where $h_q^l \in SE_q^l(n)$ is the relative position of $q$ with respect to $l$ and $t_q^l \in se_l(n)$ is the intrinsic twist corresponding to the motion of $q$ with respect to $l$.*

**Proof.** Define any set of coordinates $\Psi_l$ with $l = i, \ldots, m$ for the Euclidean spaces belonging to the Euclidean system. If we do this, we can associate with each $h_i^j \in SE_i^j(n)$ a matrix $H_i^j \in SE(n)$. In the chosen coordinates, the twist $t_i^k$ is represented by $\tilde{T}_i^k \in se(n)$, which we have seen defined as:

$$\tilde{T}_i^k = \dot{H}_i^k H_k^i \tag{1.27}$$

Furthermore, it is possible to see that

$$H_k^i = H_{i-1}^i H_{i-2}^{i-1} \ldots H_{k+1}^{k+2} H_k^{k+1}$$

By substituting the last equation in Eq. (1.27), we obtain:

$$\begin{aligned}\tilde{T}_i^k &= \frac{d}{dt}(H_{k+1}^k H_{k+2}^{k+1} \ldots H_{i-1}^{i-2} H_i^{i-1})(H_{i-1}^i H_{i-2}^{i-1} \ldots H_{k+1}^{k+2} H_k^{k+1}) = \\ &\dot{H}_{k+1}^k H_k^{k+1} + H_{k+1}^k (\dot{H}_{k+2}^{k+1} H_{k+1}^{k+2}) H_k^{k+1} + \ldots + H_{i-1}^k (\dot{H}_i^{i-1} H_{i-1}^i) H_k^{i-1} = \\ &\tilde{T}_{k+1}^k + H_{k+1}^k \tilde{T}_{k+2}^{k+1} H_k^{k+1} + \ldots + H_{i-1}^k \tilde{T}_i^{i-1} H_k^{i-1}\end{aligned} \tag{1.28}$$

and going to the vector representation of twists using what is presented on page 24 we obtain:

$$T_i^k = T_{k+1}^k + Ad_{H_{k+1}^k} T_{k+2}^{k+1} + Ad_{H_{k+2}^k} T_{k+3}^{k+2} + \ldots + Ad_{H_{i-1}^k} T_i^{i-1}. \tag{1.29}$$

which proves the result for any coordinate system. ■

In the following chapters, bond graphs will be largely used. The following result will turn out to be extremely useful to describe interconnections and relative motions in the bond graph framework.

**Theorem 1.3.9 (Twists reference).** *Given two Euclidean spaces $\mathcal{E}_i$ and $\mathcal{E}_j$ that are moving with respect to each other with a relative twist $t_i^j \in se_j(n)$, if we consider any other Euclidean space $\mathcal{E}_k$ the following identity holds:*

$$t_i^j = Ad_{h_k^j}(t_i^k - t_j^k). \tag{1.30}$$

**Proof.** From Theorem 1.3.8 we have that:

$$t_i^j = t_k^j + Ad_{h_k^j} t_i^k.$$

Furthermore, by the relations of Eq. (1.22) and Eq. (1.23), we have that:

$$t_i^j = -Ad_{h_k^j} t_j^k + Ad_{h_k^j} t_i^k.$$

Finally, using the linearity of the operator $Ad_{h_k^j}$, we obtain the result. ■

**Adjoint time derivative.** In order to find an expression of the adjoint time derivative for time varying

$$H_i^j = \begin{pmatrix} R_i^j & p_i^j \\ 0 & 1 \end{pmatrix}$$

we should take a look at

$$T_i^{i,j} = H_j^i \dot{H}_i^j = \begin{pmatrix} \tilde{\omega}_i^{i,j} & v_i^{i,j} \\ 0 & 0 \end{pmatrix} = \begin{pmatrix} R_j^i \dot{R}_i^j & R_j^i \dot{p}_i^j \\ 0 & 0 \end{pmatrix}$$

where we can deduce

$$\tilde{\omega}_i^{i,j} = R_j^i \dot{R}_i^j,\; v_i^{i,j} = R_j^i \dot{p}_i^j \text{ or } \tilde{v}_i^{i,j} = R_j^i \dot{\tilde{p}}_i^j R_i^j$$

By definition

$$\left(\dot{Ad}_{H_i^j}\right) = \begin{pmatrix} \dot{R}_i^j & 0 \\ \dot{\tilde{p}}_i^j R_i^j + \tilde{p}_i^j \dot{R}_i^j & \dot{R}_i^j \end{pmatrix} = \begin{pmatrix} \dot{R}_i^j & 0 \\ R_i^j \underbrace{R_j^i \dot{\tilde{p}}_i^j R_i^j}_{\tilde{v}_i^{i,j}} + \tilde{p}_i^j \dot{R}_i^j & \dot{R}_i^j \end{pmatrix} =$$

$$\begin{pmatrix} R_i^j & 0 \\ \tilde{p}_i^j R_i^j & R_i^j \end{pmatrix} \begin{pmatrix} \tilde{\omega}_i^{i,j} & 0 \\ \tilde{v}_i^{i,j} & \tilde{\omega}_i^{i,j} \end{pmatrix} = Ad_{H_i^j}\; ad_{T_i^{i,j}}$$

where

$$ad_{T_i^{k,j}} := \begin{pmatrix} \tilde{\omega}_i^{k,j} & 0 \\ \tilde{v}_i^{k,j} & \tilde{\omega}_i^{k,j} \end{pmatrix} \tag{1.31}$$

is called *the adjoint representation of the Lie algebra* [18]

$$ad_{T_i^{k,j}} := \begin{pmatrix} \tilde{\omega}_i^{k,j} & 0 \\ \tilde{v}_i^{k,j} & \tilde{\omega}_i^{k,j} \end{pmatrix}$$

Note that the matrix $ad_T$ is a function of a twist $T$ and not of a relative position $H$ as $Ad_H$.

Since we have $ad_{T_i^{k,j}}$ we could now investigate what $ad_{T_i^{l,j}}$ looks like. This can be answered using the following theorem.

**Theorem 1.3.10 (Coordinates changes of $ad_T$).** *Changes of coordinates of the adjoint time derivative due to changes of coordinates of the argument twist are related as follows:*

$$ad_{T_i^{l,j}} = Ad_{H_k^l}\; ad_{T_i^{k,j}}\; Ad_{H_l^k}$$

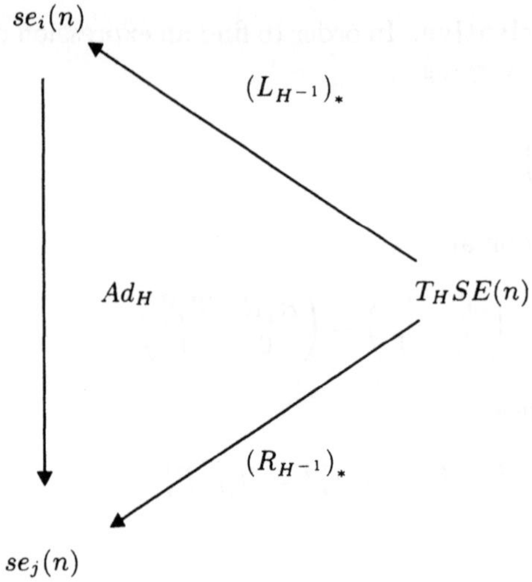


**Fig. 1.15.** The Lie group commutation diagram.

**Proof.**

$$Ad_{H_k^l}\, ad_{T_i^{k,j}}\, Ad_{H_l^k} = \begin{pmatrix} R_k^l & 0 \\ \tilde{p}_k^l R_k^l & R_k^l \end{pmatrix} ad_{T_i^{k,j}} \begin{pmatrix} R_l^k & 0 \\ \tilde{p}_l^k R_l^k & R_l^k \end{pmatrix} =$$

$$\cdots \begin{pmatrix} R_k^l \tilde{\omega}_i^{k,j} R_l^k & 0 \\ R_k^l \tilde{v}_i^{k,j} R_l^k & R_k^l \tilde{\omega}_i^{k,j} R_l^k \end{pmatrix} = ad_{T_i^{l,j}}$$

■

### 1.3.5 The link with standard Lie groups

It is possible to link the material presented as a completely coordinate-free treatment to the purely algebraical Lie group approach.

**Theorem 1.3.11.** *If we choose as a reference $r_i^j = \Psi_j^{-1}\ o\ \Psi_i$, with the coordinate functions $\Psi_k$ $k = 1, \ldots, n$ then $Ad_{r_i^j}$. $Ad_{r_j^i}$ as defined in Def. 1.2.9 will be represented by the identity, and all the maps on the right-hand side of the diagram in Figure 1.9 will have the same representations as the symmetrically corresponding maps on the left.*

**Proof.** Can be shown by using a representation with coordinates. ■

Note that the choice of $r_i^j = \Psi_j^{-1}\ o\ \Psi_i$ has a very straightforward explanation: we consider as a reference-relative position that position in which the coordinate frames $\Psi_i$ and $\Psi_j$, coincide. This choice is useful for practice but

NOT intrinsic and depends on the choice of coordinate frames for an initial relative position.

By using coordinates, we achieve that the commutation diagram of Figure 1.9 "folds" its right and left parts on one another to give the standard Lie group commutation diagram reported in Figure 1.15, where $H = \Psi_j \, o \, h_j \, o \, \Psi_j^{-1} = \Psi_i \, o \, h_i \, o \, \Psi_i^{-1} = \Psi_i \, o \, h_i^j \, o \, \Psi_j^{-1}$.

## 1.4 Wrenches

In the previous section we analyzed the relative motion of two Euclidean spaces. This resulted in a presentation of the twists as a kind of generalized, intrinsically defined, configuration-independent velocities. By considering the dual entities of velocities, we can directly define forces between two Euclidean spaces. Note that the following treatment will not look at the effect of an applied force but only at its kinematics. Hence, we do not consider dynamics at this stage.

### 1.4.1 Power and Power conjugate variables

In Sect. 1.3 we have considered relative velocities of the two Euclidean spaces $\mathcal{E}_i$ and $\mathcal{E}_j$ as elements of the tangent bundle $TSE_i^j(n)$ or $TSE_j^i(n)$. In a point $h_i^j \in SE_i^j(n)$ we can consider the co-vectors belonging to $T^*_{h_i^j} SE_i^j(n)$ which, applied to elements of $T_{h_i^j} SE_i^j(n)$, result in a scalar. This dual structure is exactly the one we need to define forces and their duality with velocities. A force can in fact be seen as a 1−form (see Sect. A.5) which, when it is applied to a vector that corresponds to a velocity, results in a scalar, which, in its turn, corresponds to the power supplied by the entity that generates the force.

It is well known that forces and velocities transform differently under coordinate changes which are not orthonormal and, therefore, they are different kinds of tensors (see Sect. A.5.1).

*Remark 1.4.1 (Geometry of a linear force).* It is obvious to visualize a velocity vector of a point mass as a vector with a direction and magnitude. Not many people realize that a force (a co-vector or one-form) can be visualized in this way as well, as only the existence of the Euclidean metric allows it. Without a metric, it would not be possible to visualize a co-vector, like a force, as a vector, it would then look like a hyperplane!

For all the reasons mentioned above, we consider a generalized force between two Euclidean spaces $\mathcal{E}_i$ and $\mathcal{E}_j$ as an element of $T^*SE_i^j(n)$ or, equivalently as an element of $T^*SE_j^i(n)$.

It is possible to consider a diagram dual to the one of Figure 1.9 and to define the dual spaces of $se_i(n)$ and $se_j(n)$, which will be indicated with

$se_i^*(n)$ and $se_j^*(n)$. We could intrinsically define elements that correspond to the dual vector spaces of $se_i(n)$ and $se_j(n)$, which are circled in Figure 1.9. Elements belonging to the intrinsic $se_i^*(n)$ and $se_j^*(n)$ will be called *wrenches.*

### 1.4.2 Numerical representation of Wrenches

Since as we already mentioned wrenches are dual to twists, if we choose for twists a six-dimensional column vector representation as

$$T = \begin{pmatrix} \omega \\ v \end{pmatrix},$$

we can choose for wrenches a six-dimensional row vector representation.

Using a dual basis for the wrenches, one thus obtains a vector representation of the wrenches:

$$W = \begin{pmatrix} m & f \end{pmatrix}$$

where $m$ represents a torque and $f$ a linear force. Thanks to the duality structure, the power may be expressed as:

$$\text{Power} = WT = m\omega + fv$$

**Changes of coordinates for Wrenches.** As the wrenches belong to the dual space of twists, the change of coordinates for twists induces the adjoint change of wrenches. Since power is a scalar and as such independent of coordinate systems, we should have

$$W^j T^{j,\cdot}_{\cdot} = W^i T^{i,\cdot}_{\cdot}.$$

Hence in the numerical representation, the transformation of wrenches is given by the transposition of the adjoint mapping:

$$(W^i)^T = Ad^T_{H_i^j}(W^j)^T$$

It is important to note, as shown in Figure 1.16, that if the mapping $Ad_{H_i^j}$ mapped twists from $\Psi_i$ to $\Psi_j$, the transposed maps wrenches in the opposite direction: from $\Psi_j$ to $\Psi_i$ ! This is because wrenches are duals of twists and they are pulled back.

## 1.5 The relation between Lie groups and Screw theory

It is now possible to formally relate Lie groups theory [17] to the the theory of Screws [12]. This section reports an extension of the results first presented in [7].

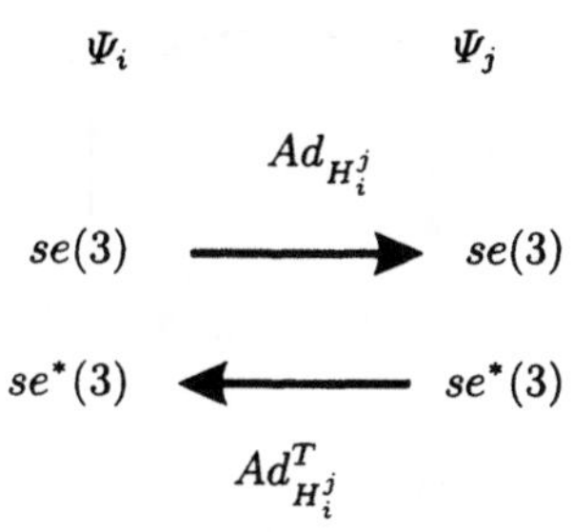


**Fig. 1.16.** The direction of the adjoint mappings

### 1.5.1 Bi-invariant forms on $se(3)$ and reciprocity of twists

It has been shown in [21] that the space of bi-invariant, and therefore intrinsic, 2-forms defined on $se_i(3)$ is a linear combination of the hyperbolic metric $\bar{\mathfrak{H}}_{ij}$ and the Killing form $\bar{\mathfrak{K}}_{ij}$. In the coordinates we used so far, their numerical representations are:

$$\bar{\mathfrak{H}} = \frac{1}{2}\begin{pmatrix} 0 & I_3 \\ I_3 & 0 \end{pmatrix} \text{ and } \bar{\mathfrak{K}} = -4\begin{pmatrix} I_3 & 0 \\ 0 & 0 \end{pmatrix}$$

And a general bi-invariant metric is given by:

$$\mathfrak{B} = h\bar{\mathfrak{H}} + k\bar{\mathfrak{K}} \qquad h, k \in \mathbb{R}.$$

Among all the possible metrics, there are two which have special properties and allow us to directly relate Lie groups to the theory of screws.

If we consider the kernel of $\mathfrak{B}$, it is different than the trivial zero element of $se_i(3)$ iff $h = 0$. A proportional element of the Killing form has therefore a first special property. Furthermore, among all the $\mathfrak{B}$s there is just one metric which is generated by an involutive map, which means that an expression of its inverse is equal to the form itself and this is obtained when $k = 0$ and $h = 2$. These involutivity property results are fundamental to relate Lie groups to screws and corresponds to the self-duality of lines.

We indicate $\mathfrak{H} = 2\bar{\mathfrak{H}}$ and $\mathfrak{K} = -\frac{1}{4}\bar{\mathfrak{K}}$. These two special 2-forms are fundamental for what follows.

### 1.5.2 Identification of twists and wrenches

Using the involutive map $\mathfrak{H}$ and the duality product on $se(3)$, one may identify the twists with the wrenches. In the sequel we shall use this identification in order to identify both twists and wrenches with screws.

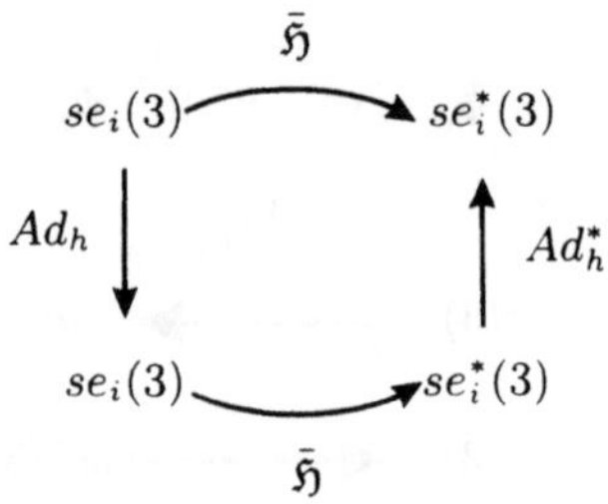


**Fig. 1.17.** The contravariant map

Therefore one may define the following map, denoted by $\mathfrak{H}^{\#}$ from twists $T \in se(3)$ to wrenches $W = \mathfrak{H}^{\#}(T)$, as follows:

$$\langle W, T' \rangle = \mathfrak{H}(T, T') \quad \forall T' \in se(3) \tag{1.32}$$

It is remarkable that, using this identification, one can identify the adjoint mapping with its contra-gradient map, in other words, the coordinate change of twists and wrenches (see Figure 1.17). Furthermore, as we can give a vector interpretation of a force as explained in Remark. 1.4.1, p.29, it is possible to give a screw interpretation of a wrench using $\mathfrak{H}$, as will be shown later.

It is even possible to define a bracket on $se_i^*(3)$ by:

$$[w_1, w_2] := [\mathfrak{H}^{-1} w_1, \mathfrak{H}^{-1} w_2] \qquad w_1, w_2 \in se_i^*(n)$$

The relation between twists (and wrenches) and lines in $\mathcal{P}^3$ is stated by the classical Chasles (and Poinsot) theorem. For expressing these theorems, some subspaces and subsets of $se_i(3)$ need to be defined using the bi-invariant forms [13].

### 1.5.3 Decomposition of $se_i(3)$

**Definition 1.5.1 (Self reciprocal sub set).** *We call self-reciprocal subset the following subset of* $se_i(3)$*:*

$$R := \{t^i \in se_i(3) \ s.t\ \mathfrak{H}_{ij} t^i t^j = 0\}.$$

**Definition 1.5.2 (Axial subspace).** *We call an axial subspace the following subspace of* $se_i(3)$*:*

$$\mathfrak{a} := \{t^i \in se_i(3) \ s.t.\ \mathfrak{K}_{ij} t^i t^j = 0\}.$$

It is the biggest subspace whose numerical representation is invariant to changes of the origin of the coordinates system.

**Definition 1.5.3 (Lines subset).** *We call a lines subspace the following subset of* $se_i(3)$*:*

$$L := R - \mathfrak{a}$$

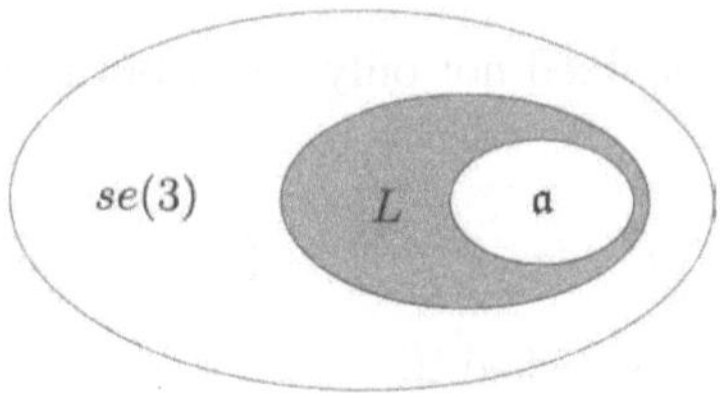


**Fig. 1.18.** The relation of the sets in $se(3)$.

Note that $R$ and $L$ are NOT subspaces, but $\mathfrak{a}$ is. Furthermore, $\mathfrak{a} \subset R$.

**Theorem 1.5.1 (Geometrical interpretation of $L$).** *Any element $\begin{pmatrix}\omega & r \wedge \omega\end{pmatrix}^T \in L$ is characterized by the fact that it leaves a line of $\mathcal{E}$ invariant in the direction $\alpha\omega$ and furthermore this line passes through $r$.*

**Proof.** Using the coordinates we introduced previously, we know that the form of an element of $L$ will be:

$$l = \begin{pmatrix}\omega \\ v\end{pmatrix} \in L \text{ where } \omega \neq 0$$

Since $\omega \neq 0$ due to the fact that $L = R - \mathfrak{a}$, we can consider in these coordinates the orthogonal complement $\omega^\perp$ of $\omega$. The fact that $l \in R$ results in the algebraic condition

$$\omega^T v = 0$$

in these coordinates. This implies that $v \in \omega^\perp$ and therefore there exist a unique $r \in \mathbb{R}^3$ such that $v = r \wedge \omega$. It is therefore possible to express ANY element of $L$ in the following form:

$$\begin{pmatrix}\omega \\ r \wedge \omega\end{pmatrix} \tag{1.33}$$

where $\omega \neq 0$ and corresponds to a line as reported in Figure 1.2. Consider now the motion of $\mathcal{E}_i$ with respect to $\mathcal{E}_j$. Take any fixed point $p \in \mathcal{E}_i$. Once we have chosen any orthonormal coordinate system $\Psi_i$ for $\mathcal{E}_i$ and $\Psi_j$ for $\mathcal{E}_j$, we can express the change of homogeneous coordinates from $\Psi_i$ to $\Psi_j$ with an homogeneous matrix $H_i^j \in SE(3)$, as usual. If we indicate the homogeneous coordinates of $p$ in $\Psi_i$ with $P^i$ and in $\Psi_j$ with $P^j$, we have:

$$\dot{P}^j = \dot{H}_i^j P^i = \tilde{T}_i^{j,j} H_i^j P^i = \tilde{T}_i^j P^j \tag{1.34}$$

where we have used the definition of twists reported in Table 1.1. Since the twist belongs to $L$ we have:

$$\tilde{T}_i^j = \begin{pmatrix}\tilde{\omega}_i^j \\ r \wedge \omega_i^j\end{pmatrix}$$

If the point $p \in \mathcal{E}_i$ is to be fixed not only for $\mathcal{E}_i$ but also for $\mathcal{E}_j$, we must have from Eq. (1.34):

$$0 = \dot{P}^j = \tilde{\omega}_i^j p^j + r \wedge \omega_i^j = \omega_i^j p^j + r \wedge \omega_i^j$$

The previous condition is satisfied if:

$$p^j = r + \alpha \omega_i^j \qquad \alpha \in \mathbb{R}$$

which is a line passing from $r$ and directed along $\omega_i^j$. ■

From the previous theorem, we can conclude that a twist $l \in L$ therefore corresponds to a rotation around this line.

*Remark 1.5.1 (Orientation of space).* From what we just stated, it follows that we can associate a pure rotation with a line because it is the set which is left invariant by a rotation. Nevertheless, this set is left invariant for any rotation around that line in both possible directions and of any magnitude. We have therefore a way to associate a rotation with a vector $\omega$ which gives the orientation of the line. Its magnitude can then be associated with the angular speed.

Nevertheless, we still have two possible choices in orienting the invariant line. These two choices correspond to the orientation of space.

Suppose that you are in front of a mirror and consider a rotation around a line which goes from our eyes to the mirror. If we consider a clockwise rotation associated with a vector directed toward the mirror, we have chosen a right-handed space; if we associate a clockwise rotation with a vector coming from the mirror toward us we have a left-handed orientation. Notice that a motion toward the mirror looked through the mirror corresponds to a motion toward us which shows that a mirror changes the orientation of space, and that is why this is called a reflection. This can be obviously associated with the usual conception of orientation: associating to the $z$ axis of a coordinate frame the orientation of the shortest rotation around $z$, bringing $x$ toward $y$.

**Theorem 1.5.2 (Geometrical interpretation of $\mathfrak{a}$).** *Any element* $\begin{pmatrix}0 & v\end{pmatrix}^T \in \mathfrak{a}$ *leaves free vectors invariant.*

**Proof.** It is easy to see that the previous twist leaves all points at infinity ($P = (a, y, z, 0)$) invariant and therefore also vectors lying on the improper hyperplane correspond to free vectors. ■

From the previous theorem, we can conclude that a twist $a \in \mathfrak{a}$ corresponds to a translation parallel to $v$.

Based on the special metrics $\mathfrak{H}$ and $\mathfrak{K}$, we can define an intrinsic surjective, but not injective operation, which is called *polar map*:

$$\mathrm{p} : se_i(3) \to \mathfrak{a} \quad ; \quad t^i \mapsto \mathfrak{H}^{jl} \mathfrak{K}_{ij} t^i$$

Note that we have inverted $\mathfrak{H}$ in order to get a proper tensor map. It is now possible to express Chasles' theorem as follows:

**Theorem 1.5.3 (Chasles).** *For any $t \in se_i(3)$, there exists a $l \in L$ and two constants $\alpha_l, \alpha_{\mathfrak{a}} \in \mathbb{R}$ such that:*

$$t = \alpha_l l + \alpha_{\mathfrak{a}} \mathrm{p}(l) \tag{1.35}$$

*In particular, when $t \in se_i(3) - \mathfrak{a}$, $l$ is unique.*

Furthermore, as a consequence of Theorem 1.5.3, we can always write a twist as:

$$T = \alpha_l \begin{pmatrix} \hat{u} \\ r \wedge \hat{u} \end{pmatrix} + \alpha_{\mathfrak{a}} \begin{pmatrix} 0 \\ \hat{u} \end{pmatrix} \tag{1.36}$$

where $\hat{u}$ is a unit vector. It is now possible to understand better what is said in Remark. 1.3.2, p.24. From the decomposition of Eq. (1.36), we can see that if we take a general twist $T = \begin{pmatrix} \omega \\ v \end{pmatrix}$, the velocity $v$ can be interpreted as a sum of two components which give the velocity of a virtual point that is passing through the coordinate system, which is used to express the twist.

The following theorem is also very important because it allows us to express a pure translation as the sum of two rotations.

**Theorem 1.5.4 (Axial decomposition).** *For any $a \in \mathfrak{a}$, there exist $l_1, l_2 \in L$ such that $a = l_1 - l_2$.*

As a direct consequence of Theorem 1.5.4, we can express any element $a \in \mathfrak{a}$ as:

$$\begin{pmatrix} 0 \\ (r_1 - r_2) \wedge \omega \end{pmatrix} = \begin{pmatrix} \omega \\ r_1 \wedge \omega \end{pmatrix} - \begin{pmatrix} \omega \\ r_2 \wedge \omega \end{pmatrix}.$$

One of the conclusions we can draw from what has been said so far is that:

$$se_i(3) = \mathrm{span}\{L\}$$

and therefore we can get any twist as a linear combination of elements of $L$ which is a known result in the context of screw theory.

It is also possible to analyse the commutativity of pure rotations and translations using the Lie algebra structure of $se(3)$. This can be also used to characterize a screw motion.

**Theorem 1.5.5 (Commutativity of Rotations and Translations).** *The sequence of a pure translation $T_{\mathfrak{a}}$ and a pure rotation $T_L$ is commutative iff the rotation is in the same direction as the translation.*

**Proof.** It is known in Lie group theory that the commutativity can be tested using the Lie algebra commutator. Consider a generic pure rotation as a twist in $L$:

$$T_L := \begin{pmatrix} \omega \\ r \wedge \omega \end{pmatrix} \Rightarrow \tilde{T}_L = \begin{pmatrix} \tilde{\omega} & r \wedge \omega \\ 0 & 0 \end{pmatrix}$$

and a pure translation as a twist belonging to $\mathfrak{a}$:

$$T_{\mathfrak{a}} := \begin{pmatrix} 0 \\ v \end{pmatrix} \Rightarrow \tilde{T}_{\mathfrak{a}} = \begin{pmatrix} 0 & v \\ 0 & 0 \end{pmatrix}$$

The order of the two motions is interchangeable iff their commutator is zero and therefore iff

$$[\tilde{T}_L, \tilde{T}_{\mathfrak{a}}] = \tilde{T}_L \tilde{T}_{\mathfrak{a}} - \tilde{T}_{\mathfrak{a}} \tilde{T}_L = 0$$

substituting we get:

$$[\tilde{T}_L, \tilde{T}_{\mathfrak{a}}] = \begin{pmatrix} \tilde{\omega} & r \wedge \omega \\ 0 & 0 \end{pmatrix} \begin{pmatrix} 0 & v \\ 0 & 0 \end{pmatrix} - \begin{pmatrix} 0 & v \\ 0 & 0 \end{pmatrix} \begin{pmatrix} \tilde{\omega} & r \wedge \omega \\ 0 & 0 \end{pmatrix} = \begin{pmatrix} \tilde{\omega} v & 0 \\ 0 & 0 \end{pmatrix}$$

The previous is clearly equal to zero iff $\tilde{\omega} v = \omega \wedge v = 0$, which is the case for nonzero vectors iff $v$ and $\omega$ are oriented in the same direction. ∎

### 1.5.4 Decomposition of $se_i^*(3)$

The decomposition which has been shown for $se_i(3)$ can be directly mapped to $se_i^*(3)$ defining $R^* = \mathfrak{H}^\#(R), \mathfrak{a}^* = \mathfrak{H}^\#(\mathfrak{a})$ and $L^* = \mathfrak{H}^\#(L)$, where $\mathfrak{H}^\#$ has been defined in Eq. (1.32). The Poinsot theorem is then expressed:

**Theorem 1.5.6 (Poinsot).** *For any $w \in se_i^*(3)$, there exist a $l^* \in L^*$ and two constants $\alpha_l, \alpha_{\mathfrak{a}} \in \mathbb{R}$ such that:*

$$w = \alpha_l l^* + \alpha_{\mathfrak{a}} \mathrm{p}^*(l^*) \tag{1.37}$$

*In particular, when $w \in se_i^*(3) - \mathfrak{a}^*$, $l^*$ is unique.*

A pure moment may be obtained as the sum of two forces.

**Theorem 1.5.7 (Dual axial decomposition).** *For any $a^* \in \mathfrak{a}^*$, there exist $l_1^*, l_2^* \in R^*$ such that $a^* = l_1^* - l_2^*$.*

One of the conclusions we can draw from what has been said so far is that:

$$se_i^*(3) = \mathrm{span}\{L^*\}$$

### 1.5.5 The space of screw vectors

We have seen that any element of $se(3)$ belonging to the lines subset $L$ has a line associated with it, namely the line which stays invariant during this rigid rotation. To do this bijectively, we need to associate with a line a direction and a magnitude which would characterize the angular velocity of the corresponding twist belonging to $L$. In this way, we loose the homogeneous character of lines, and obtain the vectors $\in L$ (or $L^*$). These vectors are called *bound line vectors* (or *rotors*). Conversely, lines (not at infinity) are members of the projectivization[11] of $L$.

Considering the whole set of projective lines, including lines at infinity, and leaving their homogeneous character, the set $R$ of self-reciprocal twists is obtained. The vectors of $R$ are called *line vectors* [13]. Those vectors that correspond to lines at infinity, member of $\mathfrak{a}$, are then called *free line vectors* [13] (or also free vectors). Conversely, projective lines are members of the projectivization of $R$.

Neither $L$ nor $R$ are vector spaces, but $L$ spans a 6-dimensional vector space that corresponds to $se(3)$.

We can now define the vector space of screws

**Definition 1.5.4 (Screw vectors).** *We define the following 6-dimensional vector space the vector space of screws in $\mathcal{E}_i$:*

$$\mathcal{S}_i^V := \{(t, w) \in se_i(3) \times se_i^*(3) \ ; \ w = \mathfrak{H}t\}$$

Note that the previous is indeed a vector space because it has inherited the vector space structure of $se_i(3)$ and/or $se_i^*(3)$. The previous space is also called the space of *motors*.

Geometrical screws, defined by a line and a pitch, are members of the projectivization of $\mathcal{S}_i^V$ [22], which therefore results in a 5-dimensional space (see Sect. 1.1).

**Definition 1.5.5 (Screws).** *We call the space of screws of the space $\mathcal{E}_i$ the following quotient space:*

$$\mathcal{S}_i := \frac{\mathcal{S}_i^V - 0}{\sim}$$

where with $\sim$ we indicate the following similitude relation:

$$s_1 \sim s_2 \Leftrightarrow \exists \alpha \in \mathbb{R} \neq 0 \ ; \ s_1 = \alpha s_2$$

[11] With projectivization is intended the process of considering elements $a, b \in L$ equivalent iff there exist a scalar $\alpha \neq 0$ such that $a = \alpha b$.

### 1.5.6 Operations on screw vectors

Operations on twists or on wrenches, or between wrenches and twists, can now be mapped to operations on screws.

Given two screws $s_1$ and $s_2$ associated with two elements $\alpha_1(t_1, \mathfrak{H}t_1)$, $\alpha_2(t_2, \mathfrak{H}t_2) \in \mathcal{S}^V$, their *reciprocal product* is defined as:

$$s_1 \bigcirc s_2 := t_1^T \bar{\mathfrak{H}} t_2$$

The reciprocal product corresponds to the duality product of twists and wrenches. Let $s_1$ and $s_2$ associated with $t_1 \in se(3)$ and $w_2 \in se(3)^*$ respectively,

$$s_1 \bigcirc s_2 := \langle t_1, w_2 \rangle$$

When $s_1 \bigcirc s_2 = 0$, the two screws are said to be *reciprocal*. Reciprocity of screws is used to express non-working condition. Line vectors are self-reciprocal; this corresponds to the self-duality of lines.

Let us consider the scalar $t_1^T \mathfrak{K} t_2$. The corresponding screw operation is called a *perpendicular product* and it is indicated with:

$$s_1 \Diamond s_2 := t_1^T \mathfrak{K} t_2$$

The vector space $se(3)$ is a Lie algebra and therefore it has an internal product called a commutator which corresponds to the Lie brackets of the associated Lie group $SE(3)$. With the previous association between screws and twists, we can define the *screw vector product* as:

$$s_1 \wedge s_2 := [t_1, t_2]$$

Using the usual coordinates we have:

$$s_1 \wedge s_2 := \left[ \begin{pmatrix} \omega_1 \\ v_1 \end{pmatrix}, \begin{pmatrix} \omega_2 \\ v_2 \end{pmatrix} \right] = \begin{pmatrix} \omega_1 \wedge \omega_2 \\ \omega_1 \wedge v_2 + \omega_2 \wedge v_1 \end{pmatrix}$$

As already mentioned, we can associate with a screw $s$ a screw corresponding to a line at infinity which is called its *polar* and will be indicated with $s'$:

$$s' = \bar{\mathfrak{H}}^{-1} \mathfrak{K} s$$

A summary of these relations is reported in Table 1.2. As screws may be decomposed as sum of bound and free line vectors; operations on screws may be reduced to operations on line vectors. Geometric distances and relations between lines may then be used very efficiently to perform calculation [13] [23]. In particular the reciprocity of line vectors is equivalent to the incidence of their lines (axes).

**Table 1.2.** The relation between Lie groups and Screw operations

| Name | Screw | Lie Group |
|---|---|---|
| Rec. product | $s_1 \bigcirc s_2$ | $t_1^T \bar{\mathfrak{H}} t_2$, $\langle t_1, w_2 \rangle$ |
| Perp. product | $s_1 \Diamond s_2$ | $t_1^T \bar{\mathfrak{K}} t_2$ |
| Screw vect. prod. | $s_1 \wedge s_2$ | $[t_1, t_2]$ |
| Polar | $s' = \mathrm{p}(s)$ | $s' = \mathfrak{H}^{-1} \mathfrak{K} s$ |

## 1.6 Conclusions

This chapter has shown that behind the use of Lie groups for rigid body motions, there are implicit assumptions that, even if more or less "natural", are not intrinsic and just depend on a choice of relative reference between spaces (see Remark. 1.2.2, p.14).

It is always important to pinpoint those assumptions which are not intrinsic: any hypothesis which is not intrinsic can in fact be considered a modeling hypothesis and should be made explicit as such.

Twists have been analyzed in detail and their mappings and relations have been shown. The intrinsic mappings using right translations are the ones which give the motion of a space with respect to an observer space directly and have been denoted with only one subscript and one superscript: $t_i^j$ or $t_j^i$.

A complete formal relation between the Lie group approach and Screw theory has then been presented. This one-to-one relation between the Lie group approach and the screw approach is due to the existence of two bi-invariant forms known as the Killing form and the Hyperbolic or Klijn form.

The space of screw is equivalent in this way to the projectivization of the Lie algebra of rigid body motions.

# 2. Kinematics of Rigid Mechanisms

The word kinematics comes from the Greek $\kappa\upsilon\nu\epsilon\mu\alpha$ (*kinema*) which means motion. This science studies the description of motion as such without considering its cause. A lot of work has been done in this field [10, 11, 13, 1]. The study of kinematics in the framework of bond graphs has been also investigated [24, 1]. The purpose of this chapter is to give a reasonably complete framework of concepts needed in the kinematics of rigid mechanisms.

In the previous chapter we treated the general description of rigid bodies with respect to each other. In order to describe the kinematics of a rigid robot or a generic rigid mechanism, we need to treat the connection between bodies. This connection is characterized by the *kinematic pairs* or *joints*. So far, we have only described impenetrability constraints (see Def. 1.2.6) which are meaningful for the description of collisions/contacts between objects. Kinematic pairs, which will be the subject of Sect. 2.1, are additional constraints representing the connections between bodies. In Sect. 2.2, we will therefore talk about the description of different kinematic structures which will be classified according to their topology. For this purpose, concepts of graph theory will be used. The necessary background is reported in Sect. A.2. For an extensive treatment on graph and matroid theory, we refer to [25].

## 2.1 Kinematic pairs

Intuitively, one would define a kinematic pair (K.P.) as an entity which physically constraints the relative motion of two objects. It is an ideal entity which completely and rigidly opposes certain relative motions between two objects. We can have two extreme cases of a K.P. : one which does not allow any relative motion between the two bodies to which it is connected, and one which does allow motion in any direction. The first one will be called a *rigid kinematic pair* and the second one a *degenerate kinematic pair*. These two cases are clearly of no particular interest since we can consider the two bodies connected to a rigid kinematic pair as one, and a degenerate K.P. can be ignored. In all other cases, we have allowed and forbidden relative motions. Actuators, springs, dampers or any other nodic element can be used to create forces which influence the motion along the allowed directions as it will be seen in Chapter 3.

It is important to realize that a K.P. is an ideal entity and that in reality we could always model the connection between two bodies by means of anisotropic[1] elastic-dissipative elements which would have high stiffness along the forbidden directions.

In this chapter we analyze the kinematics of a mechanism and do not pay any attention to the consequences of forces applied to it.

We could describe a K.P. as the set of allowed relative positions of two objects. Identifying one of the bodies $B_i$ with its Euclidian space $\mathcal{E}_i$ and another one $B_j$ with its Euclidian space $\mathcal{E}_j$, we could then define a kinematic pair as a subset of $SE_i^j(n)$. This choice only uses the structure of a set and as such it is not very powerful for analysis purposes. Furthermore, 1. certain kinds of constraints which are called nonholonomic cannot be treated in this way and therefore another approach follows, 2. we only consider *scleronomic*[2] K.P. which means entities which are not time dependent.

The starting point we choose is a differential one: we define a K.P. by associating the possible relative moving directions the K.P. allows to any relative position of the two bodies. Under certain conditions, this kind of description corresponds to a distribution defined on the manifold of isometries between the two bodies representing the relative positions. From now on we talk either about relative motion of Euclidian spaces or about bodies: when we consider a body $B_i$ we leave implicit that it has a Euclidian space $\mathcal{E}_i$ associated to it.

**Definition 2.1.1 (Kinematic Pair).** *Given two bodies $B_i$ and $B_j$, a kinematic pair between $B_i$ and $B_j$ is a set of vector fields $k_i^j := \{v_1, v_2, \ldots, v_p\}$ on the manifold $SE_i^j(n)$. For $h_i^j \in SE_i^j(n)$, the number of linear independent vectors $v_1(h_i^j), \ldots, v_p(h_i^j)$ in $T_{h_i^j} SE_i^j(n)$ is called degrees of freedom (d.o.f.) of the pair in $h_i^j$.*

Each of the vector fields gives a possible relative moving direction for the two bodies which we express as twists for the relative motion. We suppose that all the possible relative moving directions are given by a positive combination of the vector fields, as it is stated in the next assumption:

> We assume that a kinematic pair $k_i^j$ between two bodies $B_i$ and $B_j$ with relative position $h_i^j \in SE_i^j(n)$ allows a relative motion in the direction $v \in T_{h_i^j} SE_i^j(n)$ iff there exist positive real numbers $\alpha_k$ such that $v = \sum_l \alpha_l v_l$ with $v_l \in k_i^j$.

*Remark 2.1.1.* Note that, the previous definition allows that we could have a kinematic pair permitting a motion in a direction $v \in T_{h_i^j} SE_i^j(n)$ and NOT

[1] Isotropic comes from the Greek *isos* (equal) and *tropos* (rotation) which means direction independent, anisotropic means direction dependent.

[2] Scleronomic comes from the Greek and means rigid. They are given an attribute of "rigid" because the configuration manifold defined later that describes them is not time dependent and therefore the configuration manifold is a "rigid surface".

**Fig. 2.1.** An holonomic joint in a unilateral position.

**Fig. 2.2.** A two dimensional lower pair in a bilateral position.

**Fig. 2.3.** A position dependent kinematic pair.

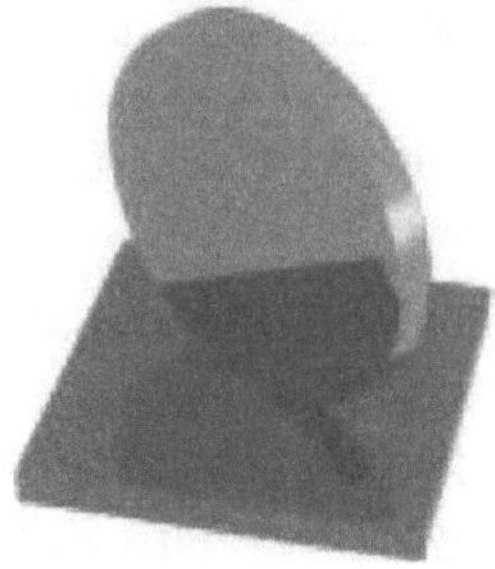

**Fig. 2.4.** A nonholonomic kinematic pair.

in the opposite direction $-v$: the K.P. cannot be characterized in general by a subspace. We call such a configuration $h_i^j$ *unilateral* and *bilateral* if allowing $v$ implies allowing $-v$.

We can graphically represent the possible moving directions as a vectorial convex hull. In general, a unilateral constraint corresponds to the contact between two bodies: penetration of two objects is not possible. Such a unilateral kinematic pair could therefore model the situation reported in Def. 1.2.6. An example of a K.P. in a unilateral configuration is shown in Figure 2.1: the moving ball is touching the socket and cannot move further in that direction. The same joint in a bilateral configuration is reported in Figure 2.2: the ball can move in any direction. Almost all the K.P.s encountered in practice will be bilateral in an open subset $O \subset SE_i^j(n)$ and unilateral in the boundary points of $O$ which will be called *end-stops*. This is what happens for the joint shown in Figure 2.1 and Figure 2.2.

For a bilateral configuration we can associate a complete subspace of $T_{h_i^j} SE_i^j(n)$ to a point $h_i^j \in SE_i^j(n)$, instead of a set of vectors in $T_{h_i^j} SE_i^j(n)$ corresponding to a convex hull. In this case, we can use a distribution which associates to a point a subspace of the tangent space of $SE_i^j(n)$ in that point.

**Definition 2.1.2 (Regular K.P.).** *We say that a kinematic pair $k_i^j$ is regular on $O$, an open set of $SE_i^j(n)$, iff in each point of $O$, $k_i^j$ is bilateral and the corresponding distribution $\Delta_{k_i^j}$ is constant dimensional.*

If we have a regular kinematic pair $k_i^j$ on $O$, then we can analyse involutivity of the corresponding defining distribution $\Delta_{k_i^j}$.

**Definition 2.1.3 (Holonomic and Nonholonomic K.P.).** *We say that a regular kinematic pair on $O$ is holonomic iff its corresponding distribution is involutive and otherwise we say that the regular K.P. is nonholonomic.*

The K.P.s of Figure 2.1, Figure 2.2 and Figure 2.3 are holonomic, but the one of Figure 2.4 is not. This because the coin of Figure 2.4 can rotate around the axis which is orthogonal to the base plane and passes through the contact point and rotate around its symmetry axis, but it cannot slide sidewise; nevertheless, any position can be reached by properly using these two degrees of freedom.

**Theorem 2.1.1 (Kinematic Pair Configuration).** *If there is a regular kinematic pair $k_i^j$ on $O \subset SE_i^j(n)$, then a unique connected sub-manifold $Q_i^j \subset O$ passing through a point in $O$ exists that corresponds to the integral of the involutive closure of the distribution $\Delta_{k_i^j}$ associated to $k_i^j$.*

**Proof.** This is a direct consequence of Frobenius theorem. See e.g. [16, 26]. ∎

Furthermore, if the involutive closure of $\Delta_{k_i^j}$ is of constant dimension $k$, this connected sub-manifold will be of dimension $k$. The largest sub-manifold of $SE_i^j(n)$ that satisfies the previous theorem and passes through the point $h_i^j$ corresponding to the current relative position, corresponds to the relative configuration manifold. We denote it as $Q_i^j(k_i^j)$ or sometimes as $Q_i^j$. For construction, the current relative position $h_i^j(0)$ at the current time instant $t = 0$ should belong to $Q_i^j(k_i^j)$. If there is a degenerate kinematic pair as defined at the beginning of this section, we have $Q_i^j(k_i^j) = SE_i^j(n)$.

*Remark 2.1.2.* If a K.P. is holonomic, the d.o.f. (dimension of the distribution $\Delta_{k_i^j}$) is the same as the dimension of the configuration manifold. In the nonholonomic case, the d.o.f. is LESS than the dimension of the configuration manifold. Consider for example the pair reported in Figure 2.4: the d.o.f. is two, but the configuration manifold is four-dimensional. This is indeed a characterizing feature of nonholonomic pairs.

### 2.1.1 Kinematic pairs actuation

A mechanism can be actuated by imposing wrenches between the bodies of which it is composed. Physically, this is done by putting actuators such as electrical motors or hydraulic pistons between pairs of links. The relative motion of these two links can be constrained by a kinematic pair. The wrenches generated by the actuators work, in general, only in certain directions. We could have actuators generating wrenches not in all the directions allowed by the kinematic pair. Depending on how many directions we can actuate, we can classify the kinematic pair. In order to do this we need to set up a proper formalism which will here be treated.

We can associate to a kinematic pair $k_i^j$ a co-distribution $\breve{\Delta}^*_{k_i^j}$ defined on $SE_i^j(n)$ which represents the possible relative forces that we can apply between bodies $i$ and $j$, by means of actuation. To each relative configuration $h_i^j \in \mathcal{Q}_i^j$, this co-distribution associates a subspace of $T^*_{h_i^j} SE_i^j(n)$. This subspace can then be mapped to $se_i^*(n)$ and $se_j^*(n)$ by the techniques shown in Chapter 1. In general these two subspaces depend on the relative configuration $h_i^j$.

If we consider a regular K.P. (see Def. 2.1.2), we can decompose $\breve{\Delta}^*_{k_i^j}$ in the following way:

$$\breve{\Delta}^*_{k_i^j} = \bar{\Delta}^*_{k_i^j} \oplus \hat{\Delta}^*_{k_i^j}$$

where $\hat{\Delta}^*_{k_i^j}$ is the maximum co-distribution contained in $\breve{\Delta}^*_{k_i^j}$ such that $\hat{\Delta}^*_{k_i^j} \Delta_{k_i^j} = 0$, $\Delta_{k_i^j}$ is a distribution defining the regular K.P. (see Def. 2.1.2) and $\bar{\Delta}^*_{k_i^j}$ is a co-distribution representing part of the applied forces.

The co-distribution $\hat{\Delta}^*_{k_i^j}$ represents those forces which are counteracted by the K.P. and therefore have no effect on the relative position of the pair. If, for example, we consider an ideal kinematic pair allowing only one relative rotation between two bodies, the force applied by any actuator trying to translate the two bodies will not have any effect. These are the forces represented by the co-distribution $\hat{\Delta}^*_{k_i^j}$. On the other hand, any vector that belongs to $\bar{\Delta}^*_{k_i^j}$ and is different than zero can correspond to an exchange of energy between the bodies connected by the pair and the element generating the force. This implies an influence on the relative motion. These forces can exchange energy because, by construction, if the K.P. is not rigid and has degrees of freedom, there is at least one allowed motion which is not in the kernel of $\bar{\Delta}^*_{k_i^j}$.

We can now consider the following distribution[3]:

$$\bar{\Delta}_{k_i^j} := \Delta_{k_i^j} - ((\bar{\Delta}^*_{k_i^j})^{\perp} \cap \Delta_{k_i^j}).$$

The distribution $\bar{\Delta}_{k_i^j}$ is by construction a sub-distribution of $\Delta_{k_i^j}$ and represents those directions which can be actuated to effect the motion. We can then classify kinematic pairs into three types according to their actuation:

[3] Note that $\bar{\Delta}_{k_i^j}$ is NOT unique.

- $\bar{\Delta}_{k_i^j} = 0$: this means there are no forces that can be actuated such that they have effect on the relative position of the kinematic pair. Such a kinematic pair is said to be *non actuated* or *passive*.
- $\bar{\Delta}_{k_i^j} \subset \Delta_{k_i^j}$: where the subset relation is strict and excludes the coincidence of the two distributions. This means that only certain directions can be actuated at the kinematic pair. Such a kinematic pair is said to be *partially actuated*.
- $\bar{\Delta}_{k_i^j} = \Delta_{k_i^j}$: this means that a force in any direction can be actuated at the kinematic pair. Such a kinematic pair is said to be *fully actuated*.

If the distribution $\bar{\Delta}_{k_i^j}$ is integrable, it is possible to define an actuation space that corresponds to the configuration of the actuators.

*Remark 2.1.3.* It is often common in robotics to consider the configuration space corresponding to the actuation space. In a more general setting, this is not correct since we can have redundantly actuated systems.

Suppose that $\Delta_{k_i^j} - \bar{\Delta}_{k_i^j}$ is involutive. We can first consider the sub-manifolds of $\mathcal{Q}_i^j$ consisting of the integral leaves of $\Delta_{k_i^j} - \bar{\Delta}_{k_i^j}$, and then we can define an equivalent relation on $\mathcal{Q}_i^j$.

**Definition 2.1.4.** *We say that $p, q \in \mathcal{Q}_i^j$ are equivalent and denote it by $p \sim q$, if they belong to the same integral leaf of $\Delta_{k_i^j} - \bar{\Delta}_{k_i^j}$.*

It is now possible to define the actuation space for the considered case.

**Definition 2.1.5 (Kinematic Pair Actuation Space).** *The quotient manifold*

$$\mathcal{A}_i^j := \mathcal{Q}_i^j / \sim$$

*is called the actuation space for an holonomic kinematic pair for which $\Delta_{k_i^j} - \bar{\Delta}_{k_i^j}$ is involutive,*

*Remark 2.1.4.* In the very common case of a one-dimensional and fully actuated K.P. we have $\mathcal{A}_i^j = \mathcal{Q}_i^j$, as in normal cases.

### 2.1.2 Position-independent Kinematic pairs: lower pairs.

Let us now consider a special case of bilateral kinematic pairs which are called *lower pairs*. These kinematic pairs have the feature of being somehow independent from the current relative position $h_i^j$ of the two bodies.

In Chapter 1, we have seen that we can intrinsically and uniquely associate to each $v \in T_{h_i^j} SE_i^j(n)$ two vectors $t_j^i$ and $t_i^j$ belonging respectively to the Lie algebras $se_i(n)$ and $se_j(n)$. These Lie algebras are vector spaces and as such they are independent of the relative current position $h_i^j$ of the two bodies. We can therefore consider those bilateral regular kinematic pairs whose defining free directions mapped to $se_i(n)$ or to $se_j(n)$ are constant subspaces of the two Lie algebras for the varying configuration $h_i^j$, as is explained in the following definition:

**Definition 2.1.6 (Lower Pairs).** *A holonomic, regular kinematic pair $k_i^j$ on $O \subset SE_i^j(n)$ is said to be a lower pair iff $\forall h_i^j \in O$, the vector subspace $\pi_{h_i^j}^j(\Delta_{k_i^j}(h_i^j)) \subset se_j(n)$ is constant for variation of $h_i^j$ allowed by $\Delta_{k_i^j}$.*

An important result which justifies the previous definition is that if an involutive distribution $\Delta_{k_i^j}$ on $SE_i^j(n)$ maps to a constant subspace of $se_j(n)$, then the inverse distribution $\Delta_{k_j^i}$ on $SE_j^i(n)$ (see Remark. 1.3.1, p.16) also maps to a constant equally dimensional subspace of $se_i(n)$. This is very important since it means that lower pairs are intrinsic entities.

**Theorem 2.1.2.** *If the map of an involutive distribution $\Delta_{k_i^j}$ on $SE_i^j(n)$ to $se_j(n)$ by means of the intrinsic mapping $\pi_{h_i^j}^j$ of Figure 1.9 is a constant subspace of $se_j(n)$ for all configurations $h_i^j \in SE_i^j(n)$ belonging to an integral leaf of $\Delta_{k_i^j}$, then the mapping of its inverse $\Delta_{k_j^i} \in SE_j^i(n)$ to $se_i(n)$ through $\pi_{h_j^i}^i$ is also a constant subspace of $se_i(n)$ of the same dimension.*

**Proof.** For simplicity let us consider two Cartesian coordinates $\Psi_i$ and $\Psi_j$ for the Euclidian spaces $\mathcal{E}_i$ and $\mathcal{E}_j$ respectively. It has been shown in Sect. 1.2.1 that an element of $h_i^j \in SE_i^j(n)$ can be represented by a matrix $H_i^j \in SE(n)$. At the same time, elements corresponding to the vectors of $se_i(n)$ or $se_j(n)$ will correspond to matrices of the form reported in Theorem 1.3.7. For consistency the notation $H_j^i = (H_i^j)^{-1}$ is used.

By hypothesis, the involutive distribution $\Delta_{k_i^j}$ maps to a constant subspace of $se_j(n)$. Suppose that this subspace is of dimension $k$, where $k$ corresponds to the degrees of freedom of the kinematic pair represented by the lower pair. Then there are $k$ constant matrices $\Lambda_l$ with $l = 1, \ldots, k$ representing a basis of the constant subspace of $se_j(n)$. Any possible moving direction can therefore be expressed as a linear combination of these matrices.

It was shown in Eq. (1.26) that the adjoint mapping $a = Ad_h b$ is represented in matrix form by $A = HBH^{-1}$. Using this, we can consider $\Phi_l := -H_j^i \Lambda_l H_i^j$ corresponding to the elements mapped by $-Ad_{h_j^i}$ to $se_i(n)$ in the following way:

$$t_j^i = -Ad_{h_j^i} t_i^j$$

Since by assumption $\Lambda_i$ is a base of the subspace and refers to the matrix representation of twists, for any allowed motion $v_j \in se_j(n)$ there will be $k$ scalars $\lambda_p$ $p = 1, \ldots, k$ such that the matrix representation $T_j$ of $v_j$ can be expressed as:

$$T_j = \dot{H}_i^j H_j^i = \sum_{p=1}^{k} \lambda_p \Lambda_p$$

which implies that:

$$\dot{H}_i^j = \left( \sum_{p=1}^{k} \lambda_p \Lambda_p \right) H_i^j.$$

Consider now the time derivative of $\Phi_l$ for any of the allowed motions corresponding to the distribution. Since the $\Lambda_l$ are by assumption constant, we have:

$$\dot{\Phi}_l = -\dot{H}_j^i \Lambda_l H_i^j - H_j^i \Lambda_l \dot{H}_i^j = \sum_{p=1}^{k} \lambda_l (H_j^i \Lambda_p \Lambda_l H_i^j - H_j^i \Lambda_l \Lambda_p H_i^j) =$$
$$H_j^i \left( \sum_{p=1}^{k} \lambda_p [\Lambda_p, \Lambda_l] \right) H_i^j \quad (2.1)$$

where $[,]$ is the matrix commutator (see Sect. A.6) representing the Lie bracket of the Lie algebra $se_j(n)$.

Since the considered distribution is involutive, the constant subspace of $se_j(n)$ representing the possible directions is closed under the operation of Lie brackets and therefore some structural constants $c_{qr}^m$ exist such that:

$$\dot{\Phi}_l = H_j^i \left( \sum_{p=1}^{k} \lambda_p \left( \sum_{m=1}^{k} c_{pl}^m \Lambda_m \right) \right) H_i^j = H_j^i \left( \sum_{p=1}^{k} \sum_{m=1}^{k} \lambda_p c_{pl}^m \Lambda_m \right) H_i^j =$$
$$H_j^i \sum_{m=1}^{k} \lambda_m^* \Lambda_m H_i^j \quad (2.2)$$

where $\lambda_m^* := \sum_{p=1}^{k} \lambda_p c_{pl}^m$.

Eventually, since $\Phi_m = -H_j^i \Lambda_m H_i^j$, we have:

$$\dot{\Phi}_l = - \sum_{m=1}^{k} \lambda_m^* \Phi_m$$

which means that the subspace of $se_i(n)$ spanned by the vectors corresponding to the $\Phi_l$ $l = 1, \ldots, k$ is invariant and therefore constant. ∎

*Remark 2.1.5.* It is important to notice that, due to the hypothesis on involutivity, this result only holds for holonomic kinematic pairs.

The latter result is related very closely to a well known result of Lie group theory which is also reported by [21], which says that the configuration manifold of a lower pair, after certain choices of reference, corresponds to a a subgroup of $SE(3)$. The followed approach is considered more intrinsic from a conceptual point of view since it does not assume any reference positions (see Remark. 1.2.2, p.14).

**Decomposition of algebras induced by a lower pair.** From the previous section, we conclude first that a lower pair is characterized by a constant subspace in $se_i(n)$ and $se_j(n)$. These two subspaces are called the *allowed twists* for the pair, in space $i$ and in space $j$ respectively, and are indicated with $\mathcal{T}_{i,j}^A$ and $\mathcal{T}_{j,i}^A$. We can then consider the complements of these subspaces. It is important to note that since there is not an intrinsic[4] positive definite metric defined on $se_i(n)$ [21], we cannot talk about orthonormal complements and therefore complements are not unique. A proper complement subspace of the allowed twists is called *forbidden twists* subspace and is denoted by $\mathcal{T}_{i,j}^F$ and $\mathcal{T}_{j,i}^F$. By definition the complements should be such that the two Lie algebras are the direct sum (see Def. A.5.10) of the free and forbidden twists:

$$se_i(n) = \mathcal{T}_{i,j}^A \oplus \mathcal{T}_{i,j}^F \quad , \quad se_j(n) = \mathcal{T}_{j,i}^A \oplus \mathcal{T}_{j,i}^F. \tag{2.3}$$

The subspaces $\mathcal{T}_{i,j}^A$ or $\mathcal{T}_{j,i}^A$ characterize the relative motion. By choosing any base for $\mathcal{T}_{i,j}^A$ or $\mathcal{T}_{j,i}^A$, we can easily express the relative velocities of the two bodies numerically.
If we consider the dual algebras $se_i^*(n)$ and $se_j^*(n)$, we can intrinsically define the subspace $\mathcal{W}_{i,j}^C$ of $se_i^*(n)$ dual to $\mathcal{T}_{i,j}^A$:

$$\mathcal{W}_{i,j}^C := \{w_j^i \in se_i^*(n) \text{ s.t. } \langle w_j^i, t_j^i \rangle = 0 \ \ \forall t_j^i \in \mathcal{T}_{i,j}^A\}$$

where $\langle , \rangle$ denotes the duality product. We can do the same to define $\mathcal{W}_{j,i}^C$. The subspace $\mathcal{W}_{i,j}^C$ is called set of *constraint wrenches.* We can define a *actuation subspace* $\mathcal{W}_{i,j}^A$ as a possible complement of the constraint wrenches such that:

$$se_i^*(n) = \mathcal{W}_{i,j}^A \oplus \mathcal{W}_{i,j}^C \quad , \quad se_j^*(n) = \mathcal{W}_{j,i}^A \oplus \mathcal{W}_{j,i}^C. \tag{2.4}$$

Once again the complements $\mathcal{W}_{i,j}^A$ and $\mathcal{W}_{j,i}^A$ are NOT unique and represent all the wrenches we can use to create relative motion. A base of the actuation wrenches subspace can be used to define controlling inputs. We can really first choose the complements $\mathcal{W}_{i,j}^A$ and $\mathcal{W}_{j,i}^A$ and then take their unique duals complements (or perp) to define $\mathcal{T}_{i,j}^F$ and $\mathcal{T}_{j,i}^F$.

**Exponential coordinates for a lower pair.** We have already seen in Theorem 2.1.1 that for a general holonomic kinematic pair $k_i^j$ between body $i$

[4] The only two intrinsic 2-covariant tensors are the Killing form which is singular and the hyperbolic metric which is not singular but also not positive definite.

and body $j$, it is possible to define a configuration sub-manifold $\mathcal{Q}_i^j \subset SE_i^j(n)$. If we are dealing with a lower pair, it is possible to find a very convenient chart for this sub-manifold around any point $r_i^j \in \mathcal{Q}_i^j$. This chart is obtained by means of the exponential map from a Lie sub-algebra to its Lie sub-group. To understand this, consider the point $r_i^j \in \mathcal{Q}_i^j$ around which we want to have a local chart. We can then define:

$$\mathcal{Q}_j := (R_{r_i^j})^{-1} \mathcal{Q}_i^j$$

where $R_{r_i^j}$ is the hybrid right translation as defined in Def. 1.2.7. By means of this position, the point $r_i^j$ will be mapped to the identity of the Lie group $SE_j(n)$. Furthermore, it is well known in literature [21] that, for a lower pair, $\mathcal{Q}_j$ is a subgroup of $SE_j(n)$.

Let us now consider a base of the constant, $k$ dimensional subspace $\mathcal{T}_{j,i}^A$ and indicate these base vectors with $\psi_j^1, \psi_j^2, \ldots, \psi_j^k$.

We can then define the exponential mapping around $r_i^j$:

**Definition 2.1.7.** *Given a lower pair $k_i^j$ with configuration sub-manifold $\mathcal{Q}_i^j(k_i^j)$ and any point $r_i^j \in \mathcal{Q}_i^j(k_i^j)$, we call exponential mapping in $j$ the following:*

$$e_i^j : \mathbb{R}^k \to \mathcal{Q}_i^j(k_i^j) \; ; \; (x^1, \ldots, x^k) \mapsto R_{r_i^j}\left(e^{\sum_{l=1}^k x^l \psi_j^l}\right)$$

*where the exponential indicates the group exponential from the Lie sub-algebra $\mathcal{T}_{j,i}^A$ to the Lie sub-group $\mathcal{Q}_j$.*

*Remark 2.1.6.* It is possible to see that the previous mapping is surjective on any compact neighborhood of $r_i^j$. It is in general NOT injective due to the topology of $SE_j(n)$.

If we consider the biggest subset $\tilde{\mathcal{Q}}_i^j(r_i^j) \subset \mathcal{Q}_i^j(k_i^j)$ for which the exponential mapping around $r_i^j$ becomes injective on $\tilde{\mathcal{Q}}_i^j(r_i^j)$ then we can define on this subset a chart around $r_i^j$ which we call the *logarithmic chart*:

$$\log_i^j(r_i^j) : \tilde{\mathcal{Q}}_i^j(r_i^j) \to \mathbb{R}^k : q_i^j \mapsto (e_i^j)^{-1}(q_i^j).$$

Using these coordinates, we can arrive to very convenient matrix forms for the kinematics. A nice treatment which uses coordinates can be found in [10].

## 2.2 Mechanisms Topology

In this section, we analyze general mechanisms. This allows us to define and calculate degrees of freedom and minimal configuration spaces. We use concepts of graph theory in our analysis. [25] reviews graph and matroid theory excellently. A summary of the needed concepts can be found in Sect. A.2. We can now give a formal definition of a *pure mechanism*. A pure mechanism describes how bodies are interconnected by means of kinematic pairs. Other elements like springs or dampers, will be introduced later.

**Definition 2.2.1 (Pure Rigid Mechanism).** *We call a pure rigid mechanism a pair of the form* $(S^m(n), K)$, *where* $S^m(n)$ *is a Euclidian system and* $K$ *is a set of kinematic pairs defined on elements of the Euclidian system* $S^m(n)$.

We represent a pure mechanism with an oriented graph in which its vertices correspond to elements in $S^m(n)$ and each edge corresponds to an element of $K$ whose adjacent vertices correspond to the elements of $S^m(n)$ on which the kinematic pair is defined. The orientation of an edge defines the reference chosen for the kinematic pair. For example, if we consider the kinematic pair $k_i^j$ connecting $\mathcal{E}_i$ to $\mathcal{E}_j$ and we use as a configuration space $\mathcal{Q}_i^j$ for this pair, we denote it in the graph with an arrow from $i$ to $j$. This also implies that when we describe the twist of the pair, we consider $t_i^j$ and NOT $t_j^i$. Nevertheless, it is always possible to invert the direction of the edge. In this case we have the configuration $\mathcal{Q}_j^i$ and the twist $t_j^i = -Ad_{h_j^i} t_i^j$ with $h_j^i \in \mathcal{Q}_j^i$.

Suppose that a real body ($B_i \neq \emptyset$) corresponds to each element of $S^m(n)$ and that there are neither fixed nor degenerate kinematic pairs in the set $K$. By analyzing the constructed graph, we can classify the mechanism and find generalized coordinates. We can now give the following definition:

**Definition 2.2.2 (Links types).** *Given a mechanism with corresponding graph* $G$, *a link (or body) can be classified on the basis of the degree of the corresponding vertex* $v$ *in* $G$ *(see Sect. A.2):*

- $\deg v = 1$ *The link will be called an extremity.*
- $\deg v = 2$ *The link will be called a connecting link.*
- $\deg v > 2$ *The link will be called a forking link.*

The base or the end effector of a robotic linkage are therefore called extremities. All the links between the extremities of a kinematic linkage are named connecting links. An example of a forking link can be the palm of a robotic hand to which more than one finger is attached.

### 2.2.1 Kinematic pairs and bond graphs

In this work, bond graphs are used extensively as a power consistent tool which can represent physical interconnection of elements.

Consider two bodies moving with respect to each other. Graphically, we can represent such a situation as in Figure 2.5 or Figure 2.6. The difference in the orientation represents a difference in the direction of the chosen mapping: in Figure 2.5, $\mathcal{Q}_i^j$ has been used as a configuration of the relative position and $t_i^j$ as a twist for the relative motion. On the other hand, in Figure 2.6,

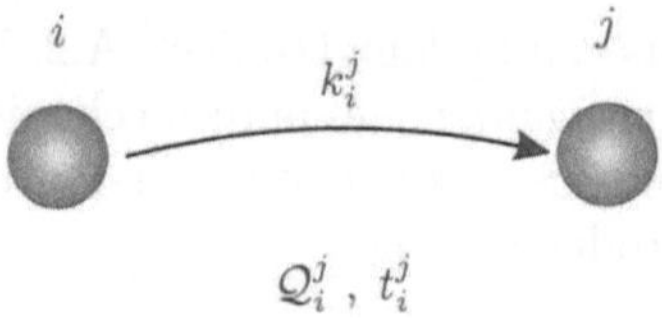


**Fig. 2.5.** Graph of a kinematic pair $k_i^j$.

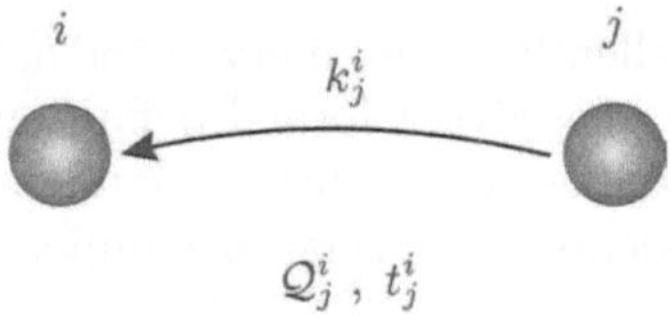


**Fig. 2.6.** Graph of a kinematic pair $k_j^i$.

$Q_j^i$ has been used as a configuration of the relative position and $t_j^i$ as a twist for the relative motion.

In order to give an equivalent representation of the relative motion in bond graphs, we must consider a few important points:

1. In bond graphs, the so called 1-junctions (see Sect. B.8) represent a velocity. A velocity is a quantity which is defined between two objects: it does not make any sense to say that an object has a certain velocity if it is not at least implicit respective to which other object this velocity is referenced. This concept typically characterizes those variables which are usually called *across variables* in network theory.
2. Bond graph junctions can be considered as operating on elements belonging to a vector space. The flows of all the bonds attached to a 0 junction should be "summable" in a vector sense and the efforts should be equal in order to express the constraints of the junction. Dually, the efforts of all the bonds attached to a 1-junction should be again "summable" and the flows equal. It can be seen that this characterizes a vector space structure with its dual for the bonds.
3. All the efforts and flows of the bonds directly connected to junctions of a simple junction structure only composed of 1 and 0 junctions and bonds with no transformers should be elements belonging to the same vector space (flows) and to its dual (efforts) respectively.

In our case, the vector spaces used are vector algebras of the Euclidian spaces in the Euclidian system. This approach clearly resembles that of the screw-vector bond graphs [24, 27].

From the previous considerations. 1. implies that we need to choose a unique reference space $\mathcal{E}_k$ when interconnections are represented by a simple junction structure, and 2. and 3. imply that in a simple junction structure, the vector space $se_k(n)$ corresponding to the Lie algebra of the reference Euclidian space is used to algebraically support the 0-junctions operation and that the dual space $se_k^*(n)$ is the vector space where operations on efforts needed by 1-junctions are algebraically executed. This means. for example. that if the bonds attached to a 0-junction represent efforts belonging to $se_k^*(n)$ and flows belonging to the dual $se_k(n)$. the addition operation of the vector space $se_k(n)$ is used to express the junction constraint which requires that the algebraic vector sum of the flows of the bonds be equal to 0.

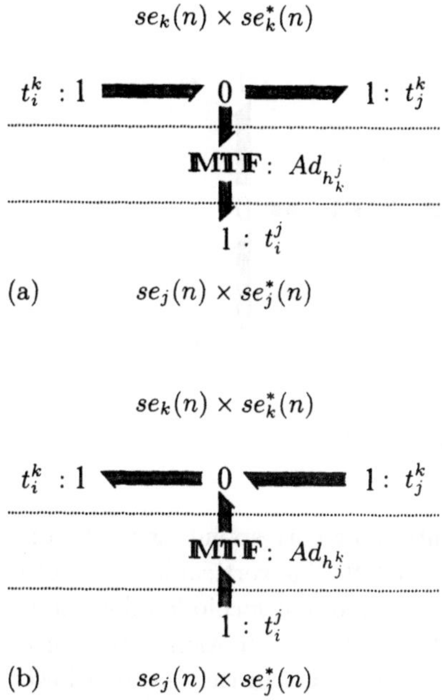


**Fig. 2.7.** Bond graphs of a relative motion for $k_i^j$.

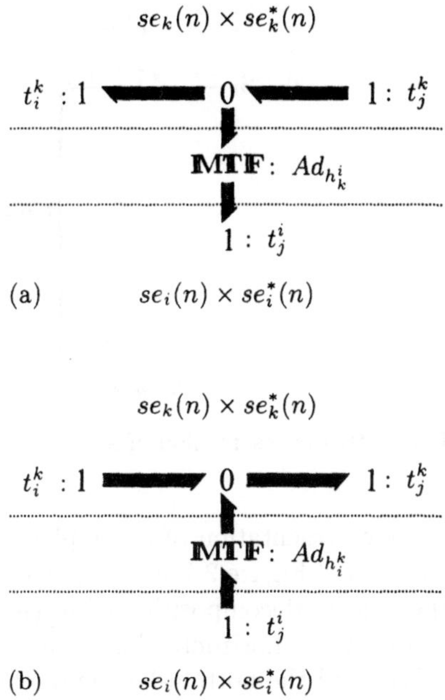


**Fig. 2.8.** Bond graphs of a relative motion for $k_j^i$.

As a consequence of the previous considerations, by using Eq. (1.30), we can give two representations of the relative motion of two bodies in bond graph terms. These representations are reported in Figure 2.7 and Figure 2.8 for the orientations of Figure 2.5 and Figure 2.6, respectively. In these representations, the 1-junctions on the top of each of the two possibilities represent the twists of body $i$ and $j$ with respect to any other reference body $k$. The 1-junction on the bottom, on the other hand, represents the relative motion of $j$ with respect to $i$ or the other way around. In Figure 2.8 (resp.Figure 2.7), the **MTF** separates simple junction structures that use the vector space structure of $se_k(n)$ and its dual from simple junction structures that use the vector space structure of $se_i(n)$ (resp.$se_j(n)$).

The top 1-junctions of each representation will be used to interconnect a complex system, whereas the bottom 1-junction will be used to connect elements attaching the two bodies $i$ and $j$ like springs, dampers or actuators, as will be seen in the following chapter.

Note that if $j$ (respectively $i$) is chosen as the reference space, $k = j$ (respectively $k = i$), we have $t_k^j = t_j^k = 0$ (respectively $t_k^i = t_i^k = 0$). This could be represented by a zero flow source on the 1-junction representing $t_j^k$ (respectively $t_i^k$) in Figure 2.7 and Figure 2.8. By applying bond graphs simplification we obtain only two 1-junctions connected by the **MTF**.

In the figures, the bonds are grayed after the notation of [24, 27].

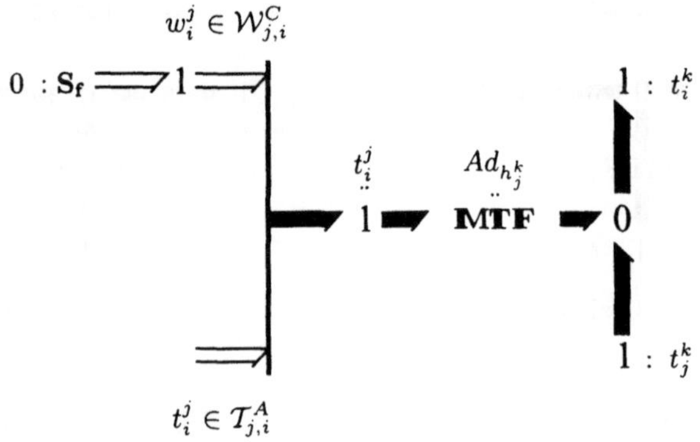


**Fig. 2.9.** Representation of a total kinematic pair $k_i^j$.

A representation of a complete kinematic pair corresponding to the orientation of Figure 2.7(b) is reported in Figure 2.9. The vertical line indicates the algebra decomposition of $se_j(n)$ in the allowed and forbidden directions. The top-left horizontal bonds represent the forbidden motion subspace of $se_j(n)$ and the bottom-left represents the allowed one. In the case of a lower pair, this decomposition is reported in Eq. (2.3) and in Eq. (2.4) for its dual and can be chosen constant by choosing constant complements. In this case, the power variables belonging to the bonds on the top-left of the figure belong to $\mathcal{T}_{j,i}^F \times \mathcal{W}_{j,i}^C$ and the bonds on the bottom-left belong to $\mathcal{T}_{j,i}^A \times \mathcal{W}_{j,i}^A$. The power flowing into the first bond is always equal to zero because the twists are forced to zero. The power flowing into the bond characterized by $\mathcal{T}_{j,i}^A \times \mathcal{W}_{j,i}^A$ will be equal to the power flowing through the **MTF**. Note that the bonds on the left side of the vertical line are NOT grayed indicating that the coordinate free decomposition is not intrinsic. This is due to the fact that the complements of $\mathcal{T}_{j,i}^A$ and $\mathcal{W}_{j,i}^C$ are NOT unique. We now discuss serial linkages to generalize our discussion to any mechanical mechanism later.

### 2.2.2 Serial linkages

The first and simplest kinematic structure we consider is a *serial kinematic linkage*. This kind of kinematic structure corresponds to a connected, cycle-free graph composed of two extremities (as defined at the beginning of Sect. 2.2) and of a certain number of connecting links. This means that there are no forking links as in a robotics hand (see Figure 2.12). A lot of common industrial robots are of this type. The graph corresponding to a SCARA manipulator is reported in Figure 2.10. We call the two extremities *base* and *tool*. For these kinds of robots, the kinematic analysis is very simple. If we suppose that the linkage is composed of $m$ bodies, and therefore $m-1$ kinematic pairs, a configuration manifold for the robot is:

$$\mathcal{Q} = \mathcal{Q}_2^1 \times \mathcal{Q}_3^2 \times \ldots \times \mathcal{Q}_m^{m-1}$$

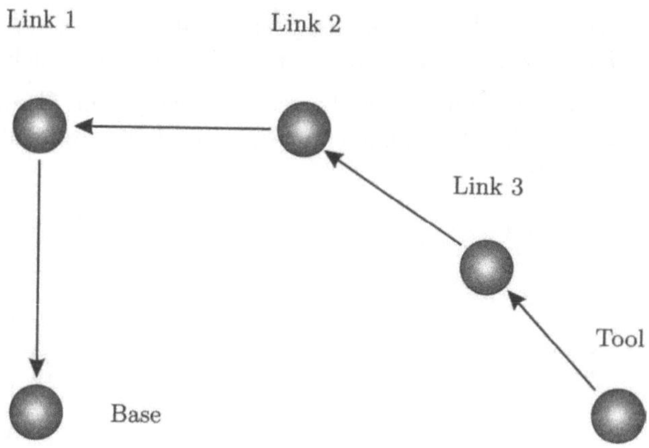


**Fig. 2.10.** The kinematics graph of a SCARA robot.

where $\mathcal{Q}_{i+1}^i$ is the configuration manifold of the kinematic pair connecting body $i$ to body $i+1$. The total configuration manifold is the Cartesian product of the kinematic pairs configuration manifolds because there are no additional constraints. The total degrees of freedom of the system will therefore be

$$\sum_{i=1}^{m-1} \dim \mathcal{Q}_{i+1}^i \quad .$$

**Forward kinematics.** Usually, in robotics, for a serial kinematic chain it is important to consider the relative position and velocity of the extremities, usually called *base* and *end effector* or *tool.* The mapping which assigns to a point in $\mathcal{Q}$ the relative position of the tool with respect to the base is called *forward kinematics.* For a serial chain composed of $m$ bodies, the forward kinematics has the following form:

$$L : \mathcal{Q} = \mathcal{Q}_2^1 \times \ldots \times \mathcal{Q}_m^{m-1} \rightarrow SE_m^1(n) \; ; \; (h_2^1, \ldots, h_m^{m-1}) \mapsto h_2^1 \circ \ldots \circ h_m^{m-1} . \tag{2.5}$$

Once coordinates and charts for the $\mathcal{Q}_i^j$ are chosen, it is possible to give a numerical expression of the previous map; this expression is called the product of exponentials [28]. A well presented treatment can be found in [10].

**Differential kinematics and the Jacobian.** It is also possible to consider the differential kinematics which maps kinematic pair velocities to the velocities between the two extremities of the linkage. Formally the differential mapping has the following form:

$$L_* : T\mathcal{Q} \rightarrow TSE_m^1(n)$$

Where the $L(\cdot)$ is the forward kinematics (see Eq. (A.3)). We can nevertheless find a more convenient form if we map the elements of the tangent bundles

$TQ$ and $TSE^1_m(n)$ to the Lie algebras $se_1(n) \times se_2(n) \times \ldots \times se_{m-1}(n)$ and $se_1(n)$, respectively, by using the mapping $\pi^j_{h^j_i}$ of Figure 1.9 on each $TSE^j_i(n)$.

This can be directly done using the result presented in Theorem 1.3.8. This theorem can be directly applied to define the following configuration-dependent linear map:

$$J(h) : \mathcal{T}^A_{1,2} \times \ldots \times \mathcal{T}^A_{m-1,m} \to se_1(n) :$$

$$(t^1_2, \ldots, t^{m-1}_m) \mapsto \begin{pmatrix} Id & Ad_{h^1_2} & Ad_{h^1_2 \circ h^2_3} & \cdots & Ad_{h^1_2 \circ \ldots \circ h^{m-2}_{m-1}} \end{pmatrix} \begin{pmatrix} t^1_2 \\ t^2_3 \\ t^3_4 \\ \vdots \\ t^{m-1}_m \end{pmatrix} \tag{2.6}$$

where $t := (t^1_2, \ldots, t^{m-1}_m)^T$ are the twists representing all the motions of the kinematic pairs expressed in $se_1(n), \ldots, se_{m-1}(n)$. $h = (h^1_2, \ldots, h^{m-1}_m)$ and $J(h)t$ is the twist of the tool with respect to the base, expressed in the twist space of the base $se_1(n)$. In the previous equation, the matrix of adjoint mappings is called the *coordinate free Jacobian matrix.*

**Mapping of wrenches.** The Jacobian $J(h)$ is a configuration dependent linear map. For any linear map we can consider what is called its adjoint map (see Def. A.5.5). This is a map which goes from the dual space of the range of the linear map to the dual space of the domain [29]. To be more specific, consider the Jacobian $J(h)$ in a fixed position $h$. Given any $t \in \mathcal{T}^A_{1,2} \times \ldots \times \mathcal{T}^A_{m-1,m}$, we can define $\tilde{t}^1_m := J(h)t$, where $\tilde{t}^1_m \in se_1(n)$. If we now take an element $\tilde{w}^1_m \in se^*_1(n)$, the dual space of $se_1(n)$, the adjoint map $J^*(h)$ maps $\tilde{w}^1_m$ to an element $w \in \mathcal{W}^A_{1,2} \times \ldots \times \mathcal{W}^A_{m-1,m}$. Furthermore, if $J^*(h)$ is to be the adjoint of $J(h)$, the following must hold:

$$\langle \tilde{w}^1_m, J(h)t \rangle = \langle J^*(h)\tilde{w}^1_m, t \rangle \tag{2.7}$$

where $\langle , \rangle$ stands for the dual product (see Def. A.5.4) of wrenches on twists. This results in their exchanged power. If we expressed this in coordinates, it would be easy to see that the matrix $\bar{J}^T(q)$ representing the adjoint $J^*(h)$ is the transpose of the matrix $\bar{J}(q)$ representing $J(h)$. This means that if the linear map is a map between twists, the adjoint map will be a map between wrenches going in the opposite direction. The application of a wrench on a twist corresponds to the power exchanged. Since at this stage we consider the mechanism ideal, no dissipation can take place. The power expressed in the two spaces should be the same and this is expressed by Eq. (2.7) where for a serial kinematic linkage, the right side of the equation is equal to the power exchanged at the kinematic pairs and the left side is equal to the power exchanged at the end effector. In bond graph terms the power-continuous

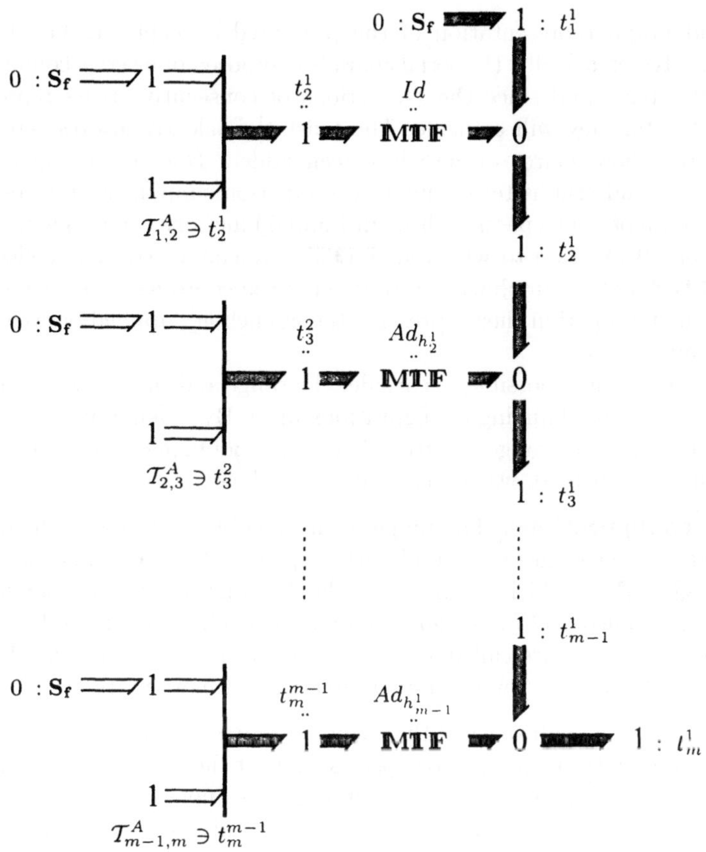


**Fig. 2.11.** A serial chain mechanism in coordinate free bond graphs.

element representing this transformation is a multidimensional transformer. The adjoint mapping of the Jacobian will then be:

$$J^*(h) : se_1^*(n) \rightarrow \mathcal{W}_{1,2}^A \times \ldots \times \mathcal{W}_{m-1,m}^A : w_m^1 \mapsto \begin{pmatrix} Id \\ Ad^*_{h_2^1} \\ Ad^*_{h_2^1 \circ h_3^2} \\ \vdots \\ Ad^*_{h_2^1 \circ \ldots \circ h_{m-1}^{m-2}} \end{pmatrix} w_m^1 \tag{2.8}$$

where $Ad^*_\bullet$ indicates the adjoint map of $Ad_\bullet$ and corresponds to the transpose matrix of its numerical representations. The adjoint mapping gives in a static situation the wrenches necessary to counteract a wrench applied between the extremities of the chain exactly.

A bond graph representation of the presented concept can be seen in Figure 2.11. Kinematically, the serial chain is composed of a series connection of kinematic pairs and since the 1-junctions of consecutive pairs represent the same twists, they will coincide. The twist $t_1^1$ is clearly always zero and that is why a flow source of zero has been added. Due to the algebra of $se_1(n)$, or in bond graphs terms due to the junction simplification rules, the vertical column of bonds passing through 1 and 0 junctions can be substituted with just one 0 junction to which all **MTF**s are connected. The collection of these **MTF**s after simplification represents the coordinate-free Jacobian. The simplified 0-junction then represents the wrench $w_m^1$ applied between the two extremities.

Kinematically, it is possible to consider the range and the null space of the Jacobian matrix for changing configurations in $\mathcal{Q}$. By looking at the change in the dimension of the range, we are able to talk about singularities. Looking at the null space instead, we can consider redundancy.

**Singular configurations.** The range of the Jacobian represents the directions in which we can move the tool with respect to the base. The elements in the set $\tilde{\mathcal{Q}} \subset \mathcal{Q}$ for which the image of the Jacobian matrix has maximum dimension are called *NON singular configurations.* The elements in the complement $\mathcal{Q}^S = \mathcal{Q} - \tilde{\mathcal{Q}}$ are called *singular configurations.* This means that in a singular configuration, some possible directions of motion are "lost".

**Redundancy.** The null space of the Jacobian represents those directions we can move along in the configuration space such that the relative configuration of the two extremities of the kinematic linkage does not change.

Before we define the major entity in the study of redundancy, it is worth explaining a few things about distributions on the configuration space. A distribution on $\mathcal{Q}$ is a smooth mapping from an element $h := (h_2^1, h_3^2, \ldots, h_m^{m-1}) \in \mathcal{Q}$ to a subspace of $T_h\mathcal{Q} := T_{h_2^1}\mathcal{Q}_2^1 \times T_{h_3^2}\mathcal{Q}_3^2 \times \ldots \times T_{h_m^{m-1}}\mathcal{Q}_m^{m-1}$. Using the mapping $\pi^i_{h^i_{i-1}}$ of Figure 1.9 for each $T_{h^i_{i+1}}\mathcal{Q}^i_{i+1}$, we can then map an element of $T_h\mathcal{Q}$ to

$$\mathcal{T}_{1,2}^A \times \ldots \times \mathcal{T}_{m-1,m}^A \subset se_1(n) \times \ldots \times se_{m-1}(n).$$

We indicate this mapping with $\pi$. It is now possible to give the following definition.

**Definition 2.2.3 (Redundant Distribution).** *The maximum distribution on $\mathcal{Q}$ which projected by means of $\pi$ on $\mathcal{T}_{1,2}^A \times \ldots \times \mathcal{T}_{m-1,m}^A$ is completely contained in the kernel of the coordinate-free Jacobian is called the redundant distribution and it is denoted by $\Delta^R$.*

It is possible to see that the distribution $\Delta^R$ corresponds to the kernel of $L_*$, the tangent mapping of the forward kinematics. For this reason, the following theorem can be stated:

**Theorem 2.2.1 (Involutivity of the Redundant distribution).** *The distribution $\Delta^R$ is involutive.*

**Proof.** The result simply follows from the fact that $\ker dL$ for any function $L$ is always an involutive distribution. For an intuitive explanation, we can consider the following. For any configuration $h$, consider the subspace $\Delta^R(h) \subset TQ$. Take any two directions $v_1, v_2 \in \Delta^R(h)$. Consider to make an infinitesimal movement along $v_1$, then along $v_2$, then along $-v_1$ and then along $-v_2$. Due to the fact that these directions belong to the kernel of $L_*$, we can conclude that at the end point $h_1$, $L(h_1) = L(h)$. This means that we could have directly moved toward $h_1$ without changing the value of $L$. This reasoning applied to any direction belonging to the distribution implies involutivity. ∎

The redundant distribution associates to each configuration those local directions in which we can move without having any effect on the relative position of the two extremities of the serial linkage.

Due to Theorem 2.2.1, $\Delta^R$ is involutive and therefore defines a foliation on $\mathcal{Q}$. Each leaf on $\mathcal{Q}$ generated by $\Delta^R$ represents what are usually called *self-motion* leaves. A motion which stays on one of these leaves will not influence the relative position of the extremities of the serial chain. It is also possible to see that the quotient space of $\mathcal{Q}$ with respect to the leaves generated by $\Delta^R$ (with a similar procedure as in Def. 2.1.5) can be used as a configuration manifold for the relative position of the extremities of the linkage.

### 2.2.3 Forking linkages

A more involved kinematic structure is obtained if we relax the hypothesis of not having forking links. An example of such a mechanism is a robotic hand [30, 31] in free space. This has only one forking link corresponding to the palm.

It is important to specify in "free space" because a hand grasping something belongs to the much more complex class of closed kinematic chains, which we will analyze in the next section. This is because the contacts between the grasped object and the fingers constrain the relative fingers-object motion. These constraints can be handled as kinematic pairs and the object as an extra link. The resulting mechanisms would therefore be described by a graph with loops.

For a forking linkage, the configuration manifold is also the Cartesian product of the kinematic pair configuration manifolds and the total degrees of freedom are therefore again the sum of the degrees of freedom of the kinematic pairs.

**Forward kinematics.** The graph representing a forking linkage is a connected circuit-free graph. Such a graph is also known as a *tree* (see Def. A.2.9). Since we have a tree, in a forking linkage we have in general more than two

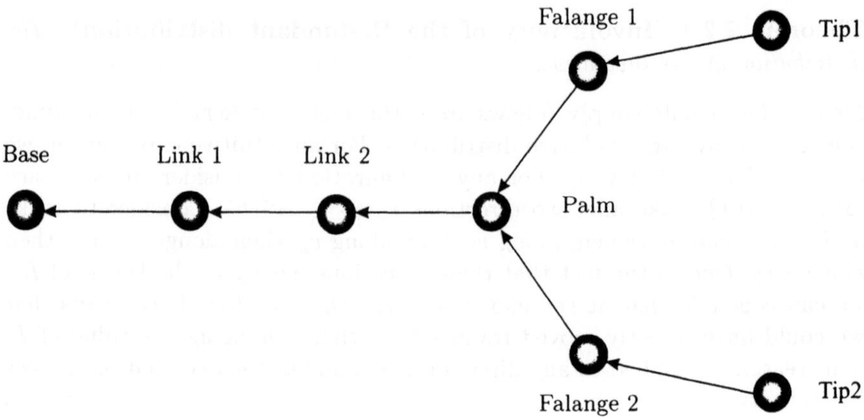


**Fig. 2.12.** The kinematics graph of a robotics hand.

extremities. Since a tree is the characterizing graph for a forking linkage, the following results are useful.

**Theorem 2.2.2 (Path uniqueness).** *For any tree $T = (V, E)$ where $V$ is the set of vertices and $E$ the set of edges, there is a unique path in $T$ which connects any two elements of $V$ that are different.*

**Proof.** This can be easily proven by contradiction: if there were two distinct paths, there would be a circuit in the tree which is not possible by definition. Note that for paths, we do not look at the orientation of the edges. ∎

**Theorem 2.2.3 (Vertices-Edges relation).** *For any tree $T = (V, E)$, if the number of elements of $V$ is $m$, the number of elements of $E$ is $m - 1$.*

**Proof.** For $m = 2$ the statement is trivial. For other $m$, let us proceed by induction and suppose we have proved the statement for the case $m = m_0$ and we want to prove it for $m = m_0 + 1$. It is easy to see that to add a vertex to the tree we need to add one and only one edge because otherwise we would create circuits and we would not have a tree anymore. ∎

The last results imply that if we have $m$ bodies in a connected forking linkage, we will have $m - 1$ kinematic pairs or joints, exactly as in a serial chain.

Suppose now that we have $n_c$ extremities. We can index the extremities with an indexing function $p(\cdot)$ in such a way that $B_{p(1)}, B_{p(2)}, \dots, B_{p(n_c)}$ correspond to the extremities bodies of the mechanism by using the notation introduced at page 8.

Another difficulty in the expression of the kinematics of a forking chain is that we do not have a one to one correspondence between bodies and twists, as in a serial chain. This is due to the fact that we can have more kinematic pairs attached to one body and therefore we need a way to express this situation. To simplify the notation, we will indicate with $l(i) \in V$ and $r(i) \in V$ the two vertices adjacent to the edge $i$. This is the same as saying that the configuration manifold of the kinematic pair corresponding to the edge $i$ in the graph is $\mathcal{Q}_{l(i)}^{r(i)}$.

We can then describe the forward kinematics as the mapping from the configuration space $\mathcal{Q}$ to the positions of the bodies $B_{p(i)}$ with respect to other bodies of the chain which we index with an indexing function $q(i)$. We require $p(i) \neq q(i)$ for a non-trivial expression.

The forward kinematics, then is a mapping with the following form:

$$L : \mathcal{Q} = \mathcal{Q}_{l(1)}^{r(1)} \times \ldots \times \mathcal{Q}_{l(m-1)}^{r(m-1)} \to SE_{p(1)}^{q(1)} \times \ldots \times SE_{p(n_c)}^{q(n_c)} ;$$

$$(h_{l(1)}^{r(1)}, \ldots, h_{l(m-1)}^{r(m-1)}) \mapsto \begin{pmatrix} h_{v_1(n_1-1)}^{q(1)} \circ \ldots \circ h_{p(1)}^{v_1(1)} \\ h_{v_2(n_2-1)}^{q(2)} \circ \ldots \circ h_{p(2)}^{v_2(1)} \\ \vdots \\ h_{v_{n_c}(n_{n_c}-1)}^{q(n_c)} \circ \ldots \circ h_{p(n_c)}^{v_{n_c}(1)} \end{pmatrix} \tag{2.9}$$

where

$$[p(i), s_i(1), v_i(1), s_i(2), \ldots, s_i(n_i-1), v_i(n_i-1), s_i(n_i), q(i)], \quad i = 1, \ldots, n_c$$

corresponds to the unique (see Theorem 2.2.2) path composed of $n_i$ edges (see Def. A.2.5) which connects $q(i)$ to the extremity $p(i)$, $v_i(\cdot)$ belongs to the set of vertices $V$ of the graph and $s_i(\cdot)$ to the set of edges $E$.

Note that by Theorem 2.2.3, the number of kinematic pairs in the system is always $m-1$ because $m$ is the number of vertices in the graph which is an equivalent of the number of bodies in the mechanism, and therefore we have a tree.

Note that, in the path from $p(i)$ to $q(i)$, we could encounter edges which are oriented in the opposite direction than the one we are following. In this case, we need to consider the inverse of the configuration of the pair corresponding to the edge.

**Differential kinematics and the Jacobian.** Once again, the differential kinematics could be defined as the tangent mapping of $L$ analogously to what said about serial kinematic chains. Nevertheless, a more suitable form can be given which expresses the mappings between the twists of the kinematic pairs. Due to the complexity of the possible paths, the expression of this Jacobian is a bit complicated. We have:

$$
\begin{aligned}
&J(h^{r(1)}_{l(1)}, \ldots, h^{r(m-1)}_{l(m-1)}) : \\
&\mathcal{T}^A_{r(1),l(1)} \times \ldots \times \mathcal{T}^A_{r(m-1),l(m-1)} \to se_{q(1)}(n) \times \ldots \times se_{q(n_c)}(n) \; ; \\
&(t^{r(1)}_{l(1)}, \ldots, t^{r(m-1}_{l(m-1)})) \mapsto \begin{pmatrix} A_1 & 0 & 0 & \ldots & 0 \\ 0 & A_2 & 0 & \ldots & \vdots \\ 0 & 0 & A_3 & \ldots & \vdots \\ \vdots & \vdots & \vdots & \ddots & \vdots \\ 0 & \ldots & \ldots & \ldots & A_{n_c} \end{pmatrix} \begin{pmatrix} S_1 \\ S_2 \\ S_3 \\ \vdots \\ S_{n_c} \end{pmatrix} \begin{pmatrix} t^{r(1)}_{l(1)} \\ t^{r(2)}_{l(2)} \\ t^{r(3)}_{l(3)} \\ \vdots \\ t^{r(m-1)}_{l(m-1)} \end{pmatrix}
\end{aligned}
\tag{2.10}
$$

where $S_i$ is a block permutation matrix interchanging the orders of the twists $(t^{r(1)}_{l(1)}, \ldots, t^{r(m-1}_{l(m-1)}))$ in the following way:

$$
\begin{pmatrix} t^{q(i)}_{v_i(n_i-1)} \\ t^{v_i(n_i-1)}_{v_i(n_i-2)} \\ \vdots \\ t^{v_i(1)}_{p(i)} \\ \times \\ \vdots \\ \times \end{pmatrix} = S_i \begin{pmatrix} t^{r(1)}_{l(1)} \\ \vdots \\ t^{r(m-1)}_{l(m-1)} \end{pmatrix}
$$

where the last $m - n_i - 1$ elements on the left vector correspond to the twists of the kinematic pairs of the edges not corresponding to $s_i(1), \ldots, s_i(n_i)$ and:

$$
A_i = \Big( \underbrace{Id \; Ad_{h^{q(i)}_{v_i(n_i-1)}} \; \ldots \; Ad_{h^{q(i)}_{v_i(n_i-1)} \circ \ldots \circ h^{v_i(2)}_{v_i(1)}}}_{n_i} \; 0 \ldots 0 \Big) .
$$

Properly speaking, $S_i$ can be more than just a permutation matrix. This is due to the fact that in the reordering of the twists for the chosen indexes, it could happen that $S_i$ maps a twist $t^u_k$ not just to itself in another position (pure permutation), but also to $t^k_u$. This would therefore also imply a mapping through $-Ad_{h^k_u}$ as explained in Eq. (1.22) and Eq. (1.23). We do not go into detail not to complicate the notation further.

**Mapping of wrenches.** With an analog procedure as used for serial chains, we can use the adjoint map of the Jacobian to map all the wrenches applied between $q(i)$ and $p(i)$ to the necessary kinematic pair wrenches $\mathcal{W}^A_{r(j),l(j)}$, where $j = 1, \ldots, (m-1)$ to balance them exactly.

The adjoint mapping is given by

$$J^*(h_{l(1)}^{r(1)}, \ldots, h_{l(m-1)}^{r(m-1)}):$$
$$se^*_{q(1)}(n) \times \ldots \times se^*_{q(n_c)}(n) \to \mathcal{W}^A_{r(1),l(1)} \times \ldots \times \mathcal{W}^A_{r(m-1),l(m-1)};$$
$$(w_{p(1)}^{q(1)}, \ldots, w_{p(n_c)}^{q(n_c)}) \mapsto$$

$$\begin{pmatrix} S_1^* & S_2^* & S_3^* & \ldots & S_{n_c}^* \end{pmatrix} \begin{pmatrix} A_1^* & 0 & 0 & \ldots & 0 \\ 0 & A_2^* & 0 & \ldots & \vdots \\ 0 & 0 & A_3^* & \ldots & \vdots \\ \vdots & \vdots & \vdots & \ddots & \vdots \\ 0 & \ldots & \ldots & \ldots & A_{n_c}^* \end{pmatrix} \begin{pmatrix} w_{p(1)}^{q(1)} \\ w_{p(2)}^{q(2)} \\ w_{p(3)}^{q(3)} \\ \vdots \\ w_{p(n_c)}^{q(n_c)} \end{pmatrix} \tag{2.11}$$

where $S_i^*$ is a block permutation matrix which corresponds to the inverse[5] permutation $S_i$ and interchanges the orders of the wrenches in the following way:

$$\begin{pmatrix} w_{l(1)}^{r(1)} \\ \vdots \\ w_{l(m-1)}^{r(m-1)} \end{pmatrix} = S_i^* \begin{pmatrix} w_{v_i(n_i-1)}^{q(i)} \\ \vdots \\ w_{p(i)}^{v_i(1)} \\ \times \\ \vdots \\ \times \end{pmatrix}$$

where the $A_i^*$ blocks are given by

$$A_i^* = \begin{pmatrix} Id \\ Ad^*_{h^{q(i)}_{v_i(n_i-1)}} \\ Ad^*_{h^{q(i)}_{v_i(n_i-1)} \circ h^{v_i(n_i-1)}_{v_i(n_i-2)}} \\ \vdots \\ Ad^*_{h^{q(i)}_{v_i(n_i-1)} \circ \ldots \circ h^{v_i(2)}_{v_i(1)}} \\ 0 \\ 0 \\ \vdots \\ 0 \end{pmatrix}.$$

This mapping gives the necessary wrenches needed at the kinematic pairs to counteract the wrenches $w_{p(1)}^{q(1)}, w_{p(2)}^{q(2)}, \ldots, w_{p(n_c)}^{q(n_c)}$ which are applied to the kinematic mechanism between each of the bodies $p(i)$ and $q(i)$ with $i = 1, \ldots, n_c$ and expressed in $se_{q(i)}(n)$.

[5] The adjoint of a permutation matrix corresponds always to the inverse permutation.

It would be possible to draw a bond graph of the mechanism exactly in the same way as it has been drawn for a serial kinematic chain. Then the topology would be more involved.

### 2.2.4 Closed kinematic chains

We now deal with the most general case of closed kinematic chains where we also relax the requirement that the graph must be circuit free. A circuit in the corresponding graph, corresponds to what is called in mechanics a kinematic loop. A kinematic loop has additional constraints on the configuration of the kinematic pairs corresponding to the edges of the circuit. A typical example of a closed chain mechanism is the Steward platform. The advantages of the use of closed chains in mechanisms are many. For example structures can be achieved that are inherently stiffer and actuators can be located all on one point in order to create a light moving structure.

Let us now briefly analyse these mechanisms by using graphs tools. For any mechanism with loops, representable by a connected graph $G = (V, E)$, consider any spanning tree of $G$ (see Def. A.2.10). This spanning tree will itself be a graph $T$ with the following form: $T = (V, \bar{C})$. Let us indicate with $C$ the set $E - \bar{C}$. If $m$ indicates the number of elements of $V$ and $n_c$ the number of elements of $C$, due to Theorem 2.2.3, the number of edges of $G$, corresponding to the number of kinematic pairs of the mechanism, will be $m + n_c - 1$. If we add to the tree $T$ one element from $C$ we obtain one and only one circuit of $G$ representing a kinematic chain of the starting mechanism. The set of $n_c$ circuits obtained in this way (by adding to $T$ each of the elements of $C$, but one at a time) is called a *fundamental system of circuits* of $G$ with respect to $T$. Each of the elements of $C$ can therefore create additional constraints in the motion which should be taken into account. Let us indicate the elements of $C$ as:

$$C = \{(p(1), q(1)), (p(2), q(2)), \ldots, (p(n_c), q(n_c))\}$$

and the elements of $\bar{C}$ as:

$$\bar{C} = \{(r(1), l(1)), (r(2), l(2)), \ldots, (r(m-1), l(m-1))\}$$

Following exactly the same line as we used for forking linkages, we can define the circuit free kinematics with respect to $T$ as the following mapping:

$$L_T : \mathcal{Q}_{l(1)}^{r(1)} \times \ldots \times \mathcal{Q}_{l(m-1)}^{r(m-1)} \to SE_{p(1)}^{q(1)} \times \ldots \times SE_{p(n_c)}^{q(n_c)} :$$

$$(h_{l(1)}^{r(1)}, \ldots, h_{l(m-1)}^{r(m-1)}) \mapsto \begin{pmatrix} h_{v_1(n_1-1)}^{q(1)} \circ \ldots \circ h_{p(1)}^{v_1(1)} \\ h_{v_2(n_2-1)}^{q(2)} \circ \ldots \circ h_{p(2)}^{v_2(1)} \\ \vdots \\ h_{v_{n_c}(n_{n_c}-1)}^{q(n_c)} \circ \ldots \circ h_{p(n_c)}^{v_{n_c}(1)} \end{pmatrix}. \quad (2.12)$$

We can then give an expression of the configuration manifold for the system as follows:

$$\mathcal{Q} := \{q \in \mathcal{Q}_{l(1)}^{r(1)} \times \ldots \times \mathcal{Q}_{l(m-1)}^{r(m-1)} \text{ s.t. } L_T(q) \in \mathcal{Q}_{p(1)}^{q(1)} \times \ldots \times \mathcal{Q}_{p(n_c)}^{q(n_c)}\} \quad .$$

If we consider the differential kinematic of $L_T$, we have an expression that is identical to Eq. (2.10). The image of the Jacobian of the simplified forking kinematics corresponding to the tree $T$. which we will indicate as $J_T(h)$, gives the hypothetical directions of motions. Unfortunately, not all of these directions are possible since the kinematic links corresponding to the edges in $C$ are also present in the mechanism. If we define:

$$\mathcal{T}_C^A := \mathcal{T}_{q(1),p(1)}^A \times \mathcal{T}_{q(2),p(2)}^A \times \ldots \times \mathcal{T}_{q(n_c),p(n_c)}^A$$

then an expression for all the possible moving directions in a configuration $h$ is:

$$\mathcal{T}^A := J_T^{-1}(h)\mathcal{T}_C^A$$

where $J_T^{-1}(h)$ is the pre-image (see Def. A.1.3) of the Jacobian corresponding to the simplified tree mechanism and $\mathcal{T}^A \subset \mathcal{T}_{r(1),l(1)}^A \times \ldots \times \mathcal{T}_{r(m-1),l(m-1)}^A$. The dimension of the subspace $\mathcal{T}^A$ gives the degrees of freedom of the complete mechanism. It could happen that $\mathcal{T}^A$ is the zero element. This means that the mechanism is *over constrained* and can not move in any direction.

It should be noted that, in general, we can also have motions of the mechanism which do not change the configuration of the kinematic pairs corresponding to the edges in $C$. These self-motions correspond to the kernel of $J_T(h)$ or, equivalently, to the pre-image of the null element through $J_T^{-1}(h)$. We could also consider the dependence of $\mathcal{T}^A$ on the configuration $h$ and define a distribution, of which the integral of the involutive closure would result in the minimal configuration manifold $\mathcal{Q}$ for the complete closed chain mechanism.

The pre-image of $J_T(h)$ could be in practice calculated with pseudo-inversion techniques.

**Internal forces of a closed chain mechanism.** One typical feature of a closed chain mechanism is that it can have *internal forces*. This means that we could apply certain equilibrated wrenches different from zero to the kinematic pairs of the structure. These wrenches would not have any effect on the motion of the structure. This corresponds to what is called *redundant actuation*.

As we have already seen in the case of forking linkages, the mapping of wrenches is represented by the adjoint of the Jacobian. In the present case we should therefore consider the adjoint of $J_T(h)$ which is a map having the following form:

$$J_T^*(h) : \mathcal{W}^A_{q(1),p(1)} \times \mathcal{W}^A_{q(2),p(2)} \times \ldots \times \mathcal{W}^A_{q(n_c),p(n_c)} \\ \rightarrow \mathcal{W}^A_{r(1),l(1)} \times \mathcal{W}^A_{r(2),l(2)} \times \ldots \times \mathcal{W}^A_{r(m-1),l(m-1)}. \quad (2.13)$$

If we denote the total set of the kinematic pairs' wrenches as:

$$\mathcal{W}^A := \mathcal{W}^A_T \times \mathcal{W}^A_C$$

where

$$\mathcal{W}^A_T := \mathcal{W}^A_{r(1),l(1)} \times \mathcal{W}^A_{r(2),l(2)} \times \ldots \times \mathcal{W}^A_{r(m-1),l(m-1)}$$

and

$$\mathcal{W}^A_C := \mathcal{W}^A_{q(1),p(1)} \times \mathcal{W}^A_{q(2),p(2)} \times \ldots \times \mathcal{W}^A_{q(n_c),p(n_c)}$$

the set of internal forces for the mechanism can be expressed as the following set:

$$\mathcal{W}^A_I := \{(w_T, w_C) \in \mathcal{W}^A \text{ s.t. } w_T = -J_T^*(h) w_C\}$$

or even better as the subspace of $\mathcal{W}^A$ represented by the kernel of the following linear map:

$$I : \mathcal{W}^A \rightarrow \mathcal{W}^A_T \; ; \; (w_T, w_C) \mapsto \left(Id \; J_T^*(h)\right) \begin{pmatrix} w_T \\ w_C \end{pmatrix}$$

If actuators exist which can generate wrenches in $\mathcal{W}^A_I$ it is possible to generate forces which do not have any noticeable effect on the motion of the chain.

An example could be the hand-object combination of a grasped object. In this case the internal forces would correspond to those forces representing the hardness of a grasp and these forces are represented in literature by means of what is called the grasp matrix [32].

In bond graphs, closed chains can be drawn in the same way as they were drawn for serial kinematic chains, namely by choosing a reference body. The bond graph will have loops which result in causal conflicts giving rise to algebraic constraints. Details of these problems can be found in [33].

## 2.3 Conclusions

In this chapter two main topics have been treated. First, the definition and description of kinematic pairs have been explained. We saw that there are kinematic pairs, called lower pairs, which have the feature of being describable independently of their configuration. These pairs are the most important ones in practice. Bond graph interpretations have been given.

Second, all the possible rigid mechanism have been analyzed and explained. It has been shown that, in general, the actuation space does not correspond with the configuration space. A typical example is a human arm: this has few degrees of freedom but a lot of muscles. These muscles could be modeled as extra actuated pairs. The resulting describing mechanism would then correspond to a closed chain mechanism. We have seen that such mechanisms can have internal forces. For a human arm these forces correspond to the contemporary contraction of opposite muscles like the biceps and triceps.

# 3. Dynamics of Rigid Mechanisms

The word dynamics comes from the Greek δύναμις (dynamis), which means force. Dynamics studies the evolution of systems as a function of time. In this chapter we therefore study how the motion of a rigid mechanism is influenced by applied forces generated by actuators, other elements like springs and dampers and gravity. Interested readers can consult [34] or [35] for an extensive review of analytic mechanical issues. More specific issues concerning the study of differential equations can be found in [16]. This chapter contributes by describing how physical elements interconnect and by deducing the corresponding equations. Related existing work can be found in [19], [36] and [1], of which this book is an extension.

In Chapter 2 we studied how the motion of mechanisms can be described, but excluded the question how motion is generated. This question will be answered in this chapter. After a general descriptive introduction to some frameworks which are used for the description of dynamic systems, the issue of decomposition and interconnection of subsystems will be briefly treated. The Lagrangian formulation will also be treated for the general case with energetic ports, although most literature on this subject only uses it for autonomous systems. The Hamiltonian description will then follow in the Poisson and Dirac framework [37, 38, 39]. The basic elements composing a mechanical system will be treated; study of their interconnections and the deduction of explicit equations will end the chapter.

## 3.1 Study of Dynamical systems

Before treating more specific issues concerning the dynamics of mechanical systems, we will reflect on the concepts behind modeling of the dynamics of systems and in particular on the concepts of physical systems.

### 3.1.1 The behavioral approach

The process of observing a system and describing its behavior, involves two entities: the system that is observed and the observer. During the process of observing, quantities of the system that is being observed are compared with the observer's own reference quantities.

In general systems, these quantities could be physically unmeaningful. In a physical system, they could be velocities, forces or other physical quantities. Most important of all these, however, is the quantity "time". It is generally accepted that time is NOT an absolute, something that people believed before the special theory of relativity, and therefore this is also a reference for the observer.

In the standard nonrelativistic study of dynamics, time is given a special role and other quantities are measured and analyzed as functions of time and called *signals*. The dynamics of the system can be described by the set of signals that the system dynamics allows.

This last approach is exactly the one taken in the behavioral framework [40, 41]. To formalize this approach, consider the set of all the possible quantities observed from the system and indicate it with $W$. In the behavioral approach, these quantities are called *manifest variables*. Furthermore, consider time as an open interval $I \in \mathbb{R}$. We can define a signal of our system as an element of $W^I$, and therefore as a function which associates with an "instant" $t \in I$, the value of the observed quantities at that instant. The dynamics can then be described as all the possible elements of $W^I$ that the system dynamics allows. This very general and interesting framework is the essence of the behavioral formalism[1].

Compared to other standard formalisms such as the state space representation, the behavioral formalism has as a major advantage its completely noncausal formulation. With noncausal we mean that no information is given about the observed quantities that indicates whether they are causes or effects. Some readers might consider this an artificial way of dealing with systems, but in fact it is the only proper way to describe some of them: causality is often only needed to simulate certain systems and it is hardly ever an intrinsic feature of a system. Consider, for example, an electrical resistor: there is no reason whatsoever to consider the current as a function of the applied voltage, or the applied voltage as a function of the current, for that matter.

The generality of the framework just explained does not make it easy to describe systems, and for this reason, the concept of *latent variables* is introduced in the behavioral framework. If we go one step further, we can require that those latent variables satisfy what is called in [40] *the axiom of state*. Intuitively, this axiom expresses that at a certain instant the value of these special latent variables should contain sufficient information about the past to be able to determine what will happen in the future. If these latent variables satisfy the axiom of state, they are called *states* and they correspond exactly to the usual notion of the state presented in any basic course on dynamical systems.

It is important to notice that, first, the description of the manifest variable is very general and defines a framework able to describe the action of

[1] Another approach to the study of dynamical systems is the differential algebraic method of Fliess [42].

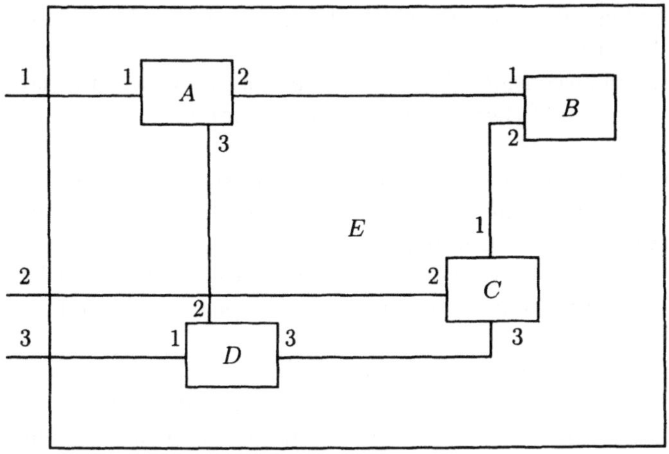


**Fig. 3.1.** Interconnection in the behavioral framework.

observation. Second, the introduction of latent variables seems at this stage an artificial step necessary to be able to work with the framework. This is because up to now we have considered only the observable signals of the system without deliberating about what could have generated them. Latent variables often result from a detailed decomposition in subparts of the system that is analyzed. The process of decomposing a system in connected subsystems is sometimes called *tearing*, *diakoptics* by Kron [43] or *hierarchical modeling* [44]. In the behavioral approach, interconnected subsystems are composed of systems which have some of their manifest variables in common equal for all times. Figure 3.1 shows an example of this approach for a system $E$ which is composed of four subsystems, namely $A, B, C, D$. The subsystems $A, C, D$, and $E$ itself have each three manifest variables. However, $B$ has only two. The interconnection means, for example, that the manifest variable 1 of system $B$ should always be equal to the manifest variable 2 of subsystem $A$, and so on. Those variables need not have any particular structure. From a physical point of view this turns out to be a shortcoming, which can be overcome by adding extra assumptions. This means that for the more specific goal of studying physical systems, extra structure and therefore constraints should be added to the behavioral approach. For the more general purpose of describing the theory of any kind of dynamical system, physical or not, the behavioral framework offers a general, formally well-posed description.

### 3.1.2 A step further: physics and bond graphs

In the previous section, we recalled very briefly concepts introduced in the behavioral framework. For the study of physical systems, we will add more structure to the model and introduce physical concepts. Probably the most important physical concept on which physics is based is the thermodynamical concept of energy [45].

Energy can be considered to have two major properties: invariance in time (the first principle of thermodynamics) and continuity in space (principle of Heaviside).

Invariance in time means that the energy in the universe is constant. This does not imply anything about the motion or the localization of energy in certain parts of space. Continuity in space, on the other hand, postulates that if energy has to move from one part of space to another, it has to travel the space between them and cannot just disappear at one point and appear at another.

Using the concept of tearing, we can then consider a system as a set of interconnected subsystems. Each of these subsystems can possess energy, which can travel from one subsystem to another by passing through an ideal boundary dividing the two. The energy that each of these subsystems contains is a function of some latent variables, which are states of the subsystem. These states can be physically defined by using thermodynamical concepts like *extensive variables* [45].

All the points explained here are precisely those used for bond graphs [46]. This framework based on physical concepts will be described later on.

## 3.2 Physical Interaction and Interconnection

We start this section with a definition of what is meant with interaction.

**Definition 3.2.1 (Interaction).** *Given two systems $A$ and $B$, we say that the systems interact if the behavior of $A$ is influenced by the presence of $B$ and/or vice versa.*

It is important to realize that a sufficient condition for interaction is therefore the existence of an exchange of energy between the two systems: there is an energy flow from $A$ to $B$ or from $B$ to $A$.

We characterize the interaction with variables that are common to both systems. In the behavioral framework, no structure is required for these common variables connecting the subsystems. We do require extra structure for the variables which we call power ports.

### 3.2.1 Power Ports

We say that the exchange of energy characterizing the interaction takes place through what are called *power ports*. Power ports model and localize the process of the exchange of energy between subsystems. To quantify this exchange of energy, we will describe a port $P$ with certain variables, for which the specific structure of a vector space $\mathcal{V}$ and its dual $\mathcal{V}^*$ are required:

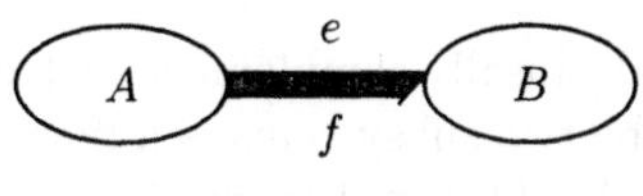


$(f, e) \in \mathcal{V} \times \mathcal{V}^*$

**Fig. 3.2.** Interaction between two systems.

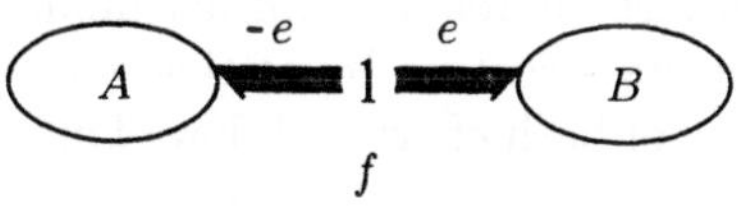


**Fig. 3.3.** Interaction between two systems in common flow form.

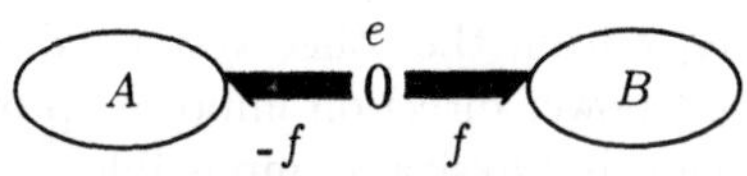


**Fig. 3.4.** Interaction between two systems in common effort form.

$$P := \mathcal{V} \times \mathcal{V}^* \tag{3.1}$$

for a suitable vector space $\mathcal{V}$. An element in $\mathcal{V}$ is called a *flow* and an element in $\mathcal{V}^*$ is called an *effort*. At each instant, the port variables are an effort and a flow. The application of an effort $e_i$ (which is a linear operator on $\mathcal{V}$) on a flow $f^i$ gives a scalar value $e_i f^i$, representing the power flowing through the described port. The interaction between two subsystems is represented in Figure 3.2, where bond graphs notation is used. The power bond direction gives the positive direction of power for $e_i f^i$: at each instant we have $e_i f^i$ power flowing from subsystem $A$ to subsystem $B$.

In Chapter 2, we used the vector space $\mathcal{V} = se_i(n)$ for an index $i$ depending on the simple junction structure used, as Sect. 2.2.1 shows.

### 3.2.2 Vector spaces and interconnections

When we want to express a power-consistent interconnection of a set of subsystems, as explained in the section above, we need a common vector space $\mathcal{V}$ and its dual $\mathcal{V}^*$.

On the other hand, each of the interconnected subsystems has its own state manifold, which differs from all the others.

It is therefore fundamental to find a way to map tangent vectors or co-vectors at a certain point of the state manifold of each subsystem to vectors and co-vectors belonging to a common vector space and its dual, where the interconnection can be expressed. This can be done using Lie group structures or Euclidean connections, depending on the specific case.

### 3.2.3 Causality

Appendix B explains general concepts about the causality of bond graphs. The major issue of causality is that two interacting systems as, for example, the ones in Figure 3.2, should causally complement each other [2]. This means that if the input of one of the two subsystems is a flow, and its output is an effort, the other subsystem should complement the first by having an effort as input and a flow as output. The situation of the first subsystem is called *impedance causality* for the interaction port, whereas the situation of the second is called *admittance causality* for the port.

The fact that a system cannot impose both the effort and the flow on a port is easily explained: then it would in fact be able to subtract infinite energy from the other system, disregarding its structure by setting a constant power direction imposing proper values of both effort and flow. This is obviously physically impossible.

### 3.2.4 Interconnection

A physical system composed of interconnected physical subsystems can be seen as a set of subsystems interacting with each other. Each subsystem has usually more ports, each of which is characterized by a vector space $\mathcal{V}_i$. Only subsystems with compatible ports (based on the same $\mathcal{V}_i$) are able to interact directly through that port, as shown in Figure 3.2.

To structure the interconnection, we consider all the bonds directly attached to a subsystem pointing inward and therefore with an incoming power that is positive. This implies that when we interconnect two subsystems, two possibilities exist for each port:

- *Common flow* interconnection, represented with an interconnection through a 1-junction. The junction means that both the bonds have the same flow and therefore opposite effort (see Figure 3.3).
- *Common effort* interconnection, represented with an interconnection through a 0-junction. The junction means that both the bonds have the same effort and therefore opposite flow (see Figure 3.4).

In this way we can express the interconnection of systems having equivalent ports and power bonds oriented toward the subsystems. Nevertheless, the requirement of causality must also be met: any two interacting subsystems should be causally complementary.

We can formally cope with this problem by means of what is called a symplectic gyrator [45] and a symplectic interconnection.

**Symplectic interconnection.** A symplectic gyrator is the element which represents duality of energetic domains, and its importance has already been studied in depth by Breedveld [45]. For our needs, a symplectic gyrator can be considered as a two-port element which swaps the role of effort and flow between its two ports. It is clear that this is possible since the dual $\mathcal{V}^*$ of a vector space $\mathcal{V}$ is a vector space of the same dimension.

In the sequel, we will often model subsystems by means of Hamiltonians systems with ports. These subsystems all need a flow as input and an effort as

**Fig. 3.5.** Common effort, dualization of $B$.

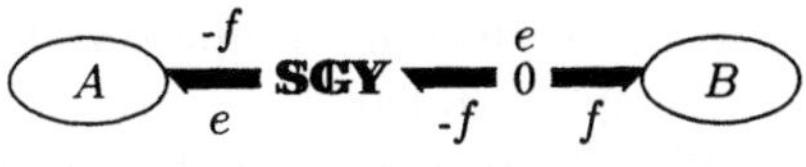


**Fig. 3.6.** Common effort, dualization of $A$.

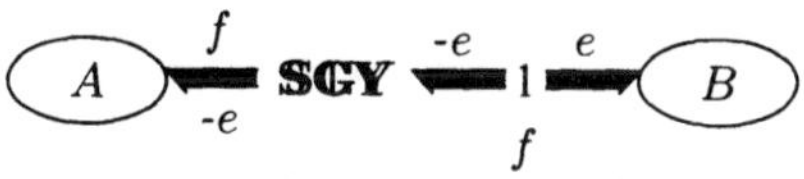


**Fig. 3.7.** Common flow, dualization of $A$.

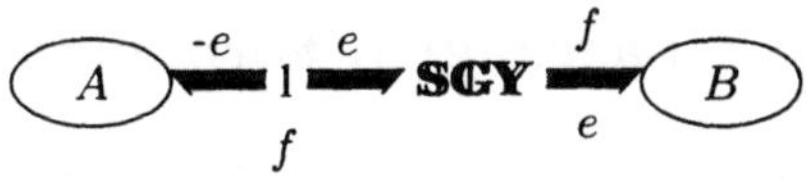


**Fig. 3.8.** Common flow, dualization of $B$.

output in the Poisson framework and as we will use them. This implies that to be able to interconnect them, we need to dualize one of the ports of the two subsystems, so that after dualization, one of the subsystems will actually formally accept an effort as input. This will not be strictly necessary if we use Dirac structures. There are only four possible symplectic connections for two systems with the same causality: these are reported in Figure 3.5, Figure 3.6, Figure 3.7, and Figure 3.8. By dualising the 0 and 1 junctions, we can see that the connection in Figure 3.5 is identical to the one in Figure 3.7 and that the connection in Figure 3.6 is identical to the one in Figure 3.8. This means that there are only two simple interconnections for systems with the same causality which correspond to:

$$\begin{bmatrix} f_2 \\ e_2 \end{bmatrix} \begin{bmatrix} 0 & I \\ -I & 0 \end{bmatrix} \begin{bmatrix} f_1 \\ e_1 \end{bmatrix} \quad \text{and} \quad \begin{bmatrix} f_2 \\ e_2 \end{bmatrix} \begin{bmatrix} 0 & -I \\ I & 0 \end{bmatrix} \begin{bmatrix} f_1 \\ e_1 \end{bmatrix}$$

where $(f_1, e_1)$ and $(f_2, e_2)$ are the flow-effort pairs of the two subsystems ports respectively.

There are two major approaches to study physical systems: the Lagrangian formulation and the Hamiltonian one. It is well known [35, 16] that the Lagrangian formulation is equivalent to the Hamiltonian one under some hyper-regularity conditions [47]. This equivalence can easily be represented by means of the Legendre transformation. Often it is not realized, however, that this equivalence between a Lagrangian and a Hamiltonian formulation is only easily expressed if the latter is represented in a canonical form.

As pointed out in [48], interconnection can be expressed explicitly by means of a Poisson manifold or a Dirac structure [49, 37, 39] formulation of Hamiltonian dynamics; Sect. 3.4 will discuss this in detail. This is not the case for a Lagrangian formulation, which is basically only based on variational methods using Hamilton's principle of least action [35].

Furthermore, the Hamiltonian formulation is more general since in its basic form, it naturally allows the description of systems which are not integrable and which have inputs with preferred impedance causality, as shown in Example 1 in [50].

We will treat both the Lagrangian and the Hamiltonian framework, but the second one will be used more extensively because it has better, more general features for the description of interconnected physical systems.

## 3.3 Lagrangian Formulation

The Lagrangian study of dynamics is based on concepts of Riemannian geometry (see Sect. A.7.6). A nice overview of modern techniques for the study of mechanical systems using the Lagrangian framework can be found in [51]. Usually, the Lagrangian formulation of dynamics is used for describing autonomous systems (without inputs). A clear-cut formulation of the first case can be found in [20], and a review including inputs is given by [47]. We will try to present the material in an intuitive and yet sufficiently precise way.

### 3.3.1 The simplest case: no potential energy and no ports

The simplest Lagrangian system we can think of is an isolated. lossless system with no potential energy. Its dynamics will be only due to the configuration dependence of what is called the inertia tensor. In a Riemannian geometric framework, this tensor represents the metric characterizing the geometry of the system.

To be more specific, consider a configuration manifold $\mathcal{Q}$ for the isolated system under consideration. With each point $q \in \mathcal{Q}$, we have associated a metric with the tangent vector space $T_q\mathcal{Q}$ which characterizes the kinetic co-energy[2]. This means that if the system finds itself in a configuration $q \in \mathcal{Q}$ with a velocity $\dot{q} \in T_q\mathcal{Q}$, the kinetic co-energy can be expressed as:

$$E_k^* : T\mathcal{Q} \rightarrow \mathbb{R}^+ \; ; \; (q, \dot{q}) \mapsto \frac{1}{2} M_{ij}(q)\dot{q}^i\dot{q}^j$$

in which we have used the tensor summing convention[3]. The manifold $\mathcal{Q}$ with the metric $M_{ij} \in T_{0,2}\mathcal{Q}$, has the structure of a Riemannian manifold. The notation $T_{0,2}\mathcal{Q}$ denotes the 0 contravariant, 2 covariant tensor bundle (see Def. A.7.14). The metric is nothing else than a position-dependent, positive definite quadratic form of the velocities in each point. We can now define what we mean by an arc.

[2] Formally speaking, the kinetic co-energy is a function of velocities and the kinetic energy of momenta.

[3] We use the implicit tensor notation here because it simplifies the expressions. A brief review is given in Sect. A.5.1.

**Definition 3.3.1 (Arc).** *An arc $\tilde{q} \subset \mathcal{Q}$ starting from $q_1 \in \mathcal{Q}$ and ending in $q_2 \in \mathcal{Q}$ is a smooth curve connecting $q_1$ to $q_2$. A parameterization of the arc is a continuous bijective function $q : [t_1, t_2] \to \tilde{q}$ from a close interval $[t_1, t_2] \subset \mathbb{R}$ to the arc, in which $q(t_1) = q_1$ and $q(t_2) = q_2$.*

We can consider an arc as a possible evolution of the system in which we consider the parameterization as the function of time associating to each instant $t \in [t_1, t_2]$ the configuration $q(t)$ of the system at that instant.

We can now define what is called *action*:

**Definition 3.3.2 (Action).** *The action of an arc $q(t)$, $t \in [t_1, t_2]$ joining $q_1 = q(t_1) \in \mathcal{Q}$ to $q_2 = q(t_2) \in \mathcal{Q}$ and a Lagrangian function $L : T\mathcal{Q} \to \mathbb{R}$ is defined as:*

$$S[q] = \int_{t_1}^{t_2} L(q(t), \dot{q}(t))dt. \tag{3.2}$$

A major principle in physics states that the arc can correspond to a possible evolution of the system $q(t)$ if it satisfies Hamilton's principle of least action [35]. This principles states that $q(t)$, to be a possible evolution of the system from $q_1$ to $q_2$, should minimize the action $S[q]$.

For our autonomous case with only kinetic co-energy, this principle should hold when $L(q, \dot{q}) = E_k^*(q, \dot{q})$. From the calculus of variations [20] it is well known that if a curve $q(t)$ is to be minimal for the integral in Eq. (3.2), it must satisfy the *Euler-Lagrange* equation:

$$\frac{d}{dt}\left(\frac{\partial L}{\partial \dot{q}}\right) - \frac{\partial L}{\partial q} = 0. \tag{3.3}$$

Note that the previous equation is a covariant equation, which means that $\partial L/\partial q$, $d/dt(\partial L/\partial \dot{q})$, $0 \in T_q^*\mathcal{Q}$.

*Remark 3.3.1 (Interpretation).* Note that if we consider $\partial L/\partial \dot{q}$ to be a "generalized momentum" $p$, Eq. (3.3) can be written as

$$\dot{p} = \frac{\partial L}{\partial q}$$

which resembles the dynamics of an inertial element with momentum $p$ connected to an elastic force $\partial L/\partial q$. For this reason a position-dependent inertia $M_{ij}(q)$ also generates an "apparent force" $\partial(M_{ij}(q)\dot{q}^i\dot{q}^j)/\partial q$ which modifies the momentum $p$. It is possible to see that if for $\mathcal{Q}$ there are coordinates $\bar{q}$ in which the inertia tensor would become constant, we will have the simplified equation $\dot{\bar{p}} = 0$. It is a standard result of Riemannian geometry that such global coordinates exist iff the curvature of the Riemannian manifold (see Sect. A.7.7) is zero everywhere. In such a case, therefore, we would obtain trivial equations with conservation of momentum.

The minimization of Eq. (3.2) with $L = E_k^*$ can be interpreted as the minimization of:

$$S[q] = \int_{t_1}^{t_2} |\dot{q}(t)|_M^2 dt \tag{3.4}$$

where $|\dot{q}(t)|_M^2$ indicates the square of the magnitude of the vector $\dot{q}(t) \in T_q\mathcal{Q}$ by using the Riemannian metric $M(q)$ in the current point $q$.

Note that the previous integral is NOT the length measured with the Riemannian metric of the arc $q(t)$ followed by the system joining $q_1$ to $q_2$, as we would expect from the theory of geodesic flow of dynamical systems. It can be shown [20] that an arc $q(t)$ minimizing Eq. (3.4), also minimizes

$$S[q] = \int_{t_1}^{t_2} |\dot{q}(t)|_M dt \tag{3.5}$$

which is now indeed the length of the arc $q(t)$ joining $q_1$ to $q_2$ for the metric $M(q)$. This implies that if we look for the arcs corresponding to the trajectories that the system evolution follows, we can consider what are called the *geodesics* of the Riemannian manifold (see Sect. A.7.7). The geodesics are intuitively those curves of minimum length joining certain points. The equation of the geodesics can be expressed by the following tensor operation of covariant differentiation (see Sect. A.7.7):

$$\nabla_{\dot{q}}\dot{q} = 0. \tag{3.6}$$

In the previous equation, the corresponding connection (see Sect. A.7.7) that defines the operation is the Levi-Civita connection (see Theorem A.7.6). This connection corresponds to the only torsion-free connection compatible with the Riemannian metric. Note that now Eq. (3.6) is a contravariant equation corresponding to Eq. (3.3) with $L = E_k^*$ multiplied by the inverse of the inertia tensor.

### 3.3.2 Adding energetic ports

In the previous section, we considered isolated systems with only kinetic co-energy. In [47], inputs for the system are defined by the use of jets. Here we use a more intuitive approach, which considers the energy exchanged with the system as a starting point. We study the case in which we can supply or subtract energy from the system through what is called an *energetic port* which localizes the exchange of energy with the rest of the universe. In this case, an energetic port is characterized by two dual vectors: a vector $(q, \dot{q}) \in T\mathcal{Q}$ corresponding to the motion of the system and a dual element $(q, \tau) \in T^*\mathcal{Q}$. These are characterized in such a way that their dual product represents the power supplied to the system through the port characterized by $(q, \dot{q})$ and $(q, \tau)$.

If $\langle \tau, \dot{q} \rangle_q$ indicates the dual product of a vector $\dot{q}$ and a co-vector $\tau$ in $q$, we need for power consistency that, indicating with $E_k^*(q, \dot{q})$ the kinetic co-energy of the Lagrangian system:

$$\left. \frac{dE_k^*}{dt} \right|_{q,\dot{q}} = \langle \tau, \dot{q} \rangle_q \tag{3.7}$$

(i.e. the rate of change in kinetic co-energy is equal to the supplied power). If we define the covariant tensor called *generalized momentum* with[4] $p_i := M_{ij}\dot{q}^j$, we have:

$$p_i = \frac{\partial E_k^*}{\partial \dot{q}_i}.$$

We can then write:

$$E_k^*(q, \dot{q}) = p_i \dot{q}^i - \frac{1}{2} M_{lj}(q) \dot{q}^l \dot{q}^j = p_i \dot{q} - E_k^*(q, \dot{q})$$

and taking the time derivative of any trajectory $q(t)$ in the previous equations, we have:

$$\frac{dE_k^*}{dt} = p_i \ddot{q}^i + \dot{p}_i \dot{q}^i - \frac{dE_k^*}{\partial q^i} \dot{q}^i - p_i \ddot{q}^i = \left( \frac{d}{dt} \left( \frac{\partial E_k^*}{\partial \dot{q}^i} \right) - \frac{\partial E_k^*}{\partial q^i} \right) \dot{q}^i \tag{3.8}$$

where more properly expressed, $(t, q, \dot{q}, \ddot{q})$ belongs to the 2-jet [16] from $\mathbb{R}$ to $\mathcal{Q}$.
For Eq. (3.8) and Eq. (3.7) to be equal for all possible $(q, \dot{q})$, it must be:

$$\frac{d}{dt} \left( \frac{\partial E_k^*}{\partial \dot{q}^i} \right) - \frac{\partial E_k^*}{\partial q^i} = \tau_i. \tag{3.9}$$

Following a similar line of argument, one can see that the previous equation in covariant differentiation terms becomes:

$$\nabla_{\dot{q}} \dot{q} = M^{ij} \tau_i \tag{3.10}$$

where $M^{ij}$ is the inverse of the inertia tensor $M_{ij}$. This last equation shows that if the Riemannian curvature corresponding to the metric $M_{ij}$ on $\mathcal{Q}$ is zero everywhere, there are coordinates $\bar{q}$ for the manifold $\mathcal{Q}$ in which the Christoffel symbols characterizing the covariant differentiation are also identically zero (see Sect. A.7.7). In that case, Eq. (3.10) takes the form:

$$\ddot{\bar{q}}^j = \bar{M}^{ij} \bar{\tau}_i \quad \text{or} \quad \bar{M}_{ij} \ddot{\bar{q}}^j = \bar{\tau}_i$$

corresponding to the most common form of Newton's law which is only valid in cases where the inertia tensor is constant.

[4] Here we use again implicit tensor notation for convenience.

### 3.3.3 Adding potential energy

So far, we have only considered the case in which the system's Lagrangian function only consists of kinetic co-energy. We now consider a situation where we also have storage of potential energy.

Potential energy is by definition only a function of the configuration; it is bounded from below and has the following form:

$$E_p : \mathcal{Q} \to \mathbb{R}.$$

Considering the differential of the potential energy, we get:

$$dE_p = \frac{\partial E_p}{\partial q} dq.$$

This implies that the change of potential energy as a function of time is:

$$\frac{dE_p}{dt} = \langle (q, \partial E_p/\partial q), (q, \dot{q}) \rangle$$

and therefore $-(q, \partial E_p/\partial q) \in T_q^* \mathcal{Q}$ corresponds to the generated elastic force[5]. If we apply this force together with one other external force which may be produced by any other external system, we get:

$$\tau_i = -\frac{\partial E_p}{\partial q^i} + \tau_i^* .$$

By substituting the last equation in Eq. (3.9), we obtain:

$$\frac{d}{dt}\left(\frac{\partial E_k^*}{\partial \dot{q}^i}\right) - \frac{\partial E_k^*}{\partial q^i} = -\frac{\partial E_p}{\partial q^i} + \tau_i^*$$

which can be written as:

$$\frac{d}{dt}\left(\frac{\partial E_k^*}{\partial \dot{q}^i}\right) - \frac{\partial (E_k^* - E_p)}{\partial q^i} = \frac{d}{dt}\left(\frac{\partial L}{\partial \dot{q}^i}\right) - \frac{\partial L}{\partial q^i} = \tau_i^*$$

where $L = E_k^* - E_p$. The latter corresponds to the well-known Euler-Lagrange equation with inputs. Note that the previous form may be given because $E_p$ does not depend on $\dot{q}$ and, therefore, $\partial E_p/\partial \dot{q}^i = 0$.

Using external additional forces would enable us to consider dissipative terms.

[5] The force corresponding to $\partial E_p/\partial q$ is the force which is applied to the element with potential energy $E_p$ and therefore, in a dual way, we can say that the opposite vector is the force that the element generates.

## 3.4 Hamiltonian Formulation

The Riemannian geometry plays the same structural role in the Lagrangian approach to dynamics as symplectic geometry in the classical Hamiltonian framework.

The most basic form of the standard Hamiltonian formalism is based on a symplectic structure [35, 34]. This structure can be directly defined if we describe a system by starting from its configuration manifold $\mathcal{Q}$, which is NOT a state manifold. We can then define a canonical 2-form on its cotangent bundle $T^*\mathcal{Q}$ which is now considered as a state manifold. Unfortunately, this approach is not general enough. This is due to the fact that a simplectic form is a 2-form and therefore can be seen as generating a mapping from a vector field to a co-vector field. What we eventually need to describe Hamiltonian dynamics is a mapping going in the opposite direction: from a co-vector representing the gradient of a Hamiltonian energy function to the corresponding Hamiltonian vector field. In a symplectic setting, this would require the "inversion" of a 2-covariant tensor and therefore this tensor cannot be degenerate. Therefore, the basic form reported in [35] is a framework that is too restrictive because it is not able to describe all the necessary physical systems. A typical example is the description of the rotational dynamics of a rigid body; this cannot be handled directly by a symplectic Hamiltonian system.

The shortcoming mentioned above can be solved by considering a 2 contravariant tensor field which could directly map a co-vector field to a vector field. This tensor field corresponds to a Poisson structure [16, 18]. This is the main approach this book follows and it is based on [48].

Unfortunately such an approach does not allow the description of implicit Hamiltonian systems. A Dirac structure is used instead to overcome this shortcoming and to describe any noncausal, implicit Hamiltonian system with ports [37, 38, 52]. This material is also treated in [53].

### 3.4.1 Port Controlled Hamiltonian Systems using Poisson structures

The basic structure we can use for an explicit Port-Controlled Hamiltonian System is a *generalized Poisson manifold.*

A Poisson manifold is a manifold $\mathcal{X}$ with a generalized Poisson bracket $\{,\}$ [37, 50].

A *generalized Poisson bracket* $\{,\}$ is a bilinear map from $C^\infty(\mathcal{X}) \times C^\infty(\mathcal{X})$ into $C^\infty(\mathcal{X})$ which satisfies the skew-symmetry:

$$\{F, G\} = -\{G, F\}$$

and the Leibniz rule:

$$\{F, G \cdot H\} = \{F, G\} \cdot H + G \cdot \{F, H\}.$$

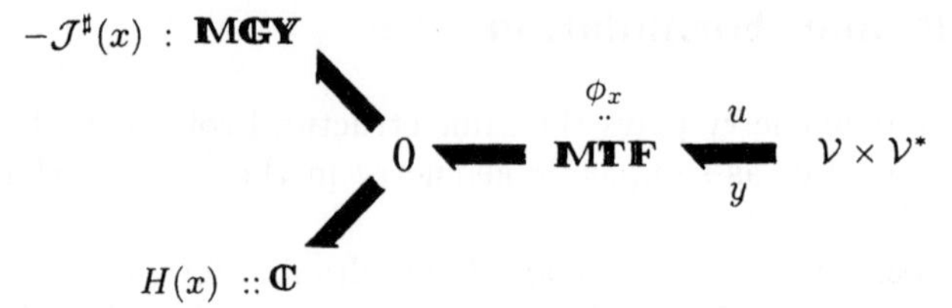


**Fig. 3.9.** The bond graph of a Hamiltonian system with Poisson structure.

It is important to say that in this case, the manifold we consider is NOT a configuration manifold as the one in the Lagrangian framework, but a state manifold: each point of the manifold corresponds to a physical state of the system.

$\mathcal{J}(x)$ denotes the skew-symmetric 2 contravariant tensor field called *generalized Poisson tensor* associated with the bracket $\{.,\}$, and defined by: $\mathcal{J}^{ij} dG_i dH_j = \{G, H\}$. $\mathcal{J}^\sharp(x)$ denotes the vector bundle morphism from $T^*\mathcal{X}$ to $T\mathcal{X}$ mapping exact 1-forms (see Sect. A.5) to Hamiltonian vector fields such that:

$$\mathcal{J}^\sharp(x)(dH)(dG) = \{G, H\}. \tag{3.11}$$

Choosing coordinates $x_1, \ldots, x_m$ in a neighborhood of a point $x$ of the manifold $\mathcal{X}$, we may represent the generalized Poisson tensor by the *structure matrix* $J(x)$:

$$J_{kl}(x) = \{x_k, x_l\}(x) \qquad k, l = 1, \ldots, m \tag{3.12}$$

where the coordinates are interpreted as coordinate functions. A *generalized 1-port controlled Hamiltonian system (PCHS)* [50, 37, 19] on the generalized Poisson manifold $\mathcal{X}$ is defined by a Hamiltonian function $H(x)$, a port interaction vector space $\mathcal{V}$, a fiber bundle linear map $\phi_x$ from $\mathcal{V} \times \mathcal{X}$ to $T_x\mathcal{X}$ and:

$$\dot{x} = \mathcal{J}^\sharp(x) dH(x) + \phi_x(u), \tag{3.13}$$

$$y = \phi_x^* dH(x), \tag{3.14}$$

where $\phi_x^*$ denotes the adjoint map of $\phi_x$ (see Def. A.5.5).

Figure 3.9 shows a bond graph representation of the previous system. The **C** element represents the energetic storage element with energy function $H(x)$. The element **MGY** represents the Poisson structure, and **MTF** represents the power-continuous element representing the fiber bundle linear map $\phi_x$ to the vector space $\mathcal{V}$ and its dual.

The drift vector field of this system is a Hamiltonian vector field defined with respect to the Poisson bracket. In Figure 3.9 this Hamiltonian vector field corresponds to the pair **MGY** and **C**. The interaction of the system with its environment is defined by the two *adjoint port interaction maps* $\phi_x$

and $\phi_x^*$ corresponding to the **MTF** in Figure 3.9 and by the *two port variables* $u \in \mathcal{V}$ and $y \in \mathcal{V}^*$. The port variables are said to be *power conjugated* because their duality product expresses the rate of change of internal energy:

$$\frac{dH}{dt} = \langle y, u \rangle. \tag{3.15}$$

The structure of $\mathcal{V} \times \mathcal{V}^*$, as in Eq. (3.1), therefore characterizes the port of the Hamiltonian system through which it exchanges energy with the rest of the universe.
If the space of port variables is $\mathbb{R}^k$, (which may correspond for instance to the choice of a basis $(v_1, \ldots, v_k)$ of $\mathcal{V}$), the interaction maps $\phi_x^i$ are defined by vector fields $\tilde{g}^i(x) = \phi_x(v_i)$. Choosing some coordinates $(x_1, \ldots, x_n)$ for the Poisson manifold, we define the port-controlled Hamiltonian system as follows [50] [37] :

$$\dot{x} = J(x)\frac{\partial H}{\partial x}(x) + \sum_{i=1}^{k} g^i(x)u_i, \tag{3.16}$$

$$y_i = g^{iT}(x)\frac{\partial H}{\partial x}(x) \tag{3.17}$$

where $\frac{\partial H}{\partial x}(x)$ denotes the gradient of the Hamiltonian function and $g^i(x)$ denotes the coordinate vectors of the vector fields $\tilde{g}^i(x)$. In a particular case the generalized Hamiltonian systems may be related to the standard Hamiltonian systems defined on symplectic manifolds, and this is when the generalized Poisson bracket satisfies the Jacobi identity:

$$\{F, \{G, H\}\} + \{G, \{H, F\}\} + \{H\{F, G\}\} = 0. \tag{3.18}$$

In this case, because of the Darboux theorem [16], the generalized Poisson bracket becomes a (true) Poisson bracket [18]. Suppose, moreover, that the structure matrix has a constant rank $2n$ in a neighborhood of a point $x_0 \in M$, then the Jacobi identity (Eq. (3.18)) ensures the existence of canonical coordinates $(q, p, r) = (q_1, \ldots, q_n, p_1, \ldots, p_n, r_1, \ldots, r_l)$ such that the structure matrix $J(q, p, r)$ becomes:

$$J(q,p,r) = \begin{pmatrix} 0_n & I_n & 0_{n\times l} \\ -I_n & 0_n & 0_{n\times l} \\ 0_{l\times n} & 0_{l\times n} & 0_{l\times l} \end{pmatrix}. \tag{3.19}$$

Using these canonical coordinates, the Hamiltonian drift vector field $\mathcal{J}(x)(dH)$ takes the following expression:

$$\begin{aligned} \dot{q}_i &= \frac{\partial H}{\partial p_i}(q, p, r) \\ \dot{p}_i &= -\frac{\partial H}{\partial q_i}(q, p, r) \qquad i = 1, \ldots, n \quad j = 1, \ldots, l. \\ \dot{r}_j &= 0 \end{aligned} \tag{3.20}$$

The latter expression of the Hamiltonian vector field shows that the coordinates may be split into two parts. First the coordinates $r_j$ are *dynamical invariants*, also called *Casimir functions* [18]. The first $2n$ equations form a standard Hamiltonian vector field parameterized by the invariants $r$.

*Remark 3.4.1.* It is important to note that the Casimir functions only depend on the Poisson bracket and remain invariant for any Hamiltonian vector field defined with respect to this bracket. The Poisson structure corresponds to the interconnection of the elements composing the system and therefore the Casimir functions are a consequence of the interconnection and not of the energy function.

It should be noted that the dynamics of mechanical systems with nonholonomic constraints leads to the introduction of a bracket on the constrained state space that satisfies the skew symmetry and the Leibniz rule, but that does not satisfy the Jacobi identity [54].

Note that the presented Hamiltonian framework is more powerful than the Lagrangian one for the following reasons:

- The Hamiltonian system is characterized by two entities: the Poisson structure and the energy function.The Lagrangian framework is only characterized by the Lagrangian function.
- The Poisson structure represents the interconnection of the system's parts and is independent of the energy function.
- The Hamiltonian function directly corresponds to the stored energy.

### 3.4.2 Port-controlled Hamiltonian Systems using Dirac Structures

An even more general approach to model Hamiltonian systems uses what is called a Dirac structure[49]. We start by introducing a Dirac structure on a vector space.

**Dirac structures on vector spaces.** Consider a $m$-dimensional real vector space $\mathcal{V}$ and denote by $\mathcal{V}^*$ its dual as usual. A *Dirac structure* [49] on the vector space $\mathcal{V}$ is an $m$-dimensional subspace $L$ of $\mathcal{V} \times \mathcal{V}^*$ such that:

$$\forall (f, e) \in L,\ \langle e, f \rangle = 0 \tag{3.21}$$

where $\langle , \rangle$ denotes the duality product. In order to give some constructive definition of different Dirac structures in the sequel, we shall define Dirac structures in terms of two linear maps: this is called a *kernel representation of a Dirac structure* [55].

Every Dirac structure $L \subset \mathcal{V} \times \mathcal{V}^*$ is uniquely defined in a basis $\mathcal{B} = \{v_1, .., v_m\}$ of $\mathcal{V}$ by the pair of real matrices $(F. E)$ called *structure matrices* and satisfying the condition:

$$E F^T + F E^T = 0 \tag{3.22}$$

by:

$$L = \{(f,e) \in \mathcal{V} \times \mathcal{V}^* \ ; \ F\bar{f} + E\bar{e} = 0\} \tag{3.23}$$

where $\bar{f}$ is the coordinate vector of $f$ in the basis $\mathcal{B}$ of $\mathcal{V}$ and $\bar{e}$ is the coordinate vector of $e$ in the dual basis to $\mathcal{B}$ of $\mathcal{V}^*$.

A Dirac structure defined on a vector space $\mathcal{V}$ encompasses the two structures of a Poisson vector space and a pre-symplectic vector space [49]. The condition of Eq. (3.22) implies Eq. (3.21).

Such Dirac structures arise in electrical circuits (or their mechanical analogue) where Kirchhoff's laws actually endows the set of currents and voltages of the circuit or the rate variables and the co-energy variables of an LC-circuit (possibly with elements in excess) with a Dirac structure [56] [57].

**Dirac structures on differentiable manifolds.** The definition of Dirac structures on vector spaces may be extended to differentiable manifolds as follows. Let $\mathcal{M}$ be a differentiable manifold: than a *Dirac structure* [49] on $\mathcal{M}$ is given by a smooth vector sub-bundle $L \subset T\mathcal{M} \times T^*\mathcal{M}$ such that the linear space $L(x) \subset T_x\mathcal{M} \times T_x^*\mathcal{M}$ is a Dirac structure on the tangent vector space $T_x\mathcal{M}$ for every $x \in \mathcal{M}$.

More intrinsic definitions in terms of duality and some characteristic distributions and co-distributions may be found in [49] [58].

In this case Dirac structures may also be locally characterized by a pair of *structure matrices* (depending now smoothly on the point $x$ in $\mathcal{M}$). Indeed, by choosing some coordinates $(x_1, .., x_m)$ in some neighborhood of a point $x \in M$, we now obtain a Dirac structure defined by a pair of matrices $F(x), E(x)$ depending smoothly on $x$ such that:

$$L(x) = \{(f,e) \in T_xM \times T_x^*M \text{ s.t. } F(x)f + E(x)e = 0\} \tag{3.24}$$

Once again, for power continuity, a condition similar to Eq. (3.22) should hold, but for a state-dependent matrix. On this local representation it may be seen that Dirac structures defined on differentiable manifolds generalize as well as (generalized) Poisson manifolds as pre-symplectic manifolds in the same way as Dirac structures do generalize pre-symplectic and Poisson vector spaces [49].

*Remark 3.4.2.* On a Dirac structure there exist also some additional closeness conditions which generalize the Jacobi identities of Poisson brackets, taking into account also the input port. They also play an essential role in the geometrical structure and representation of Dirac structures; the interested reader is referred to [49] [59] [58] [60].

**Dirac structures and Implicit Hamiltonian systems.** Using Dirac structures the generalized Hamiltonian systems may be extended to *implicit* Hamiltonian systems. Such systems were introduced by Courant [49] and are extensively used in the analysis of symmetries in evolution equations [59]. Contrary to the statement in [49], it seems that such structures also play a fundamental role in finite dimensional systems [54] [58] [56] [61] [39], as we shall see below.

An implicit Hamiltonian system with ports is defined by a state space, a differentiable $C^\infty$ manifold, denoted by $\mathcal{M}$, a vector bundle over $\mathcal{M}$ called *port interaction space* and denoted by $W_x$, a Dirac structure $L(x)$ on the vector bundle with fibers: $T_x\mathcal{M} \times W_x \times T_x^*\mathcal{M} \times W_x^*$ and a Hamiltonian function $H(x) \in C^\infty(\mathcal{M})$ by:

$$(\dot{x}, f, \frac{\partial H}{\partial x}(x), -e) \in \mathcal{D}(x), \quad x \in M \tag{3.25}$$

where $x \in \mathcal{M}$ is the state of the system, the pair $(f, e)$ are the external variables, also called *port variables* and the minus sign corresponds to the sign convention ensuring that the power going into the system is counted positively.

From the power continuity condition it follows that an implicit Hamiltonian system as the one of Eq. (3.25) satisfies the balance equation:

$$\frac{dH}{dt} = \langle e, f \rangle \tag{3.26}$$

Hence, if the Hamiltonian function is bounded from below, the implicit port-controlled Hamiltonian system is passive and lossless. Using a kernel representation of the Dirac structure $L(x)$ with structure matrices: $F_1(x), F_2(x), E_1(x), E_2(x)$, the Hamiltonian system of Eq. (3.25) becomes:

$$F_1(x)\dot{x} + F_2(x)f + E_1(x)\frac{\partial H}{\partial x}(x) - E_2(x)e = 0 \tag{3.27}$$

For a detailed exposure and analysis of such implicit port-controlled Hamiltonian systems, the reader is referred to [58] [56] [39].

## 3.5 Nodic elements

In the theory of physical systems, nodicity can be defined for an $n-$port subsystem [62]. Intuitively, we would say that a nodic element is a physical dynamical subsystem with $n > 1$ ports whose state change does NOT depend on the absolute value of its ports but on their relative values. Nodicity can only be properly defined in a coordinate-free manner for a set of ports of a subsystem which are compatible: they should be based on the same vector space (see Sect. 3.2.1).

Suppose that we have a subsystem $A$ with $n$ ports and that its ports are all compatible and based on the same space $\mathcal{V}$. We can define nodicity for such a system.

**Definition 3.5.1 (Nodic elements).** *A subsystem with n compatible ports $(f^i, e_j) \in \mathcal{V} \times \mathcal{V}^*$ based on a vector space $\mathcal{V}$ is said to be nodic with flow continuity and effort relativity (resp. effort continuity and flow relativity) iff the following properties hold:*

1. *Flow continuity (resp. effort continuity)*

$$\sum_{i=1}^{n} f^i(t) = 0 \left( \textit{resp. } \sum_{i=1}^{n} e_i(t) = 0 \right)$$

2. *Effort relativity (resp. flow relativity)*

$$f^i = \Gamma^i(e_1 - e_n, \ldots, e_{n-1} - e_n)$$
$$\left( \textit{resp. } e_i = \Gamma^i(f^1 - f^n, \ldots, f^{n-1} - f^n) \right) \quad (3.28)$$

*where $e_i \in \mathcal{V}^{*T}$ and $f_i \in \mathcal{V}^T$ are signals in $\mathcal{V}^*$ and $\mathcal{V}$ respectively.*

Note that the compatibility of the ports is necessary for summing the flows $f^i$ (resp. the efforts $e_i$).

In the electrical domain, all elements are nodic. This is due to the fact that a required reference or special set of references does not exist: the behavior of any electrical component only depends on the relative potential of its poles. Furthermore, the sum of the currents flowing in the element should always be zero.

The mechanical domain is special in this respect: it has physical elements which are NOT nodic, namely masses, which store kinetic energy. Masses are elements that are NOT nodic because their kinetic energy is only a function of their velocity with respect to a so-called inertial reference.

### 3.5.1 Inertial references

It is well known that in the expression of the kinetic energy of a mass, the motion of that mass must be considered relative to a reference space belonging to the class of inertial reference spaces. It can be seen that this latter class corresponds to an equivalence class: two Euclidean spaces are said to be *inertially equivalent* if they translate with respect to each other with a constant velocity. This intuitive idea can be formalized by using the Lie group concept of the killing form, see Sect. A.8.3. The killing form is a quadratic degenerate form defined on $se_i(n)$. Consider two spaces $\mathcal{E}_i$ and $\mathcal{E}_j$ and their relative twists $t_j^i \in se_i(n)$ and $t_i^j \in se_j(n)$. If we denote the killing form on space $i$ with:

$$\langle , \rangle_{K_i} : se_i(n) \times se_i(n) \to \mathbb{R} \; ; \; (t^{i'} . t^{i''}) \mapsto \mathrm{tr}(ad_{t^{i'}} ad_{t^{i''}})$$

where 'tr' indicates the trace operator, we can define the following equivalence relation for Euclidean spaces belonging to the same Euclidean system.

**Definition 3.5.2 (Inertial equivalence).** *We say that two Euclidean spaces $\mathcal{E}_i$ and $\mathcal{E}_j$ are inertially equivalent iff their relative motion is represented by a constant $t_j^i \in se_i(n)$ and by a constant $t_i^j \in se_j(n)$ so that:*

$$\langle t_i^j, t_i^j \rangle_{K_j} = \langle t_j^i, t_j^i \rangle_{K_i} = 0.$$

It can be proven that the previous relation is a properly defined equivalence relation.

Of all inertial equivalence classes we consider the *class of inertial systems.* Mechanical laws are invariant for changing references within this class. The kinetic energy of a body $B_i \subset \mathcal{E}_i$ is only meaningful if it is expressed as a function of the relative motion of $B_i$ with reference to one of the spaces in the class of inertial systems. Furthermore, the laws of classical mechanics are invariant for changes of reference within the class of inertial systems. This means that in the mechanical domain "special Euclidean spaces" exist and this is the reason why in this domain non-nodic elements exist, namely those masses whose energy has to be expressed as a function of a variable with reference to an inertial reference system.

The existence of non-nodic elements in mechanical systems has been given proper attention [62] and should be taken into account when studying the different physical domains.

## 3.6 Springs

Springs are those elements in the mechanical domain which store potential energy. With the name "spring" we do not mean a physical spring, but the ideal power-conserving concept of storage of potential energy.

In modeling, it is important to discriminate between *ideal elements* and *physical elements.* An ideal element represents an ideal concept like storage of energy or dissipation. A physical element can be an existing object with a predominating behavior similar to the one of an ideal element, but nevertheless it always has what are called *parasitic effects.* For example, a physical spring is never only a power-conserving element storing potential energy, but it always has some free energy dissipation and some distributed mass as well. In what follows, unless explicitly stated, we mean by the term *spring* the ideal energy-conserving concept of potential energy storage.

We consider two kinds of springs: *geometrical springs* and *winding springs.* The first kind is the one usually considered in the geometric literature as [63, 64, 10, 65, 66]. The second kind, which is often forgotten by mathematicians, has important implications for the dynamic description of systems and it is relevant in practice. These springs are elements which are able to create different wrenches for exactly the same configurations of the bodies they are attached to. To understand this, just think of two wooden sticks connected by just one nail. The sticks will only be able to rotate around the axis of the nail. If we now connect the two sticks with an elastic band in a place close to the nail, and rotate one stick 360 degrees around the nail keeping the other stick fixed, we are exactly at the same position but with a completely differently generated torque. The major attention will be given to geometrical springs.

### 3.6.1 Geometrical Springs

A spring is a nodic element and because of this, its definition always requires at least two bodies to which it is connected. In general, given $m$ bodies associated with the Euclidean spaces $\mathcal{E}_1, \dots, \mathcal{E}_m$, we define a spring by its energy storage function. The element is nodic because the energy is a function of only the relative positions of the bodies which the spring is attached to. Furthermore, any bodies can be used in the definition: these do not need to belong to any particular class of references.

**Definition 3.6.1 (Spring).** *Given m bodies contained in m Euclidean spaces $\mathcal{E}_1, \dots, \mathcal{E}_m$, we call a function of the following form:*

$$V : SE_2^1(n) \times SE_3^2(n) \times \dots \times SE_m^{m-1} \to \mathbb{R} \tag{3.29}$$

*a spring connecting these bodies.*

An element $h \in SE_2^1(n) \times SE_3^2(n) \times \dots \times SE_m^{m-1}$ is sufficient to describe all the necessary relative positions of the $m$ bodies with respect to each other.

If a spring is a passive physical element[6], it must have a finite minimum. This is clear because otherwise it would be possible to extract infinite energy from this elastic element. It is also possible to define an equivalent energy of the relative positions of all the bodies (to which the spring is attached) with respect to any Euclidean space $\mathcal{E}_j$:

$$V_j : SE_1^j(n) \times SE_2^j(n) \times \dots \times SE_m^j \to \mathbb{R}:$$
$$(h_1^j, h_2^j, \dots, h_m^j) \mapsto V(h_j^1 \circ h_2^j, h_j^2 \circ h_3^j, \dots, h_j^{m-1} \circ h_m^j, ). \tag{3.30}$$

The previous representation is useful for a direct interconnection between the spring and the masses.

Note that Eq. (3.29) is a function of $m-1$ relative positions of the elements to which the spring is connected, and in Eq. (3.30) of $m$ relative positions with respect to an additional space. It is characteristic of nodicity that the element is characterized by an energy function of $n-1$ relative positions where $n$ is the number of power ports with respect to an additional space $\mathcal{E}_j$.

[6] Intuitively, a passive element is an element which cannot supply infinite energy and therefore all energy functions of physical elements should be bounded from below.

**Characterizing equations of a spring.** To calculate the constitutive equations of a spring, we need to consider the differential of the energy function. The differential is a function of the form:

$$dV : SE_2^1(n) \times \ldots \times SE_m^{m-1}(n) \to T^*SE_2^1(n) \times \ldots \times T^*SE_m^{m-1}(n).$$

In a certain point $h = (h_2^1, \ldots, h_m^{m-1}) \in SE_2^1(n) \times \ldots \times SE_m^{m-1}(n)$, and for a velocity $\dot{h} \in T_{h_2^1}SE_2^1(n) \times \ldots \times T_{h_m^{m-1}}SE_m^{m-1}(n)$, the differential is such that $dV(h)$ applied to $(h, \dot{h})$ represents the increase of energy due to a change of relative positions $\dot{h}$. To simplify the description, we can map $dV(h)$ to $se_1^*(n) \times \ldots \times se_{m-1}^*(n)$ by means of $\chi_h^*$, which is the dual of $\chi_h := \chi_{h_2^1} \times \ldots \times \chi_{h_m^{m-1}}$, where $\chi_{h_i^j} := (\pi_{h_i^j}^j)^{-1}$ and $\pi_{h_i^j}^j$ is the intrinsic map to the Lie algebra reported in Figure 1.9. We get:

$$d\bar{V} : SE_2^1(n) \times \ldots \times SE_m^{m-1}(n) \to se_1^*(n) \times \ldots \times se_{m-1}^*(n) : \\ h := (h_2^1, \ldots, h_m^{m-1}) \mapsto \chi_h^* dV(h). \quad (3.31)$$

If we then define

$$T_V := \pi_{h_2^1}^1(\dot{h}_2^1) \times \ldots \times \pi_{h_m^{m-1}}^{m-1}(\dot{h}_m^{m-1}) = t_2^1 \times \ldots \times t_m^{m-1}$$

and

$$W_V := d\bar{V}(h)$$

the increase of energy of the spring at a certain instant of time is:

$$\dot{V} = \dot{\bar{V}} = \langle W_V, T_V \rangle = w_2^1 t_2^1 + \ldots + w_m^{m-1} t_m^{m-1} \quad (3.32)$$

where $w_i^j \in se_j^*(n)$ denotes the wrench applied by body $B_i$ to the spring connecting $B_i$ and $B_j$ and is expressed in the space of $B_j$ or $-w_i^j$, which is the wrench that the spring applies to body $B_i$ expressed in the space $B_j$.

*Remark 3.6.1.* If we would change the spaces in which we express some of the twists in Eq. (3.32) by using the transformations reported in Theorem 1.3.6, then the corresponding wrenches should change accordingly in order to keep $\dot{V}$ invariant. This implies that if we consider a simple spring which connects just two objects $i$ and $j$, Eq. (3.32) becomes $\dot{V} = w_i^j t_i^j$, and since $t_i^j = -Ad_{h_i^j} t_j^i$, in order to obtain $\dot{V} = w_j^i t_j^i$, we need $w_j^i = -Ad_{h_i^j}^* w_i^j$, which gives the intrinsic relation between the wrenches on the two sides of the spring (expressed in the same space, they should be opposite wrenches). With the same reasoning we can see that $-w_i^{i,j}$, corresponding to the wrench applied by the spring to body $i$ and expressed in the same space, is equal to $w_j^i$. This is a natural consequence of the fact that a spring is defined as an energy function of the relative positions of its hinge points. In [67, 68], these relations are proven for certain conditions of the energy function expressed with coordinates. The presented method seems more natural because it starts from the axiom that a spring is nodic and as such, its energy function should only be a function of its relative configurations. Furthermore, the presented approach is treated without using coordinates.

It is now possible to describe a general spring in Hamiltonian terms. The following equations hold:

$$\dot{h} = \chi_h T_V \qquad (3.33)$$
$$W_V = \chi_h^* \frac{\partial V}{\partial h}$$

where it can be seen that the Poisson structure is degenerate and that if the inputs are null, the conserved quantities (Casimir functions) are any energy functions $V(\cdot)$. Figure 3.10 gives a representation of the elastic element in coordinate-free bond graphs. If $\mathcal{E}_j$ is an inertial Euclidean space, the $m$ 1-junctions corresponding to $t_1^j, \ldots, t_m^j$ are the power ports to which we could directly attach inertial elements. Note that the nature of the **MTF** directly attached to the storage $\mathbb{C}$ element is different than that of the other $m-1$ **MTF**s. This is due to the fact that it is a mapping not between vector spaces, but between the vector space $se_1(n) \times \ldots \times se_{m-1}(n)$ and a vector bundle.

Given a spring function $V$, we can define a restriction of $V$ considering the relative motion of just two of the bodies $B_1, \ldots, B_m$. Suppose these two bodies to be $B_i$ and $B_j$. It is then possible to define:

$$V_i^j(h_1^j, \ldots, h_{i-1}^j, h_{i+1}^j, \ldots, h_m^j) : SE_i^j(n) \to \mathbb{R}\, ;$$
$$h_i^j \mapsto V(h_j^1 \;o\; h_2^j, \ldots, h_j^{m-1} \;o\; h_m^j).$$

By considering the differential of this function and projecting it to $se_j^*(n)$ as in Eq. (3.31), we obtain the wrench corresponding to an external bond with dual twist $t_i^j$, represented by a 1-junction which is labeled $t_i^j$ in Figure 3.10.

Consider now two Cartesian coordinates $\Psi_i$ and $\Psi_j$ for the spaces $\mathcal{E}_i$ and $\mathcal{E}_j$ in the following way. If we denote by $r_i^j \in SE_i^j(n)$ the relative position of $i$ with respect to $j$ for which the potential energy $V_i^j$ is minimum once the other given relative positions $(h_1^j, \ldots, h_{i-1}^j, h_{i+1}^j, \ldots, h_m^j)$ are fixed, then we can choose $\Psi_i$ and $\Psi_j$ in such a way that $\Psi_j \;o\; r_i^j \;o\; \Psi_i^{-1} = I$, the identity matrix. We can then study the elastic behavior and the associated twists and wrenches by using coordinates. A relative position of $i$ with respect to $j$ is then represented by an homogeneous matrix $H_i^j$ which has the form presented in Eq. (1.15).

It is now possible to follow the line of [69, 64, 70], and, using coordinates, we give an expression of the elastic wrenches as functions of the relative

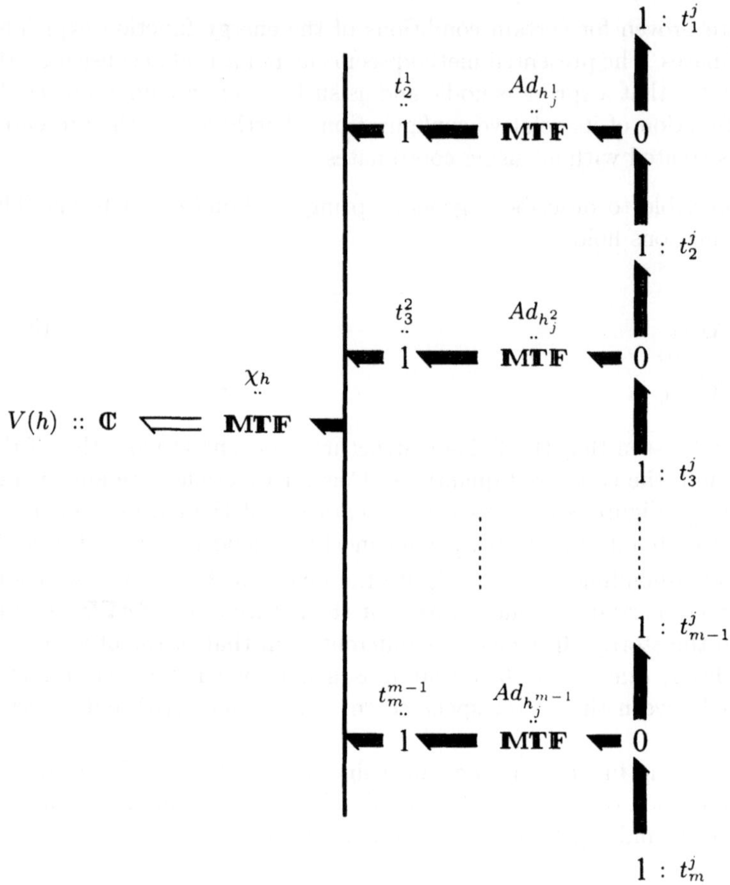


**Fig. 3.10.** The bond graph of a spring with $m$ hinge points.

positions of the objects attached to the spring. This will be done in Chapter 5 for a specific application. Furthermore, we can then consider the behavior of the port corresponding to the position of space $i$ with respect to $j$ in the case where $(h_1^j, \ldots, h_{i-1}^j, h_{i+1}^j, \ldots, h_m^j)$ are constant. This case was first studied in [21] and then in [67, 68, 70]. It is possible to introduce the concepts of center of stiffness and center of compliance[7].

**Center of Stiffness.** Suppose now that $V_i^j$ has a minimum in $r_i^j$ and consider the following two energy functions[8]:

$$\breve{V}_i(h_i) : SE_i \to \mathbb{R} \; ; \; h_i \mapsto V_i^j(r_i^j \, o \, h_i^{-1}) \tag{3.34}$$

[7] It has been shown in [21] that in general the center of stiffness and the center of compliance do NOT coincide.

[8] We do not specify the dependence of $V_i^j$ on the other configurations here, in order to simplify the notation.

and

$$\breve{V}_j(h_j) : SE_j \to \mathbb{R} \,;\; h_j \mapsto V_i^j(h_j \, o \, r_i^j). \tag{3.35}$$

By construction, $\breve{V}_i(\cdot)$ and $\breve{V}_j(\cdot)$ have a minimum at the identity of $SE_i(n)$ and $SE_j(n)$ respectively and we have $\breve{V}_i(h_i^{-1}) = \breve{V}_j(K_{r_i^j}(h_i))$, where $K_{r_i^j}$ is the hybrid conjugation map of Def. 1.2.8.

We can consider $d\breve{V}_l : SE_l(n) \to T^*SE_l(n)$ with $l = i, j$. The *stiffness map* $K_l$ is then defined as the linearization of $d\breve{V}_l$ around the identity $e_l \in SE_l(n)$. These mappings have the following form:

$$K_l : se_l(n) \to se_l^*(n) \,;\; t_l \mapsto \frac{d}{dt}\left(d\breve{V}(e^{t_l t})\right)\Big|_{t=0} \qquad l = i, j.$$

As is well known in tensor calculus, it is possible to see that the linearization of $d\breve{V}_l(h_l)$ transforms tensorially only in points $h_l$ such that $d\breve{V}_l(h_l) = 0$, which is the case for $K_l$.

With the previous definitions, it is possible to prove that:

$$K_j = Ad^*_{r_j^i} K_i Ad_{r_j^i} \tag{3.36}$$

which relates the stiffness in the two spaces with respect to one another. Now consider two Cartesian coordinates $\Psi_i$ and $\Psi_j$ for the space $\mathcal{E}_i$ and $\mathcal{E}_j$, respectively. The representations of $\breve{V}_i(\cdot)$ and $\breve{V}_j(\cdot)$ in these coordinates have the form:

$$\bar{V}_l : SE(n) \to \mathbb{R} \,;\; H \mapsto \breve{V}_l(\Psi_l \, o \, H \, o \, \Psi_l^{-1}) \quad l = i, j$$

and in a similar way, we denote the numerical representations of $K_i$ and $K_j$ we found using the coordinates $\Psi_i$ and $\Psi_j$, so by $\bar{K}_i$ and $\bar{K}_j$.

A proper treatment of the center of stiffness (COS) is reported in [21]. The COS is a point in space defined such that the numerical representation of the stiffness takes a special form, namely one in which there is minimum coupling between translation and rotation. It is possible to state the following symmetry property –also called *port-indifference* in [67]– which seems more natural in the presented framework.

**Theorem 3.6.1 (Stiffness Symmetry).** *If there is a point $p_i$ in space $\mathcal{E}_i$ such that, using Cartesian coordinates $\Psi_i$ for which $\Psi_i(p_i) = 0$, the stiffness matrix numerical representation $\bar{K}_i$ takes a special form, e.g. the one in which $p_i$ corresponds to the center of stiffness, then a point $p_j$ in $\mathcal{E}_j$ and coordinates $\Psi_j$ exist for which $\Psi_j(p_j) = 0$ such that $\bar{K}_j = \bar{K}_i$ and $p_j = r_i^j(p_i)$.*

**Proof.** This is a direct consequence of the application of Eq. (3.36) considering that, as shown in Sect. 1.3.5, the numerical representation of $Ad_{r_i^j}$ is the identity, where $r_i^j$ is the minimum energy-relative position in which the frames corresponding to $\Psi_i$ and $\Psi_j$ coincide. ∎

*Remark 3.6.2 (Port Indifference).* When the spring is at its minimal energy $r_i^j$, the previous theorem proves that the center of stiffness, if existent, is a unique point in space. For $\mathcal{E}_i$ and $\mathcal{E}_j$ this point coincides in the relative position equal to $r_i^j$. The same result was found in [67], but here it is presented as a direct consequence of spring nodicity.

A lot of interesting work has been done on the synthesis of spatial springs using simple linear or screw springs [65, 66].

### 3.6.2 Winding Springs

The winding spring is a kind of spring often overlooked by mathematicians, yet it is the only rotational spring that can be easily implemented. Intuitively, if there is a geometrical spring between just two bodies and we rotate one of the two bodies by 360 degrees around a fixed axis, we are back to the previous configuration and have the same forces between the bodies. Such a spring can be easily implemented for control purposes, but in mechanical terms it is not easily realizable.

We only define this type of spring between two bodies $B_i$ and $B_j$ which are connected by a kinematic pair with configuration manifold $\mathcal{Q}_i^j \subset SE_i^j(n)$.

**Definition 3.6.2 (Winding configuration).** *Consider two bodies $B_i$ and $B_j$ which are connected by a kinematic pair with configuration manifold $\mathcal{Q}_i^j \subset SE_i^j(n)$. A winding configuration is a connected manifold $\tilde{\mathcal{Q}}_i^j$ strictly greater than $\mathcal{Q}_i^j$ with an equivalence relation $\sim$ defined for $\tilde{\mathcal{Q}}_i^j$. This equivalence relation should be such that equivalent points are isolated and $\mathcal{Q}_i^j = \tilde{\mathcal{Q}}_i^j / \sim$.*

It is now possible to define a winding spring

**Definition 3.6.3 (Winding spring).** *A winding spring between two bodies $B_i$ and $B_j$ is a spring with a lower bounded energy function having the form:*

$$V_w : \tilde{\mathcal{Q}}_i^j \to \mathbb{R}$$

*such that $q_1 \sim q_2$, $q_1 \neq q_2$ implies $V_w(q_1) \neq V_w(q_2)$.*

The easiest example of a winding spring is a rubber band connecting two sticks that can just rotate with respect to each other around one axis. Clearly the configuration manifold of the kinematic pair describing the relative position would be a circle. Nevertheless, the winding configuration manifold is homeomorphic to a straight line the equivalent points of which represent the periodicity of $2\pi$ for the circle.

*Remark 3.6.3.* It is very important to realize that, if one wants to describe systems with winding springs, it is NOT sufficient to consider only the kinematic configuration of a mechanism. Instead, for each separate winding spring, the configuration $\tilde{\mathcal{Q}}_i^j$ should be used. This is necessary in order to cope with different elastic forces for equal configurations in $\mathcal{Q}_i^j$.

## 3.7 Inertias

In this section we analyse the dynamics of a rigid body in detail. In the previous section we described springs. We saw that it is not important which reference we use to describe the spring since it is a nodic element. Masses are very different and they are in fact the only non-nodic mechanical elements. This is due to the fact that kinetic energy must be a function of 'across quantities' which must be referenced to a subclass of possible references, namely to the class of inertial references treated in Sect. 3.5.1.

### 3.7.1 Kinetic co-energy and energy

Kinetic energy must be expressed as a function of the relative motion of a body with respect to an inertial reference. From now on we will consider a Euclidean space $\mathcal{E}_0$ as belonging to the inertial class. We reference everything with respect to $\mathcal{E}_0$.

Even basic courses of mechanics teach that it is well known that the co-energy with respect to $\mathcal{E}_0$ of a point of mass $m$ and velocity $v$ with respect to an inertial reference is:

$$E_k^*(v) = \frac{1}{2}mv^2.$$

One could wonder why the previous quantity is called co-energy. Formally speaking, $E_k^*$ is a function of a velocity, and as a function it differs from:

$$E_k(p) = \frac{1}{2m}p^2$$

which is called kinetic energy of the mass, where $p$ denotes the momentum. For more details on the difference between energy and co-energy see Appendix B or [46, 45].

We can now consider a space $\mathcal{E}_i$ with which a mass distribution function $\rho : \mathcal{E}_i \to \mathbb{R}^+$ is associated. This function associates a mass density with each point in $\mathcal{E}_i$. We can then formally define a body subset $B_i \subset \mathcal{E}_i$ as:

$$B_i := \{x \in \mathcal{E}_i \ ; \ \rho(x) > 0\}.$$

Furthermore, each Euclidean space has an $n-$form $\Omega_i$ corresponding to the volume form. The mass of a body $B_i$ can thus be defined as:

$$m_i = \int_{\mathcal{E}_i} \rho_i(p)\Omega_i(p). \tag{3.37}$$

Now consider a relative motion of $\mathcal{E}_i$ with respect to $\mathcal{E}_0$. The kinetic co-energy of an infinitesimal volume element $dV$ in a certain point $p \in \mathcal{E}_i$ which has velocity $v$ with respect to $\mathcal{E}_0$ is:

$$dE_k^* = \frac{1}{2}\rho(p)v^2 dV.$$

The total energy is nothing else than the integral over all the space. To handle the energy easily, we look for an expression of the velocity $v$ which is a function of $t_i^{i,0} \in se_i(n)$, the twist of space $\mathcal{E}_i$ with respect to $\mathcal{E}_0$. Consider for this purpose any Cartesian coordinates $\Psi_i, \Psi_0$ and express the relative position of $\mathcal{E}_i$ with respect to $\mathcal{E}_0$ as an element:

$$\Psi_0 \, o \, h_i^0 \, o \, \Psi_i^{-1} = \begin{pmatrix} R & p \\ 0 & 1 \end{pmatrix}$$

where $R \in SO(n)$, $p \in \mathbb{R}^n$. To any fixed point $p_i \in \mathcal{E}_i$, the point $p_0 = h_i^0(p_i)$ corresponds in the inertial space $\mathcal{E}_0$. Denoting with capital letters the numerical representation of a point, we obtain, using coordinates:

$$\begin{pmatrix} \Psi_0(p_0) \\ 1 \end{pmatrix} =: \begin{pmatrix} P_0 \\ 1 \end{pmatrix} = \begin{pmatrix} R & p \\ 0 & 1 \end{pmatrix} \begin{pmatrix} P_i \\ 1 \end{pmatrix}.$$

This implies that the velocity of the fixed point $p_i \in \mathcal{E}_i$ ($\dot{P}_i = 0$) for the space $\mathcal{E}_0$ in the chosen coordinates is:

$$\begin{pmatrix} \dot{P}_0 \\ 0 \end{pmatrix} = \begin{pmatrix} \dot{R} & \dot{p} \\ 0 & 0 \end{pmatrix} \begin{pmatrix} P_i \\ 1 \end{pmatrix} = \begin{pmatrix} R & p \\ 0 & 1 \end{pmatrix} \begin{pmatrix} \Omega & v \\ 0 & 0 \end{pmatrix} \begin{pmatrix} P_i \\ 1 \end{pmatrix}$$

where the matrix containing $\Omega$ and $v$ is a numerical representation of $t_i^{i,0}$, the twist of the relative motion of $\mathcal{E}_i$ with respect to $\mathcal{E}_0$ expressed in $se_i(n)$. It is possible to see that, after few manipulations:

$$\dot{P}_0 = (-R\tilde{P}_i \; R) \begin{pmatrix} \omega \\ v \end{pmatrix}$$

where, by $\tilde{x}$ we denote the antisymmetric matrix such that $\tilde{x}y = x \wedge y \;\; \forall y$ and $\Omega = \tilde{\omega}$. In other words, $\tilde{x}$ corresponds to an antisymmetric matrix which represents the vector product of the vector $x$ with another vector of the same dimension.

We have shown therefore that for any fixed point $p_i \in \mathcal{E}_i$ there is a linear map from $se_i(n)$ to the velocity of the point $p_i$ with respect to the inertial reference $\mathcal{E}_0$. This also implies that the square of the magnitude of $\dot{P}_0$ can be expressed as a quadratic form:

$$||\dot{P}_0||^2 = T_i^T \bar{M}(P_i) T_i$$

where

$$\bar{M}(P_i) := \begin{pmatrix} \tilde{P}_i^T R^T R \tilde{P}_i & \tilde{P}_i R^T R \\ -R^T R \tilde{P}_i & R^T R \end{pmatrix} = \begin{pmatrix} \tilde{P}_i^T \tilde{P}_i & \tilde{P}_i \\ \tilde{P}_i^T & I_n \end{pmatrix}$$

and $T_i = \left(\omega^T\ v^T\right)^T$. The matrix $\bar{M}(P_i)$ can be singular for a single point $P_i$, but the sum of $\bar{M}(P_i)$ for a sufficient number of noncollinear $P_i$ gives a nonsingular matrix. A consequence of this is that the integral of this matrix over a nonzero volume would always give a nonsingular matrix.

Hence the total kinetic co-energy is:

$$E_k^*(T_i) = \frac{1}{2} T_i^T \bar{I} T_i$$

where

$$\bar{I} := \int_{\mathcal{E}_i} \bar{M}(P_i)\rho(P_i)\Omega_i(P_i)$$

Because of the previous arguments concerning the non-singularity of the integral of $\bar{M}(\cdot)$ and because of physical reasons, $\bar{I}$ is always a positive definite matrix. The physical reason is that if we had a singular $\bar{I}$, a non-zero twist would exist with no energy associated with it. For an object with a finite volume, this is clearly impossible.

It can be seen that $\bar{I}$ represents a quadratic form defined in the Lie algebra $se_i(n)$ which will be called the *inertia tensor* in body space. If we indicate with $I$ this quadratic form on $se_i(n)$, of which $\bar{I}$ is a numerical representation, we can define the following:

$$\langle , \rangle_I : se_i(n) \times se_i(n) \to \mathbb{R} \; ; \; (t^1, t^2) \mapsto \langle t^1, t^2 \rangle_I . \tag{3.38}$$

In a coordinate-free form, we can then see that the co-energy is equal to:

$$E_k^*(t_i^{i,0}) = \frac{1}{2} \langle t_i^{i,0}, t_i^{i,0} \rangle_I .$$

The quadratic form $I$ uniquely defines a bijection between elements of $se_i(n)$ and elements of $se_i^*(n)$ such that $m^i \in se_i^*(n)$ corresponds to $t_i^{i,0} \in se_i(n)$ iff:

$$m^i(t_i^*) = \langle t_i^{i,0}, t_i^* \rangle_I \quad \forall t_i^* \in se_i(n).$$

The quantity $m^i$, corresponding through the inertia tensor $I$ to $t_i^{i,0}$, is called the *momentum*, and it corresponds to the twist $t_i^{i,0}$.

*Remark 3.7.1.* Note that the quadratic form $I$ in the Lie algebra is a constant quadratic form, independent of the relative position of $\mathcal{E}_i$ with respect to any other space. Therefore it is a quantity intrinsically associated with the space $\mathcal{E}_i$ to which the body belongs. It is possible to see that, by the group operation of left transport, this quadratic form defines a left invariant Riemannian metric in the Lie group $SE_i(n)$.

Since $I$ is nonsingular, we can define the following function of $m^i$, which will be called *kinetic energy* of the body:

$$E_k(m^i) = \frac{1}{2}\langle m^i, m^i \rangle_Y. \tag{3.39}$$

Where $\langle , \rangle_Y$ denotes the quadratic form on $se^*(n)$ corresponding to $Y = I^{-1}$. Numerically, we have for corresponding $m^i$ and $t_i^{i,0}$:

$$E_k(m^i) = E_k^*(t_i^{i,0}) = \frac{1}{2} m^i t_i^{i,0}.$$

### 3.7.2 The second law of dynamics

Newton's second law of dynamics states that the variation of the momentum of a body has to be equal to the applied wrench for an inertial observer. With the notation used, if we consider the space $\mathcal{E}_0$ as our inertial frame, the second law of dynamics takes the following form:

$$\dot{m}^{0,i} = w^0 \tag{3.40}$$

where $m^{0,i} := Ad^*_{h_0^i} m^i$ is the momentum of a body for the observer $\mathcal{E}_0$, $w^0$ is the wrench applied to the body expressed in the inertial space and $Ad^*_{h_0^i}$ is a mapping from $se_i^*(n)$ to $se_0^*(n)$ at $h_0^i$. Since the momentum and the wrench are elements of $se_0^*(n)$, we have:

$$w^0 = Ad^*_{h_0^i} w^i.$$

The second law of dynamics expressed in body space becomes:

$$\frac{d}{dt}\left(Ad^*_{h_0^i(t)} m^i\right) = Ad^*_{h_0^i(t)} w^i.$$

We therefore have to calculate the time derivative of $Ad^*_{h_0^i(t)} m^i$. To do so, consider $h_0^i(t)$ as the following function of time:

$$h_i^0(t) = h_i^0(0) \; o \; h_i(t) = h_i^0(0) \; o \; e^{t_i^{i,0} t}$$

where we have chosen $h_i(t) = e^{t_i^{i,0} t}$ to have a relative twist equal to $t_i^{i,0}$ at time $t = 0$. The used exponential is the group exponential reviewed in Sect. A.8.6. The relative motion $h_i^0(t)$ represents the relative motion of space $\mathcal{E}_i$ with respect to $\mathcal{E}_0$ and passes from $h_i^0(0)$ at time $t = 0$ with a velocity of which the corresponding twist in $\mathcal{E}_i$ space is $t_i^{i,0} \in se_i(n)$. We therefore get:

$$h_0^i(t) = (h_i^0(t))^{-1} = e^{-t_i^{i,0} t} \; o \; h_0^i(0).$$

Furthermore, it is possible to see that:

$$Ad_{\left(e^{-t_i^{i,0} t} \; o \; h_0^i(0)\right)} = Ad_{e^{-t_i^{i,0} t}} \; o \; Ad_{h_0^i(0)}$$

and therefore:

$$Ad^*_{\left(e^{-t_i^{i,0}t} \, o \, h_0^i(0)\right)} = Ad^*_{h_0^i(0)} \; o \; Ad^*_{e^{-t_i^{i,0}t}}$$

where it is important to note that the adjoint map at $e^{-t_i^{i,0}t}$ is the usual group adjoint defined in Def. A.8.5. Finally, we use Def. A.8.7 for the dual of $ad^*_{t_i^{i,0}}$ and since $Ad^*_{h_0^i(0)}$ is constant, we obtain:

$$\frac{d}{dt}\left(Ad^*_{h_0^i(t)}\right)\bigg|_{t=0} = -Ad^*_{h_0^i(0)} \; o \; ad^*_{t_i^{i,0}}$$

and since $Ad^*_{h_0^i}$ is a linear map we obtain:

$$Ad^*_{h_0^i} w^i = \frac{d}{dt}\left(Ad^*_{h_0^i(t)} m^i(t)\right)\bigg|_{t=0} = -Ad^*_{h_0^i(0)} ad^*_{t_i^{i,0}} m^i + Ad^*_{h_0^i(0)} \dot{m}^i .$$

Furthermore, since $Ad^*_{h_0^i(0)}$ is a full rank linear map, we obtain:

$$\dot{m}^i = ad^*_{t_i^{i,0}} m^i + w^i \tag{3.41}$$

which is an expression for Newton's second law in body space. When we use the numerical representation of $ad$ introduced in Eq. (1.31) and the corresponding coordinates for twists, the previous equation takes the following form:

$$(\dot{M}^i)^T = ad^T_{T_i^{i,0}} (M^i)^T + (W^i)^T . \tag{3.42}$$

Factorizing the matrices, this can be written as:

$$\begin{pmatrix} (\dot{M}^i_\omega)^T \\ (\dot{M}^i_v)^T \end{pmatrix} = \begin{pmatrix} -\tilde{\omega}_k^{k,0} & -\tilde{v}_k^{k,0} \\ 0 & -\tilde{\omega}_k^{k,0} \end{pmatrix} \begin{pmatrix} (M^i_\omega)^T \\ (M^i_v)^T \end{pmatrix} + (W^i)^T . \tag{3.43}$$

*Remark 3.7.2 (Group structure).* It is important to stress that the correspondence between Eq. (3.41) and Eq. (3.40) is only a consequence of the description of the relative motion, and this description is true for any co-vector mapping, independent of inertial parameters. It is a characterization of the group structure and it is independent of the value that the inertia tensor $\bar{I}_i$ has for the body $i$.

### 3.7.3 A Hamiltonian formulation of a rigid body motion: the Lie-Poisson bracket

In the general Hamiltonian formulation we have seen that any Hamiltonian system with ports is characterized by a Poisson bracket, an energy function and input vector fields corresponding to an input transformer in bond graphs terms. The Poisson bracket represents the interconnection and is independent of the energy function.

If for the moment we ignore potential energy and thus concentrate our study on the motion of a free rigid body floating in a space without gravitation, then the Hamiltonian corresponds to the following kinetic energy:

$$H(m^i) = E_k(m^i) = \frac{1}{2}\langle m^i, m^i \rangle_Y$$

as reported already in Eq. (3.39). Since the state is represented by the momenta $m^i \in se_i^*(n)$, the Poisson bracket we are looking for should be defined on functions of $se_i^*(n)$.

*Remark 3.7.3.* It is important to realize that the quadratic form $I$ (and therefore $Y$) expressed in body coordinates is constant and independent of the body configuration. The Hamiltonian function is therefore ONLY a function of $m^i$. This characterizes what is called a left-invariant metric on $SE_i(n)$, and implies that $\nabla H = Ym^i = t_i^{i,0}$ is the twist of the relative motion of $\mathcal{E}_i$ with respect to $\mathcal{E}_0$ expressed in space $\mathcal{E}_i$, where the notation $\nabla H$ denotes the gradient of $H$.

Since $\nabla H = t_i^{i,0}$, we can write Eq. (3.41) as:

$$\dot{m}^i = ad^*_{\nabla H} m^i + w^i.$$

This means that the Hamiltonian vector field describing the dynamics of an input equal to zero ($w^i = 0$) is $ad^*_{\nabla H} m^i$. To find the corresponding Poisson bracket, we need to consider the application of this vector field to a general function $F : se_i^*(n) \to \mathbb{R}$. The application of this vector field to $F$ gives a new function which corresponds to the Poisson bracket $\{F, H\}$. This function, being the application of the vector field to a generic function $F$, can be expressed as:

$$\langle ad^*_{\nabla H} m^i, \nabla F \rangle$$

Where $\langle , \rangle$ indicates the dual product of a co-vector and a vector. By the definition of adjoint we get:

$$\langle ad^*_{\nabla H} m^i, \nabla F \rangle = \langle m^i, ad_{\nabla H} \nabla F \rangle = \langle m^i, [\nabla H, \nabla F] \rangle$$

where $[,]$ corresponds to the algebra commutator of $se_i(n)$ and the result reported in Theorem A.8.1 has been used. Note that since $F(\cdot), H(\cdot)$ are functions of elements of $se_i^*(n)$, their gradients $\nabla F(\cdot), \nabla H(\cdot)$ are vectors belonging to $se_i(n)$. We therefore define as Poisson bracket:

$$\{F, H\}(m^i) := \langle m^i, [\nabla F(m^i), \nabla H(m^i)] \rangle$$

The previous bracket is a well-known bracket called the *Lie-Poisson bracket.* If we use $X_H(m^i)$ to denote the Hamiltonian vector field with respect to this Poisson bracket for any Hamiltonian function $H : se_i^*(n) \to \mathbb{R}$, then Eq. (3.41) takes the following form:

$$\dot{m}^i = X_H(m^i) + w^i.$$

*Remark 3.7.4.* The reasoning followed makes clear that the Lie-Poisson bracket satisfies all the properties a Poisson bracket requires and that it is defined on generic functions of the dual of any Lie algebra. This bracket characterizes the group geodesic flow of any left-invariant Riemann metric of a general Lie group $G$. In general this Poisson bracket is singular. The symplectic structure, defined by the restriction of this Poisson bracket to the current leaf which is defined as a level set of the Casimir functions of the bracket, is called the Kirillov form [35].

The mapping $se_i(n) \to se_i^*(n)$ ; $t_i^{i,0} \mapsto ad^*_{t_i^{i,0}} m^i$ is linear due to the bilinearity property of the algebra commutator and the mapping is clearly a function of $m^i$. We denote this linear mapping with $m^i\wedge$. Using this notation and adding the external applied wrench, we can write:

$$\dot{m}^i = (m^i\wedge)t_i^{i,0} + w_i. \tag{3.44}$$

It can easily be seen that the previous equation expressed in coordinates, can be obtained directly by manipulating Eq. (3.43) in order to get:

$$(\dot{M}^i)^T = \begin{pmatrix} \tilde{M}^i_\omega & \tilde{M}^i_v \\ \tilde{M}^i_v & 0 \end{pmatrix} \begin{pmatrix} \omega_i^{i,0} \\ v_i^{i,0} \end{pmatrix} + (W^i)^T \tag{3.45}$$

and therefore a numerical representation of $m^i\wedge$ is:

$$(M^i\wedge) := \begin{pmatrix} \tilde{M}^i_\omega & \tilde{M}^i_v \\ \tilde{M}^i_v & 0 \end{pmatrix}$$

Eq. (3.44) shows that the momentum dynamics in body space is independent of the configuration of the body. The position of $\mathcal{E}_i$ relative to $\mathcal{E}_0$ does not play any role in principle. Nevertheless, if we associate a potential energy depending on the configuration of space $\mathcal{E}_i$ with respect to $\mathcal{E}_0$, we must somehow account for it. Consider a potential energy $V : SE_i^0(n) \to \mathbb{R}$. We have seen the constitutive equations of a spring in Eq. (3.33). In this case, where we only deal with two bodies and the force applied to the mass is of opposite sign, we have the following equations, similar to the Lie group representation in [19, 18]:

$$\begin{pmatrix} \dot{h}_i^0 \\ \dot{m}^i \end{pmatrix} = \begin{pmatrix} 0 & \chi_{h_i^0} \; o \; Ad_{h_i^0} \\ -Ad^*_{h_i^0} \; o \; \chi^*_{h_i^0} & m^i\wedge \end{pmatrix} \begin{pmatrix} \frac{\partial H}{\partial h_i^0} \\ \frac{\partial H}{\partial m^i} \end{pmatrix} + \begin{pmatrix} 0 \\ Ad^*_{h_i^0} \end{pmatrix} w_i^0 \tag{3.46}$$

$$t_i^0 = (0 \; Ad_{h_i^0}) \begin{pmatrix} \frac{\partial H}{\partial h_i^0} \\ \frac{\partial H}{\partial m^i} \end{pmatrix}$$

where $H(h_i^0, m^i) = V(h_i^0) + \frac{1}{2}\langle m^i, m^i\rangle_Y$ and

$$\chi_{h_i^j} := \left(\pi^j_{h_i^j}\right)^{-1} \qquad \chi^*_{h_i^j} := \left(\pi^{j\,*}_{h_i^j}\right)^{-1}. \tag{3.47}$$

The map $\pi^j_{h_i^j}$ maps a velocity belonging to the tangent bundle of $SE_i^j(n)$ to an intrinsic twist in $se_j(n)$ (see Figure 1.9), and the map $\chi_{h_i^j}$ is the inverse map.

## 3.8 Dissipation elements

By the first principle of thermodynamics, energy cannot be destroyed but only transformed. This is what is meant by "energy dissipation". In this case a reconvertible energy is irreversibly transformed into thermal energy. It is important to realize [45] that for a lot of applications, the environmental temperature can be considered constant and the concept of *free energy* is implicitly used instead of energy. In mathematical terms, free energy is defined as the Legendre transformation of energy with respect to temperature. One could say that free energy is destroyed when by an isothermical transformation real energy is converted to the thermal domain.

After this short discussion we can now treat dissipating elements of a mechanical system. In the framework previously explained, a dissipative element can be seen as a subsystem with two energy ports: a mechanical one and a thermal one. Energy will always flow toward this dissipating element through the mechanical port. When we are interested in free energy, the thermal port is usually not considered. In bond graphs, such a dissipative element is represented by the bond graph element **IR**.

We will now analyse the structure of a mechanical port of a dissipating element.

### 3.8.1 Irreversibility

As we have stated before, the dissipation of free energy represents an irreversible transfer of energy. If we represent a purely dissipating element as a sub-model **IR** with one energetic port, this implies that the positive flow of energy only goes towards the sub-model **IR**.

This means that if the vector space characterization of the power port of the element is $\mathcal{V}$, we have for the positive convention of inward power:

$$f^i e_i > 0 \,,\ \forall (f^i, e_i) \in \mathcal{V} \times \mathcal{V}^*.$$

**Definition 3.8.1 (Impedance Dissipating map).** *A dissipative element in impedance form between the Euclidean spaces $\mathcal{E}_i$ and $\mathcal{E}_j$ is a tensor field with the following form:*

$$R_I : SE_i^j(n) \to \mathrm{sym}(T_{0,2}SE_i^j(n))$$

*which associates a two covariant, symmetric, semi-positive definite tensor with each relative position of the two bodies attached to the dissipating element.*

**Definition 3.8.2 (Admittance Dissipating map).** *A dissipative element in admittance form between the Euclidean spaces $\mathcal{E}_i$ and $\mathcal{E}_j$ is a tensor field with the following form:*

$$R_A : SE_i^j(n) \to \mathrm{sym}(T_{2,0}SE_i^j(n))$$

*which associates a two contravariant, symmetric, semi-positive definite tensor with each relative position of the two bodies attached to the dissipating element.*

A mapping of the form $R_I$ is said to have an impedance form because it can be used as a local mapping from "generalized velocities" belonging to $T_{h_i^j}SE_i^j(n)$ to "generalized forces" in $T^*_{h_i^j}SE_i^j(n)$. The admittance form on the other hand, can be used to map from "generalized forces" to "generalized velocities".

If $R_I(\cdot)$ (resp. $R_A(\cdot)$) is nonsingular everywhere, it is obviously possible to associate a mapping $R_A(\cdot)$ (resp. $R_I(\cdot)$) corresponding to its inverse at each point $h_i^j \in SE_i^j(n)$ in a unique way.

### 3.8.2 Algebra representation

To describe the interconnection of dissipative elements in a more convenient way than as $R_I(\cdot)$ or $R_A(\cdot)$, we express them as an operation in the algebra $se_i(n)$ or equivalently $se_j(n)$. We can then consider the following mappings:

$$\tilde{R}_I(h_i^j) : se_j(n) \to se_j^* \; ; \; t_j \mapsto (\chi^*_{h_i^j} \; o \; R_I(h_i^j) \; o \; \chi_{h_i^j})(t_j) \quad \text{or} \tag{3.48}$$

$$\tilde{R}_A(h_i^j) : se_j^*(n) \to se_j \; ; \; w_j \mapsto (\pi_{h_i^j} \; o \; R_A(h_i^j) \; o \; \pi^*_{h_i^j})(w_j). \tag{3.49}$$

Due to the semi-positive definiteness of the original operators $R_A$ and $R_I$, we have $\tilde{R}_I(h_i^j)(t_i)(t_i) \geq 0$ and $\tilde{R}_A(h_i^j)(w_i)(w_i) \geq 0$. Furthermore, when the operators are non-singular, they are also strictly positive definite.

## 3.9 Actuators and Interaction ports

Actuators are ideal elements generating wrenches between two bodies. We can associate an actuator with each kinematic pair of which the wrenches belonging to $\mathcal{W}^A_{j,i}$, or to a subspace of it can be set to any value. These elements are ideal and are used to interact with the system.

Furthermore, we also need to express the above for a system which can interact with the environment in a less structural way than through kinematic pairs.

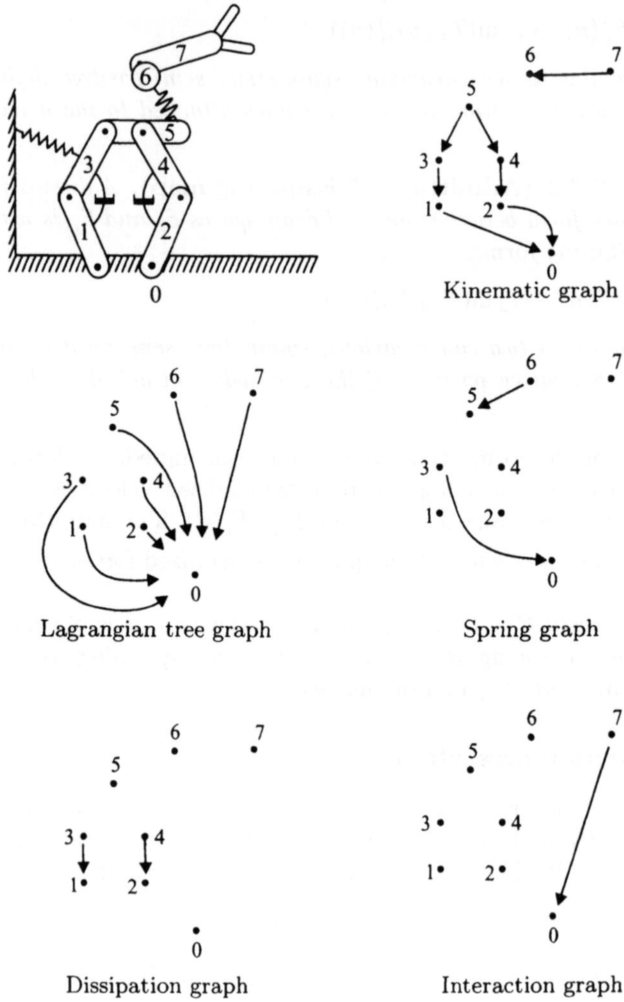


**Fig. 3.11.** The various graphs.

## 3.10 Elements Interconnection

Having studied all the elements which can be present in a physical mechanical lumped system, we can now consider how to interconnect those elements. We have already given some considerations on the topology of interconnections of bodies by means of kinematic pairs in Sect. 2.2. We now extend what has been presented for the interconnection to all the separate elements. Before doing so, some remarks are called for:

1. Any physical element has a certain number of power ports through which it exchanges energy with the rest of the universe.

2. The dynamic description of the element is based on equations which relate its port variables: efforts and flows.
3. The flow at each of its ports corresponds to a twist which is ALWAYS a quantity that relates two different spaces.
4. For nodic elements these two spaces can be anything, but for the masses one of these spaces MUST be inertial.

Point 3 implies that we can describe a general interconnection of physical elements as an oriented graph in which its vertices correspond to all the considered spaces; for this reason the vertex set $V$ corresponds to the Euclidean system, and each of the dynamic elements is represented by one or by a set of edges in the case of a multi-port element as the geometric springs presented. The total graph $G_T = (V_T, E_T)$, for which $V_T = V_B \cup \{0\}$ with $\{0\}$ is the vertex corresponding to the inertial space $\mathcal{E}_0$, and $E_T = E_K \cup E_B \cup E_S \cup E_D \cup E_I$, representing the complete interconnection, is therefore composed of the following subgraphs:

- **Kinematic graph** $G_K := (V_B \cup \{0\}, E_K)$
  This graph corresponds to the graph we presented in Chapter 2, where the vertex set $V_B$ corresponds to the set of bodies $V_B$ (and associated spaces) plus in some cases $\{0\}$, and the set of edges $E_K$ corresponds to the ideal kinematic pairs (see Sect. 2.2).
- **Lagrangian tree graph** $G_B := (V_B \cup \{0\}, E_B)$
  If we consider only one common inertial reference, we can, as a consequence of point 4, represent all the inertial elements in the system by what is called a *Lagrangian tree* [19]. The set of vertices is equal to the set of bodies plus an extra vertex representing an inertial Euclidean reference space which we denote with the index 0. The set $E_B$ is composed of edges pointing from each vertex corresponding to a body (each element of $V_B$) to the inertial vertex reference $\{0\}$[9] (see Sect. 3.7).
- **Springs graph** $G_S = (V_B \cup \{0\}, E_S)$
  To each spring we can associate a set of edges. The union of these edges will correspond to $E_S$. To be physically meaningful, each spring should be attached to bodies with non-zero mass (see Sect. 3.6).
- **Dissipation graph** $G_D = (V_B \cup \{0\}, E_D)$
  Each edge in $E_D$ will represent a part in the graph where free energy can be dissipated by means of $\mathbb{R}$ elements (see Sect. 3.8).
- **Interaction graph** $G_I(V_B \cup \{0\}, E_I)$
  The edges $E_I$ indicate the place where an external system is able to interact with the mechanism. This graph is useful to define a work space configuration $\mathcal{X}$ in a general way. If an edge $(i, j) \in E_I$, this means that the environment can apply a wrench between space $\mathcal{E}_i$ and $\mathcal{E}_j$. Each edge is

[9] Note that the orientation chosen in [19] is the opposite of the one proposed here. This orientation has been chosen because, to intrinsically consider the twist of a body with respect to an inertial frame, we need to consider the positive isometry from the Euclidean space corresponding to the body to the inertial space.

therefore a representative of a power port through which the environment can apply a wrench. As a consequence, if the corresponding twist is different from zero, it exchanges energy with the system. For a robotic space hand, which is only used to implement tip grasps of objects floating in space, we could consider the interaction graph as composed of edges connecting the vertices corresponding to the finger tips of the hand. If the robotic hand is connected to a fixed base on earth, edges connecting each vertex representation of the tips should also connect to the fixed base so wrenches can be generated between them.

The graph $G_P = (V_B \cup \{0\}, E_S \cup E_K \cup E_D \cup E_I)$ is called the *primary graph* and the graph $G_C = (V_B \cup \{0\}, E_S \cup E_K \cup E_D \cup E_I \cup E_B)$ is called the *port connection graph* in [19, 46].

A simple example of the introduced graphs is given in Figure 3.11. We considered the joint connecting body 5 and 7 as elastic. To model this, body 6 is added. Between body 6 and 7 there is a kinematic pair where a motor is attached and body 5 and 6 are connected with an elastic element representing the elasticity of the joint. Furthermore, we suppose that the robot can interact with the environment only by means of body 7 and 0 between which the environment can apply forces.

### 3.10.1 Kirchhoff's laws and power continuous interconnection.

For the total graph $G_T$, both generalizations of Kirchhoff's laws for electrical circuits hold. First, it is possible to see that the sum of the twists along any cycle of the graph vanishes. This is a direct consequence of Theorem 1.3.8, which yields:

$$\sum_{i \in C} Ad_{h^k_{r(i)}} t^{r(i)}_{l(i)} = 0 \tag{3.50}$$

where $C$ indicates the set of edges belonging to a cycle in $G_T$, $r(i)$ the vertex towards which the edge is pointing and $l(i)$ the other vertex of each edge $i$ belonging to the considered cycle. In the example of Figure 3.11, an element of $C$ could, for example, be composed of $(3,0), (1,0) \in G_B$ and $(3,1) \in G_k$. The vertex $k$ could be any, but to obtain the expression in Theorem 1.3.8, it should be a vertex belonging to the cycle. Considering the just mentioned cycle in the example, we could choose $k = 0$ and the generalized Kirchhoff's law would become:

$$t^0_3 + t^0_1 + Ad_{h^0_1} t^1_3 = 0$$

For the second Kirchhoff's law, the existence of the Lagrangian tree is fundamental. This law corresponds to the third law of dynamics on action and reaction. In graph terms, it states that the sum of the wrenches along any co-cycle, which is also called the cut-set, vanishes in the graph $G_T$.

This means that if we consider the set of edges adjacent to each vertex of the graph, the sum of the wrenches corresponding to the edges will be zero. This law can be analytically expressed as:

$$\sum_{i \in \bar{C}_k} Ad^*_{h_k^{r(i)}} w_{l(i)}^{r(i)} = 0 \tag{3.51}$$

where $\bar{C}_k$ indicates the set of edges adjacent to a vertex $k$, and $w_{l(i)}^{r(i)}$ the wrench corresponding to an adjacent edge oriented from $l(i)$ to $r(i)$, where either $r(i) = k$ or $l(i) = k$. This law must be satisfied in order to have power continuity in the interconnection, as will be seen later.

These two laws make possible a direct correspondence between the total graph $G_T$ and a bond graph junction structure. The bond graph junction structure is created by substituting a 1-junction for each vertex and either the top or the bottom representations of Figure 2.7 and Figure 2.8 for each edge. Once these substitutions have been made, the previous laws correspond to basic junction laws, as expected.

As already said, we consider the Lagrangian tree to be a spanning tree (see Def. A.2.10) for the total graph $G_T$. This means that each vertex except the one corresponding to the common inertial reference corresponds to a Euclidean space $\mathcal{E}_i$ with a nonzero mass $m_i$ associated with it as defined in Eq. (3.37).

To each graph $G$, we can associate a matrix $\mathbf{B} \in \{-1, 0, 1\}^{v(G) \times e(G)}$ called the *incidence matrix*, where $v(G)$ is the number of vertices of $G$ and $e(G)$ the number of edges. Each column of $\mathbf{B}$ corresponds to an edge of $e(G)$ and is composed of one 1, one $-1$ and all other elements are zero. The row corresponding to the position of the 1 corresponds to the vertex to which the edge is pointing and the row with $-1$ corresponds to the vertex on the other side of the edge.

We denote by $\mathbf{B}_k$, $\mathbf{B}_b$, $\mathbf{B}_s$, $\mathbf{B}_d$ and $\mathbf{B}_i$ the incidence matrices of the graphs $G_K$, $G_B$, $G_S$, $G_D$ and $G_I$, respectively, to which we also add rows of zeros, if necessary, in such a way that row $i$ corresponds to vertex $i$ for each $i \in V_B$.

Due to the definition of the Lagrangian tree, in $\mathbf{B}_b$ the line corresponding to the vertex $\{0\}$ will be a line of ones.

For convenience of notation, we define $\tilde{t}_i^k := Ad_{h_{r(i)}^k} t_{l(i)}^{r(i)}$, which corresponds to the twist associated with edge $i$ and expressed in space $k$. We can then group the edges for the various graphs and denote them by:

$$\tilde{t}_K^k := \left(\tilde{t}_{k(1)}^k, \tilde{t}_{k(2)}^k, \ldots, \tilde{t}_{k(e(G_K))}^k\right)^T \tag{3.52}$$

$$\tilde{t}_B^k := \left(\tilde{t}_{b(1)}^k, \tilde{t}_{b(2)}^k, \ldots, \tilde{t}_{b(e(G_B))}^k\right)^T \tag{3.53}$$

$$\tilde{t}_S^k := \left(\tilde{t}_{s(1)}^k, \tilde{t}_{s(2)}^k, \ldots, \tilde{t}_{s(e(G_S))}^k\right)^T \tag{3.54}$$

$$\tilde{t}_D^k := \left(\tilde{t}_{d(1)}^k, \tilde{t}_{d(2)}^k, \ldots, \tilde{t}_{d(e(G_D))}^k\right)^T \tag{3.55}$$

$$\tilde{t}_I^k := \left(\tilde{t}_{i(1)}^k, \tilde{t}_{i(2)}^k, \ldots, \tilde{t}_{i(e(G_I))}^k\right)^T \tag{3.56}$$

where $k(\cdot), b(\cdot), s(\cdot)$, $d(\cdot)$ and $i(\cdot)$ are indexing functions for the edges of the kinematic, Lagrangian, spring, dissipation and interaction graphs, respectively.

Due to the spanning-tree hypothesis of $G_B$, we can consider for each edge $e$ in $E_S, E_K, E_D, E_I$ the cycles composed of $e$ and the other two edges belonging to the Lagrangian tree that connect the inertial reference to the vertices adjacent to $e$. The set of these cycles is known as the *fundamental cycles* for the graph $G_T$ associated with the spanning tree $G_B$ (see [25]).

If we arrange the order of the rows of $\mathbf{B}_k, \mathbf{B}_s, \mathbf{B}_d$ and $\mathbf{B}_i$ in such a way that their row $i$ corresponds to the vertex $b(i)$, where $b(\cdot)$ is the indexing function for the edges of the Lagrangian tree, and if we apply Eq. (3.50) to each of these cycles, we obtain the following equations:

$$(\mathbf{B}_k)^T \tilde{t}_B^k + \tilde{t}_K^k = 0 \tag{3.57}$$

$$(\mathbf{B}_s)^T \tilde{t}_B^k + \tilde{t}_S^k = 0 \tag{3.58}$$

$$(\mathbf{B}_d)^T \tilde{t}_B^k + \tilde{t}_D^k = 0 \tag{3.59}$$

$$(\mathbf{B}_i)^T \tilde{t}_B^k + \tilde{t}_I^k = 0 \tag{3.60}$$

which can be written as:

$$\mathbf{C}\tilde{t}^k = 0$$

with $\tilde{t}^k = \left((\tilde{t}_B^k)^T \ (\tilde{t}_K^k)^T \ (\tilde{t}_S^k)^T \ (\tilde{t}_D^k)^T \ (\tilde{t}_I^k)^T\right)^T$ and

$$\mathbf{C} = \begin{pmatrix} (\mathbf{B}_K)^T & I_{e(G_K)} & 0 & 0 & 0 \\ (\mathbf{B}_S)^T & 0 & I_{e(G_S)} & 0 & 0 \\ (\mathbf{B}_D)^T & 0 & 0 & I_{e(G_D)} & 0 \\ (\mathbf{B}_I)^T & 0 & 0 & 0 & I_{e(G_I)} \end{pmatrix} = \left(\mathbf{B}^T \ I_{e(G_K)+e(G_S)+e(G_D)+e(G_I)}\right) \tag{3.61}$$

where $\mathbf{B} := \left(\mathbf{B}_K \ \mathbf{B}_S \ \mathbf{B}_D \ \mathbf{B}_I\right)$ is the incidence matrix of the total graph.

In graphic theoretic terms, $\mathbf{C}$ is called the *cycle* or *loop matrix* for the graph $G_T$ associated with the spanning tree $G_B$.

For the example presented in Figure 3.11, we would get:

$$\mathbf{B}_K = \begin{pmatrix} 1 & 1 & 0 & 0 & 0 & 0 & 0 \\ -1 & 0 & 1 & 0 & 0 & 0 & 0 \\ 0 & -1 & 0 & 1 & 0 & 0 & 0 \\ 0 & 0 & -1 & 0 & 1 & 0 & 0 \\ 0 & 0 & 0 & -1 & 0 & 1 & 0 \\ 0 & 0 & 0 & 0 & -1 & -1 & 0 \\ 0 & 0 & 0 & 0 & 0 & 0 & 1 \\ 0 & 0 & 0 & 0 & 0 & 0 & -1 \end{pmatrix}$$

where the $i$-th row of the matrix corresponds to vertex $i$ of the kinematic graph of the figure and each column corresponds to one of its edges.

We can arrive at a similar expression for a representation of Eq. (3.51). If we denote by $\tilde{w}_i^k := Ad^*_{h^k_{r(i)}} w^{r(i)}_{l(i)}$ and in a similar way we define:

$$\tilde{w}_K^k := \left(\tilde{w}^k_{k(1)}, \tilde{w}^k_{k(2)}, \ldots, \tilde{w}^k_{k(e(G_K))}\right)^T \tag{3.62}$$

$$\tilde{w}_B^k := \left(\tilde{w}^k_{b(1)}, \tilde{w}^k_{b(2)}, \ldots, \tilde{w}^k_{b(e(G_B))}\right)^T \tag{3.63}$$

$$\tilde{w}_S^k := \left(\tilde{w}^k_{s(1)}, \tilde{wt}^k_{s(2)}, \ldots, \tilde{w}^k_{s(e(G_S))}\right)^T \tag{3.64}$$

$$\tilde{w}_D^k := \left(\tilde{w}^k_{d(1)}, \tilde{w}^k_{d(2)}, \ldots, \tilde{w}^k_{d(e(G_D))}\right)^T \tag{3.65}$$

$$\tilde{w}_I^k := \left(\tilde{w}^k_{i(1)}, \tilde{w}^k_{i(2)}, \ldots, \tilde{w}^k_{i(e(G_I))}\right)^T \tag{3.66}$$

and $\tilde{w}^k = \left((\tilde{w}^k_B)^T \; (\tilde{w}^k_K)^T \; (\tilde{w}^k_S)^T \; (\tilde{w}^k_D)^T \; (\tilde{w}^k_I)^T\right)^T$, the result is that Eq. (3.51) corresponds to the following equation:

$$\tilde{w}^k_B - \mathbf{B}_K \tilde{w}^k_K - \mathbf{B}_S \tilde{w}^k_S - \mathbf{B}_D \tilde{w}^k_D - \mathbf{B}_I \tilde{w}^k_I = 0 \tag{3.67}$$

where the difference in sign is due to the fact that the edges belonging to the Lagrangian tree are pointing outwards from each vertex. This last equation can be written as:

$$\mathbf{Q}\tilde{w}^k = 0$$

where

$$\mathbf{Q} = \left(I_{e(G_B)} \; -\mathbf{B}\right) = \left(I_{e(G_B)} \; -\mathbf{B}_K \; -\mathbf{B}_S \; -\mathbf{B}_D \; -\mathbf{B}_I\right) \tag{3.68}$$

is called the *fundamental cut-set matrix* associated with the spanning tree $G_B$. From the previous definitions one can observe that $\mathbf{C}\mathbf{Q}^T = 0$ for any possible primary graph. This is important for the following theorem:

**Theorem 3.10.1 (Interconnection power continuity).** *The interconnection represented by any* $\mathbf{B}$ *is always power continuous.*

**Proof.** Denote by $\mathcal{W} := \ker \mathbf{Q}$ the set of wrenches and by $\mathcal{T} := \ker \mathbf{C}$ the set of twists satisfying the Kirchhoff's laws. A result of basic algebra is that the perp of $\mathcal{W}$ indicated with $\mathcal{W}^{\perp}$ and corresponding to the subsets of dual twists that are not producing energy if applied to wrenches in $\mathcal{W}$ is $\mathcal{W}^{\perp} = \operatorname{Im} \mathbf{Q}^T$. Since $\mathbf{C}\mathbf{Q}^T = 0$, we have $\mathcal{W}^{\perp} \subset \mathcal{T}$ and since it can be seen that $\dim(\mathcal{W}^{\perp}) = \dim(\mathcal{T})$, it follows that $\mathcal{W}^{\perp} = \mathcal{T}$, proving power continuity for the interconnection. ■

## 3.11 Hamiltonian interconnection in the Poisson framework

We now present two methods to get an expression of the equations of the complete interconnected system. The first method is a review of the procedure presented in [19] with some modifications. This method yields an implicit Hamiltonian with Lagrangian multipliers. Its advantage is that the interconnection structure is directly used, but the disadvantage of the method is that we are left with an implicit system which should be brought to an explicit form.

### 3.11.1 Inertias equations

In Sect. 3.7, we analyzed in detail the equations describing a rigid body. These have resulted in Eq. (3.46). We can group these equations for each body and get an expression with the form:

$$\dot{x}_B = J_B(x_B)\frac{\partial H_B(x_B)}{\partial x_B} + \phi_B(x_B)\tilde{w}_B^k \tag{3.69}$$

$$\tilde{t}_B^k = \phi_B^*(x_B)\frac{\partial H_B(x_B)}{\partial x_B} \tag{3.70}$$

where, if we denote by $l(\cdot)$ an indexing function for the bodies, we have:

$$x_B = (h^0_{l(1)}, m^{l(1)}, \ldots, h^0_{l(e(G_B))}, m^{l(e(G_B))}) \text{ and}$$

$$H_B(x_B) = \sum_{i=1}^{e(G_B)} H_{l(i)}(h^0_{l(i)}, m^{l(i)}) \tag{3.71}$$

in which $H_{l(i)}(h^0_{l(i)}, m^{l(i)})$ represents the Hamiltonian function associated with body $l(i)$ from which the edge $i$ of $G_B$ is departing. We recall that the function $e()$ indicates the number of edges of the graph given as its argument. Furthermore, in the previous equations we have

$$J_B(x_B) = \begin{pmatrix} j_B(1) & 0 & 0 \\ 0 & \ddots & 0 \\ 0 & 0 & j_B(e(G_B)) \end{pmatrix}$$

where

$$\phi_B(x_B) = \begin{pmatrix} 0 \\ Ad^*_{h^k_{l(1)}} \\ \vdots \\ 0 \\ Ad^*_{h^k_{l(e(G_B))}} \end{pmatrix}, j_B(i) := \begin{pmatrix} 0 & \chi_{h^0_{l(i)}} \; o \; Ad_{h^0_{l(i)}} \\ -Ad^*_{h^0_{l(i)}} \; o \; \chi^*_{h^0_{l(i)}} & m^{l(i)} \wedge \end{pmatrix}.$$

### 3.11.2 Springs equations

In Sect. 3.6, we analyzed in detail the equations describing elastic elements. These have resulted in Eq. (3.33). We can group these equations for each of the $n_S$ springs and get an expression like:

$$\dot{x}_S = \phi_S(x_S)\tilde{t}^k_S \tag{3.72}$$

$$\tilde{w}^k_S = \phi^*_S(x_S)\frac{\partial H_S(x_S)}{\partial x_S} \tag{3.73}$$

where $x_S = (h_{s(1)}, \dots, h_{s(n_S)})$, $H_S(x_S) = \sum_{i=1}^{n_S} V_{s(i)}(h_{s(i)})$, in which the function $V_{s(i)}(h_{s(i)})$ represents the potential energy associated with spring $i$ in a configuration $h_{s(i)}$. Furthermore, $\phi_S(x_S)$ is a suitable map from $\tilde{t}^k_S$ to $(T^1_V, \dots, T^{n_S}_V)$, where $T^i_V$ corresponds to the $T_V$ in Eq. (3.33) for each spring $i$.

### 3.11.3 Dissipative elements equations

We can collect all the equations of the dissipative elements in the following form:

$$\tilde{w}^k_D = \tilde{R}_I(h)\tilde{t}^k_D \tag{3.74}$$

for which $h = (h^{r(d(1))}_{l(d(1))}, \dots, h^{l(d(e(G_D)))}_{l(d(e(G_D)))})$ and

$$\tilde{R}_I(h) = \begin{pmatrix} \tilde{R}^{d(1)}_I(h^{r(d(1))}_{l(d(1))}) & 0 & 0 \\ 0 & \ddots & 0 \\ 0 & 0 & \tilde{R}^{d(e(G_D))}_I(h^{r(d(e(G_D)))}_{l(d(e(G_D)))}) \end{pmatrix}$$

and $\tilde{R}^{d(i)}_I$ corresponds to the impedance map for the edge $i$ of $G_D$.

### 3.11.4 The interconnected system equations

It is now possible to collect the equations of the elements and their interconnection.

If we rewrite Eq. (3.69) and Eq. (3.70) using the twists' Kirchhoff interconnection relations of Eq. (3.57), Eq. (3.58), Eq. (3.59), and Eq. (3.60), and similarly, the wrenches' interconnection relations of Eq. (3.67), we obtain:

$$\dot{x}_B = J_B(x_B)\frac{\partial H_B(x_B)}{\partial x_B} + \phi_B(x_B)\mathbf{B}\begin{pmatrix}\tilde{w}_K^k\\ \tilde{w}_S^k\\ \tilde{w}_D^k\\ \tilde{w}_I^k\end{pmatrix} \tag{3.75}$$

$$\begin{pmatrix}-\tilde{t}_K^k\\ -\tilde{t}_S^k\\ -\tilde{t}_D^k\\ -\tilde{t}_I^k\end{pmatrix} = \mathbf{B}^T\phi_B^*(x_B)\frac{\partial H_B(x_B)}{\partial x_B} \tag{3.76}$$

Combining the latter equations with Eq. (3.72) and Eq. (3.73), we obtain:

$$\begin{pmatrix}\dot{x}_B\\ \dot{x}_S\end{pmatrix} = \begin{pmatrix}J_B(x_B) & 0\\ 0 & 0\end{pmatrix}\begin{pmatrix}\frac{\partial H_B(x_B)}{\partial x_B}\\ \frac{\partial H_S(x_S)}{\partial x_S}\end{pmatrix} + \begin{pmatrix}\phi_B(x_B)\mathbf{B} & 0\\ 0 & \phi_S(x_S)\end{pmatrix}\begin{pmatrix}\tilde{w}_K^k\\ \tilde{w}_S^k\\ \tilde{w}_D^k\\ \tilde{w}_I^k\\ \tilde{t}_S^k\end{pmatrix} \tag{3.77}$$

$$\begin{pmatrix}-\tilde{t}_K^k\\ -\tilde{t}_S^k\\ -\tilde{t}_D^k\\ -\tilde{t}_I^k\\ \tilde{w}_S^k\end{pmatrix} = \begin{pmatrix}\mathbf{B}^T\phi_B^*(x_B) & 0\\ 0 & \phi_S^*(x_S)\end{pmatrix}\begin{pmatrix}\frac{\partial H_B(x_B)}{\partial x_B}\\ \frac{\partial H_S(x_S)}{\partial x_S}\end{pmatrix}. \tag{3.78}$$

If we now consider a symplectic interconnection with a common wrench $\tilde{w}_S^k$ as seen in Figure 3.4, we obtain:

$$\begin{aligned}\begin{pmatrix}\dot{x}_B\\ \dot{x}_S\end{pmatrix} &= \begin{pmatrix}J_B(x_B) & \phi_B(x_B)\mathbf{B}_S\phi_S^*(x_S)\\ -\phi_S(x_S)\mathbf{B}_S^T\phi_B^*(x_B) & 0\end{pmatrix}\begin{pmatrix}\frac{\partial H_B(x_B)}{\partial x_B}\\ \frac{\partial H_S(x_S)}{\partial x_S}\end{pmatrix}\\ &+ \begin{pmatrix}\phi_B(x_B)\mathbf{B}_K & \phi_B(x_B)\mathbf{B}_D & \phi_B(x_B)\mathbf{B}_I\\ 0 & 0 & 0\end{pmatrix}\begin{pmatrix}\tilde{w}_K^k\\ \tilde{w}_D^k\\ \tilde{w}_I^k\end{pmatrix}\end{aligned} \tag{3.79}$$

$$\begin{pmatrix}-\tilde{t}_K^k\\ -\tilde{t}_D^k\\ -\tilde{t}_I^k\end{pmatrix} = \begin{pmatrix}\mathbf{B}_K^T\phi_B^*(x_B) & 0\\ \mathbf{B}_D^T\phi_B^*(x_B) & 0\\ \mathbf{B}_I^T\phi_B^*(x_B) & 0\end{pmatrix}\begin{pmatrix}\frac{\partial H_B(x_B)}{\partial x_B}\\ \frac{\partial H_S(x_S)}{\partial x_S}\end{pmatrix}. \tag{3.80}$$

If we substitute the dissipative elements of Eq. (3.74), we obtain:

$$\begin{pmatrix} \dot{x}_B \\ \dot{x}_S \end{pmatrix} = \bar{J}(x) \begin{pmatrix} \frac{\partial H_B(x_B)}{\partial x_B} \\ \frac{\partial H_S(x_S)}{\partial x_S} \end{pmatrix} + \begin{pmatrix} \phi_B(x_B)\mathbf{B}_K & \phi_B(x_B)\mathbf{B}_I \\ 0 & 0 \end{pmatrix} \begin{pmatrix} \tilde{w}_K^k \\ \tilde{w}_I^k \end{pmatrix} \tag{3.81}$$

$$\begin{pmatrix} -\tilde{t}_K^k \\ -\tilde{t}_I^k \end{pmatrix} = \begin{pmatrix} \mathbf{B}_K^T \phi_B^*(x_B) & 0 \\ \mathbf{B}_I^T \phi_B^*(x_B) & 0 \end{pmatrix} \begin{pmatrix} \frac{\partial H_B(x_B)}{\partial x_B} \\ \frac{\partial H_S(x_S)}{\partial x_S} \end{pmatrix} \tag{3.82}$$

with

$$\bar{J}(x) := \left( \begin{pmatrix} J_B(x_B) & \phi_B(x_B)\mathbf{B}_S \phi_S^*(x_S) \\ -\phi_S(x_S)\mathbf{B}_S^T \phi_B^*(x_B) & 0 \end{pmatrix} - \begin{pmatrix} \phi_B(x_B)\mathbf{B}_D \tilde{R}_I(h)\mathbf{B}_D^T \phi_B^*(x_B) & 0 \\ 0 & 0 \end{pmatrix} \right) \tag{3.83}$$

where $\bar{J}(x)$ is composed of an antisymmetric power-conservative part minus a symmetric semi-positive part representing irreversible transformation of free energy to heat.

The last step to be taken is to satisfy the constraints implemented by the kinematic pairs. As seen in Sect. 2.1.2, for each kinematic pair, we can decompose $se_k(n)$ and its dual as the direct sum of two subspaces. Denote by $A_i$ the projection operator[10] which projects a twist $t^k \in se_k(n)$ to the allowed subspace $\mathcal{T}^A_{r(i),l(i)}$ for the kinematic pair indexed by $i$. If we group all these projection operators, we have $A := (A_{k(1)}, \ldots, A_{k(e(G_K))})$. Then, to satisfy the constraints of the kinematic pairs, we obtain:

$$\begin{pmatrix} \dot{x}_B \\ \dot{x}_S \end{pmatrix} = \bar{J}(x) \begin{pmatrix} \frac{\partial H_B(x_B)}{\partial x_B} \\ \frac{\partial H_S(x_S)}{\partial x_S} \end{pmatrix} + \begin{pmatrix} \phi_B(x_B)\mathbf{B}_K(Id - A)^* & \phi_B(x_B)\mathbf{B}_K A^* & \phi_B(x_B)\mathbf{B}_I \\ 0 & 0 \end{pmatrix} \begin{pmatrix} \lambda \\ W \\ W_I \end{pmatrix} \tag{3.84}$$

$$\begin{pmatrix} 0 \\ T \\ T_I \end{pmatrix} = \begin{pmatrix} (Id - A)\mathbf{B}_K^T \phi_B^*(x_B) & 0 \\ A\mathbf{B}_K^T \phi_B^*(x_B) & 0 \\ \mathbf{B}_I^T \phi_B^*(x_B) & 0 \end{pmatrix} \begin{pmatrix} \frac{\partial H_B(x_B)}{\partial x_B} \\ \frac{\partial H_S(x_S)}{\partial x_S} \end{pmatrix} \tag{3.85}$$

where $\lambda$ belongs to the constrained space and corresponds to constraint forces, $W$ and $W_I$ are, respectively, the actuation wrenches applied to the kinematic pairs and the external forces applied by the environment on the mechanisms between the bodies specified in the interconnection graph $G_I$. The twists' duals of $\lambda$ should be 0 to satisfy the constraints.

As we can see, the obtained equations are not in explicit form even if the mechanism does not have closed kinematic chains. This can be a shortcoming if explicit equations are needed. The following section presents a method which yields explicit equations in common cases.

---

[10] Note that here we can talk about projection since $se_k(n)$ is the direct sum of $\mathcal{T}^A_{i,j} \oplus \mathcal{T}^F_{i,j}$.

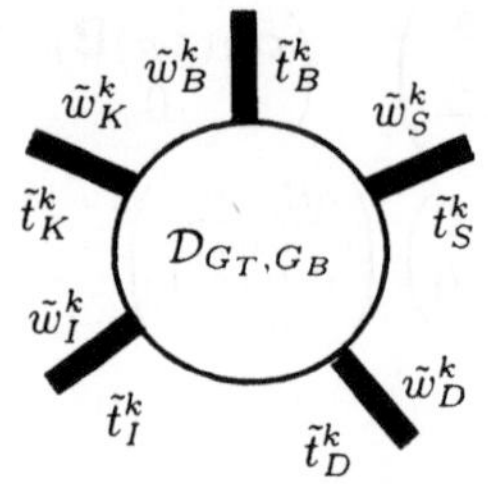


**Fig. 3.12.** The Dirac structure of the mechanism's topology.

## 3.12 Hamiltonian interconnection in the Dirac framework

It is possible to directly associate a Dirac structure with the total graph $G_T$ of Sect. 3.10 using what is presented in [8]. The Dirac structure can be directly expressed using the kernel representation introduced in Eq. (3.23). The $(n+q) \times (n+q)$ structure matrices are:

$$F = \begin{pmatrix} \mathbf{C} \\ 0_{q\times(q+n)} \end{pmatrix} \qquad \text{and} \qquad E = \begin{pmatrix} 0_{n\times(q+n)} \\ \mathbf{Q} \end{pmatrix}$$

where $q = e(G_B)$ is the number of edges of the Lagrangian tree and corresponds to the number of rigid bodies, $n = e(G_P)$ is the number of edges of the primary graph, $\mathbf{C}$ is the fundamental loop matrix associated with the spanning-tree $G_B$ introduced in Eq. (3.61) and $\mathbf{Q}$ is the fundamental cut-set matrix associated with the spanning-tree $G_B$ introduced in Eq. (3.68).

Kirchhoff's laws are then equivalent to defining the set of admissible twists and wrenches as:

$$\mathcal{D}_{G_T,G_B} = \{(\tilde{t}^k, \tilde{w}^k) \in (se(3)_k)^{n+q} \times (se(3)_k^*)^{n+q} : F\tilde{t}^k + E\tilde{w}^k = 0\} \tag{3.86}$$

From Theorem 3.10.1 it follows that $\ker \mathbf{C}$ and $\ker \mathbf{Q}$ are dual spaces and it can be seen that $\dim \mathcal{D}_{G_T,G_B} = 6(n+q)$; this is exactly Tellegen's theorem. Hence the admissible set of twists and wrenches associated with the constraint relation of the port connection graph form a constant Dirac structure on $(se(3)_k)^{n+q}$. A graphical representation of the Dirac structure is reported in Figure 3.12 where the space on which the structure is defined is drawn using power ports on the composing parts defined in Sect. 3.10.1.

The previous Dirac structure does not take into account the constraints of each of the kinematic pairs. This can be done defining a Dirac structure on a differentiable manifold, which is actually a submanifold of the state manifold $(SE_i^0(3) \times se_i^*(3))^q$ corresponding to the $q$ rigid bodies from which the mechanism is composed (see Eq. (3.46)). This submanifold $\mathcal{K}$ is actually:

$$\mathcal{K} := SE_{l(1)}^{r(1)}(3) \times \ldots \times SE_{l(n)}^{r(n)}(3)$$

where $n = e(G_K)$ is the number of kinematic pairs also used in Eq. (3.52) and $l(i), r(i)$ index the vertices in $G_K$ and correspond to the two rigid bodies which the $i$-th kinematic pair connects (see Sect. 3.10). We can then define a diffeomorphism between $T_h\mathcal{K}$ and $(se_k(3))^n$ where $h := \left(h_{l(1)}^{r(1)}, \ldots, h_{l(1)}^{r(1)}\right)$ in the following way:

$$\begin{aligned}\pi_K := T_h\mathcal{K} \to (se_k(3))^n \; ; \; (h, \dot{h}) \mapsto \\ \left(\left(Ad_{h^k_{r(1)}} \, o \, \pi^{r(1)}_{h^{r(1)}_{l(1)}}\right)\left(h_{l(1)}^{r(1)}, \dot{h}_{l(1)}^{r(1)}\right), \ldots, \right. \\ \left. \left(Ad_{h^k_{r(n)}} \, o \, \pi^{r(n)}_{h^{r(n)}_{l(n)}}\right)\left(h_{l(n)}^{r(n)}, \dot{h}_{l(n)}^{r(n)}\right)\right)\end{aligned} \tag{3.87}$$

where $\pi^j_{h^j_i}$ is defined in Figure 1.9.

If for simplicity we restrict our study to the case of regular kinematic pairs as defined in Def. 2.1.2, it is possible to associate with each kinematic pair $i$ two subspaces of $se_k(3)$ and two subspaces of $se^*_k(3)$ as done for lower pairs in Eq. (2.3) and Eq. (2.4) by mapping the distribution and co-distribution defining each kinematic pair using respectively $Ad_{h^k_r} \, o \, \pi^r_{h^r_l}$ and $((\pi^r_{h^r_l})^* \, o \, Ad^*_{h^k_r})^{-1}$. It is important to note that the achieved subspaces are now *configuration dependent* and not constant:

$$se_k(n) = \mathcal{T}^A_{K_i}\left(h^{l(i)}_{r(i)}\right) \oplus \mathcal{T}^F_{K_i}\left(h^{l(i)}_{r(i)}\right) \tag{3.88}$$

for the twists of kinematic pair $i$ and

$$se^*_k(n) = \mathcal{W}^A_{K_i}\left(h^{l(i)}_{r(i)}\right) \oplus \mathcal{W}^C_{K_i}\left(h^{l(i)}_{r(i)}\right) \tag{3.89}$$

for the wrenches of kinematic pair $i$. If the kinematic pairs are regular, they are characterized by a constant number of degrees of freedom. If we denote with $\mathrm{dof}_i$ the number of degrees of freedom of kinematic pair $i$, we have that:

$$\dim\left(\mathcal{T}^A_{K_i}(\cdot)\right) = \mathrm{dof}_i \qquad \dim\left(\mathcal{T}^F_{K_i}(\cdot)\right) = 6 - \mathrm{dof}_i, \tag{3.90}$$

and

$$\dim\left(\mathcal{W}^A_{K_i}(\cdot)\right) = \mathrm{dof}_i \qquad \dim\left(\mathcal{W}^C_{K_i}(\cdot)\right) = 6 - \mathrm{dof}_i. \tag{3.91}$$

We can therefore find for each kinematic pair $i$ full rank matrices $T_{Fi}\left(h^{r(i)}_{l(i)}\right)$, $T_{Ai}\left(h^{r(i)}_{l(i)}\right)$, $W_{Ci}\left(h^{r(i)}_{l(i)}\right)$ and $W_{Ai}\left(h^{r(i)}_{l(i)}\right)$ such that the rows of

$$\begin{pmatrix} T_{Ai} \\ T_{Fi} \end{pmatrix} \begin{matrix} \text{dof} \\ \text{6-dof} \end{matrix}$$

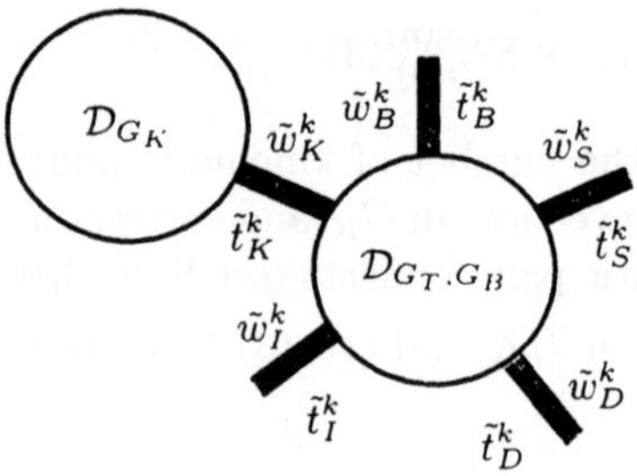


**Fig. 3.13.** The interconnection with the kinematic pairs.

are a basis of $se_k(3)$, the rows of $T_{Ai}$ are a basis of $\mathcal{T}_{K_i}^A\left(h_{l(i)}^{r(i)}\right)$, the rows of $T_{Fi}$ are a basis of $\mathcal{T}_{K_i}^F\left(h_{l(i)}^{r(i)}\right)$, the rows of

$$\begin{pmatrix} W_{Ai} \\ W_{Ci} \end{pmatrix} \begin{matrix} \text{dof} \\ 6\text{-dof} \end{matrix}$$

are a basis of $se_k^*(3)$, the rows of $W_{Ai}$ are a basis of $\mathcal{W}_{K_i}^A\left(h_{l(i)}^{r(i)}\right)$, the rows of $W_{Ci}$ are a basis of $\mathcal{W}_{K_i}^C\left(h_{l(i)}^{r(i)}\right)$ and furthermore the two bases of $se_k(3)$ and $se_k^*(3)$ are dual of each other. All these matrices are functions of $h$, but this has been omitted for notational convenience. We therefore clearly have that:

$$\ker W_{Ai}\left(h_{l(i)}^{r(i)}\right) = \mathcal{T}_{K_i}^F\left(h_{l(i)}^{r(i)}\right) \quad \text{and} \quad \ker T_{Fi}\left(h_{l(i)}^{r(i)}\right) = \mathcal{W}_{K_i}^A\left(h_{l(i)}^{r(i)}\right). \tag{3.92}$$

Because of power conservation we must have that

$$\mathcal{T}_{K_i}^F\left(h_{l(i)}^{r(i)}\right) = \left(\mathcal{W}_{K_i}^A\left(h_{l(i)}^{r(i)}\right)\right)^{\perp} \tag{3.93}$$

and

$$\mathcal{T}_{K_i}^A\left(h_{l(i)}^{r(i)}\right) = \left(\mathcal{W}_{K_i}^C\left(h_{l(i)}^{r(i)}\right)\right)^{\perp}. \tag{3.94}$$

We have therefore:

$$\ker W_{Ai} = \operatorname{range} T_{Fi}^T \qquad \text{and} \qquad \ker T_{Fi} = \operatorname{range} W_{Ai}^T \tag{3.95}$$

and we can use as structural matrices to define the Dirac structure associated with the $i$-th kinematic pair:

$$F_i\left(h_{l(i)}^{r(i)}\right) = \begin{pmatrix} W_{Ai}(h_{l(i)}^{r(i)}) \\ 0_{(6-\text{dof}_i)\times 6} \end{pmatrix} \quad \text{and} \quad E_i\left(h_{l(i)}^{r(i)}\right) = \begin{pmatrix} 0_{\text{dof}_i\times 6} \\ T_{Fi}(h_{l(i)}^{r(i)}) \end{pmatrix}. \tag{3.96}$$

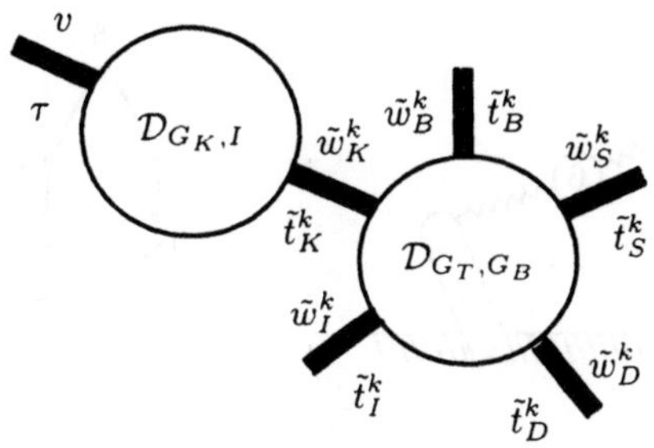


**Fig. 3.14.** The interconnection with the controllable kinematic pairs.

It is now possible to define the Dirac structure associated with all kinematic pairs as:

$$\mathcal{D}_{G_K}(h) = \Big\{(f,e) \in T_h\mathcal{K} \times T_h^*\mathcal{K}\ ;\ E(h)\tilde{w}_K^k + F(h)\tilde{t}_K^k = 0\ ;$$
$$\tilde{t}_K^k := \pi_K(f), \tilde{k}_K^k := (\pi_K^*)^{-1}(e)\Big\} \quad (3.97)$$

where

$$F(h) := \operatorname{diag}\left(F_1\left(h_{l(1)}^{r(1)}\right), \ldots, F_n\left(h_{l(n)}^{r(n)}\right)\right) \quad (3.98)$$

,

$$E(h) := \operatorname{diag}\left(E_1\left(h_{l(1)}^{r(1)}\right), \ldots, E_n\left(h_{l(n)}^{r(n)}\right)\right). \quad (3.99)$$

and $\tilde{t}_K^k$ and $\tilde{w}_K^k$ are defined respectively in Eq. (3.52) and Eq. (3.62). It is also shown in [71] that the interconnection of Dirac structures is again a Dirac structure and therefore the obtained Dirac structure can be graphically represented as in Figure 3.13, where the interconnection is implemented using the structure of the power port $(\tilde{t}_K^k, \tilde{w}_K^k)$.

It is also possible to consider kinematic pairs with which we can interact through the degrees of freedom. This can be done using the following structural matrices :

$$F_i\left(h_{l(i)}^{r(i)}\right) = \begin{pmatrix} W_{Ai}(h_{l(i)}^{r(i)}) & -I_{\mathrm{dof}_i} \\ 0 & 0 \\ W_{Ci}(h_{l(i)}^{r(i)}) & 0 \end{pmatrix} \quad (3.100)$$

and

$$E_i\left(h_{l(i)}^{r(i)}\right) = \begin{pmatrix} 0 & 0 \\ T_{Ai}(h_{l(i)}^{r(i)}) & I_{\mathrm{dof}_i} \\ 0 & 0 \end{pmatrix}. \quad (3.101)$$

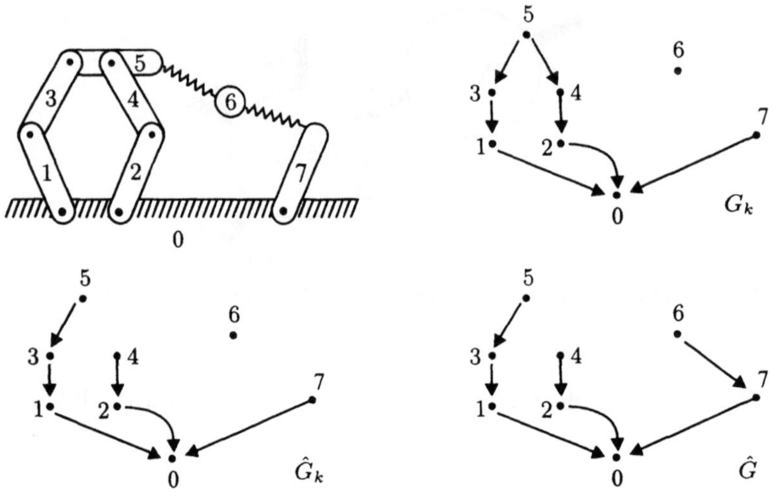


**Fig. 3.15.** A simple example.

The new port that is defined will be a numerical representation of elements in $\mathcal{T}^A \times \mathcal{W}^A$:

$$\begin{aligned}\mathcal{D}_{G_K,I}(h) := \{((f,v),(e,\tau)) \in (T_h\mathcal{K}\times\mathbb{R}^m)\times(T_h^*\mathcal{K}\times\mathbb{R}^m)\ ;\\ E(h)\begin{pmatrix}\tilde{w}_K^k\\ \tau\end{pmatrix} + F(h)\begin{pmatrix}\tilde{t}_K^k\\ v\end{pmatrix} = 0\ ;\ \tilde{t}_K^k := \pi_K(f), \tilde{k}_K^k := (\pi_K^*)^{-1}(e)\Big\}\end{aligned} \tag{3.102}$$

where $m = \sum_{i=1}^n \text{dof}_i$ and $E(h), F(h)$ are defined as a diagonal composition of the matrices defined in Eq. (3.100) and Eq. (3.101) as done previously. It would then be possible to interconnect the obtained Dirac structure shown in Figure 3.14 to the energy storage elements corresponding to inertias and springs and the dissipating elements in order to obtain a complete implicit, dynamic model of the system. This could be done by expressing Eq. (3.69), Eq. (3.70), Eq. (3.72) and Eq. (3.73) as the interconnection of simple Dirac structures and pure storage elements.

## 3.13 Direct Lagrangian equations

In this section we describe a way to achieve explicit dynamic equations for mechanisms having no kinematic loops and only holonomic kinematic pairs.

We first consider a spanning forest (see Def. A.2.12) $\hat{G}_K = (V_B \cup \{0\}, \hat{E}_K)$ of the kinematic graph $G_K$. In general $\hat{G}_K$ is NOT a connected graph since we could have inertial elements (vertices in $V_B$) not constrained to other masses by means of kinematic pairs as the body indexed by 6 in Figure 3.15.

Let us now consider a tree $\hat{G} = (V_B \cup \{0\}, E)$ such that $\hat{E}_K \subset E$. If $V_B$ has $m$ elements (there are $m$ bodies in the system), $E$ will have $m$ edges[11]. In the example of Figure 3.15, $m = 7$.

If we indicate with $l(i)$ and $r(i)$ the departing and the destination vertices for an edge $i$, and if we index the edges in $\hat{E}_K$ with $1, \ldots, \hat{n}_k$ where $\hat{n}_k := e(\hat{G}_K)$, the number of edges in $\hat{G}_K$, and the remaining edges in $E$ with $\hat{n}_k + 1, \ldots, m$, where $m = e(\hat{G})$, then we can define:

$$\mathcal{Q} := \mathcal{Q}_{l(1)}^{r(1)} \times \ldots \mathcal{Q}_{l(\hat{n}_k)}^{r(\hat{n}_k)} \times SE_{l(\hat{n}_k+1)}^{r(\hat{n}_k+1)}(n) \times \ldots SE_{l(m)}^{r(m)}(n). \tag{3.103}$$

Because $\hat{G}$ is a spanning tree by construction, a point in $\mathcal{Q}$ will uniquely identify the relative position of any body in the mechanism with respect to any other body and it will identify in particular the relative position of any body with respect to the inertial reference space $\mathcal{E}_0$ represented by the vertex $\{0\}$.

### 3.13.1 Total kinetic co-energy

We now calculate a Riemannian metric on $\mathcal{Q}$ corresponding to the total kinetic co-energy of the system.

First, however, we must calculate a mapping from an element belonging to $T\mathcal{Q}$ to the various twists of the bodies with respect to the inertial frame and expressed in their own space. Once this is done, the total inertia tensor can be pulled back directly to $T\mathcal{Q}$ (see Def. A.7.20).

**The needed mapping.** Each element $(q, \dot{q}) \in T\mathcal{Q}$ is such that $q = (h_{l(1)}^{r(1)}, \ldots, h_{l(m)}^{r(m)})$ and $\dot{q} = (\dot{h}_{l(1)}^{r(1)}, \ldots, \dot{h}_{l(m)}^{r(m)})$. Realizing this, we can define the following map:

$$\tilde{J}(q) : T_q\mathcal{Q} \to se_1(n) \times \ldots \times se_m(n) \; ; \; \dot{q} \mapsto A(q)J(q)\Pi(q) \begin{pmatrix} \dot{h}_{l(1)}^{r(1)} \\ \vdots \\ \dot{h}_{l(m)}^{r(m)} \end{pmatrix} \tag{3.104}$$

where $J(q)$ is a Jacobian similar to the Jacobian in Eq. (2.10), where instead of $m-1$ we have $m$, $\mathcal{T}_{r(i),l(i)}^{A} = se_{r(i)}(n)$, $\forall i = \hat{n}_k + 1, \ldots, m$, $n_c = m$, $q(i) = 0$ and $p(i) = i$, $\forall i$,

$$A(q) := \begin{pmatrix} Ad_{h_0^1} & 0 & \ldots & 0 \\ 0 & \ddots & & \vdots \\ \vdots & & \ddots & 0 \\ 0 & \ldots & 0 & Ad_{h_0^m} \end{pmatrix} \text{ and } \Pi(q) := \begin{pmatrix} \pi^{r(1)}_{h_{l(1)}^{r(1)}} & 0 & \ldots & 0 \\ 0 & \ddots & & \vdots \\ \vdots & & \ddots & 0 \\ 0 & \ldots & 0 & \pi^{r(m)}_{h_{l(m)}^{r(m)}} \end{pmatrix}$$

[11] Recall that a tree with $m+1$ vertices always has $m$ edges.

This map gives the twist of each body with respect to the inertial frame and expressed in its own space as a function of the configuration velocities.

**The total inertia tensor and the Riemannian metric.** The total kinetic co-energy of the system can be expressed as:

$$E_k^*(t_1^{1,0}, \dots, t_m^{m,0}) = \sum_{i=1}^{m} \frac{1}{2} \langle t_i^{i,0}, t_i^{i,0} \rangle_{I_i}$$

where $t_i^{i,0} = Ad_{h_0^i} t_i^0 \in se_i(n)$ is the twist of the body $i$ with respect to the inertial frame and expressed in the body $i$ space and $\langle . \rangle_{I_i}$ is the quadratic form in $se_i(n)$ corresponding to the kinetic co-energy defined in Eq. (3.38).

We can therefore express the total energy as a quadratic form in $se_1(n) \times \dots \times se_m(n)$ such that:

$$E_k^*(t) = \frac{1}{2} t^T I_T t$$

where

$$I_T := \begin{pmatrix} I_1 & 0 & \dots & 0 \\ 0 & \ddots & & \vdots \\ \vdots & & \ddots & 0 \\ 0 & \dots & 0 & I_m \end{pmatrix} \qquad t := \begin{pmatrix} t_1^{1,0} \\ \vdots \\ t_m^{m,0} \end{pmatrix}$$

and $I_i$ is the inertia tensor of body $i$ as introduced in Sect. 3.3.1.

We can now define a Riemannian metric on $\mathcal{Q}$ which gives the total kinetic co-energy by means of a pullback:

$$E_k^* : T\mathcal{Q} \to \mathbb{R}^+ \ ; \ (q, \dot{q}) \mapsto \frac{1}{2} \dot{q}^T \tilde{J}^*(q) I_T \tilde{J}(q) \dot{q}$$

where $\tilde{J}^*(q)$ is the adjoint of $\tilde{J}(q)$. The Riemannian metric is therefore equal to $\tilde{J}^*(q) I_T \tilde{J}(q)$ in a position $q$.

### 3.13.2 The potential energy and the Lagrangian

The total potential energy is the sum of the potential energies of all the springs represented by the edges of $G_S$. Since a point in $\mathcal{Q}$ uniquely identifies the relative positions of all the bodies and the reference space, it is possible to restrict the energy function of each spring to $\mathcal{Q}$ in order to get:

$$E_p : \mathcal{Q} \to \mathbb{R}^+ \ ; \ q \mapsto \sum_{i=1}^{n_S} V^i(q)$$

where $V^i(\cdot)$ corresponds to the potential energy function corresponding to spring $i$ as given in Eq. (3.29).

We can now define the Lagrangian function for the system as

$$L : T\mathcal{Q} \to \mathbb{R} \ ; \ (q, \dot{q}) \mapsto E_k^*(q, \dot{q}) - E_p(q)$$

in the same way as in Sect. 3.3.

### 3.13.3 The Constraints

If we have kinematic loops or nonholonomic kinematic pairs, we have to treat them explicitly. To start with, we can consider a map

$$L_K : \mathcal{Q} \to \bar{\mathcal{Q}}$$

where $\bar{\mathcal{Q}} := \mathcal{Q}_{l(p(1))}^{r(p(1))} \times \ldots \times \mathcal{Q}_{l(p(n_k))}^{r(p(n_k))}$, $p(i)$ is the $i$–th edge in the kinematic graph and[12] $n_k := e(G_K)$. This map should associate with each point in $\mathcal{Q}$ the relative configurations of all the kinematic pairs present in the mechanism and the map can be expressed with methods similar to those presented in Chapter 2. We can then consider its Jacobian, which is defined as:

$$J_K(q) : T_q\mathcal{Q} \to T_{L_K(q)}\bar{\mathcal{Q}} \; ; \; \dot{q} \mapsto D_{L_K}(q)\dot{q}$$

where $D_{L_K}(q) \in \mathbb{R}^{m \times \hat{n}_k}$ is the differential of $L_K$ in $q \in \mathbb{R}^m$.

If all the kinematic pairs are holonomic and if there are no kinematic loops ($G_K$ is circuit free), the motion of each kinematic pair is allowed. If there are nonholonomic pairs or kinematic loops, only some directions are allowed, namely those expressed by the distribution $\Delta_{k_{l(p(i))}^{r(p(i))}}$ describing the $i$-th pair and defined in Def. 2.1.2. If we denote by $\Delta^C_{k_{l(p(i))}^{r(p(i))}}$ a complementary distribution to $\Delta_{k_{l(p(i))}^{r(p(i))}}$ such that

$$T_{h_{l(p(i))}^{r(p(i))}} \mathcal{Q}_{l(p(i))}^{r(p(i))} = \Delta^C_{k_{l(p(i))}^{r(p(i))}}(h_{l(p(i))}^{r(p(i))}) \oplus \Delta_{k_{l(p(i))}^{r(p(i))}}(h_{l(p(i))}^{r(p(i))})$$

we can define the projection $P_i$ of an element belonging to $T_{h_{l(p(i))}^{r(p(i))}} \mathcal{Q}_{l(p(i))}^{r(p(i))}$ on $\Delta^C_{k_{l(p(i))}^{r(p(i))}}(h_{l(p(i))}^{r(p(i))})$ for each kinematic pair $i$ at each position $h_{l(p(i))}^{r(p(i))}$:

$$P_i(h_{l(p(i))}^{r(p(i))}) : T_{h_{l(p(i))}^{r(p(i))}} \mathcal{Q}_{l(p(i))}^{r(p(i))} \to \Delta^C_{k_{l(p(i))}^{r(p(i))}}(h_{l(p(i))}^{r(p(i))})$$

We can then consider:

$$P(q) := \begin{pmatrix} P_1(h_{l(p(1))}^{r(p(1))}) & 0 & \ldots & 0 \\ 0 & \ddots & & \vdots \\ \vdots & & \ddots & 0 \\ 0 & \ldots & 0 & P_{n_k}(h_{l(p(n_k))}^{r(p(n_k))}) \end{pmatrix}$$

and express the constraints, which should be satisfied as:

$$P(q)J_K(q)\dot{q} = 0$$

*Remark 3.13.1.* Note that for a holonomic kinematic pair $i$, we have $T_{h_{l(p(i))}^{r(p(i))}} \mathcal{Q} = \Delta_{k_{l(p(i))}^{r(p(i))}}(h_{l(p(i))}^{r(p(i))})$ and therefore the operator $P_i(\cdot)$ is identically zero.

[12] Note that $n_k \geq \hat{n}_k$ since the graph $\hat{G}_k$ is defined deleting edges of $G_k$.

### 3.13.4 Damping elements

Following the same reasoning as in the previous section, we could consider a map

$$L_D : \mathcal{Q} \to \mathcal{Q}_D$$

where $\mathcal{Q}_D := SE_{l(d(1))}^{r(d(1))}(n) \times \ldots \times SE_{l(d(n_d))}^{r(d(n_d))}(n)$, $d(i)$ is the $i$–th edge in the dissipation graph associated with the mechanism and $n_d := e(G_D)$. This map should associate with each point in $\mathcal{Q}$ the relative configurations of all the relative spaces adjacent to the edges of $G_D$. We can then consider its Jacobian, which is defined as:

$$J_D(q) : T_q\mathcal{Q} \to T_{L_D(q)}\mathcal{Q}_D \; ; \; \dot{q} \mapsto D_{L_D}(q)\dot{q}$$

where $D_{L_D}(q)$ is the differential of $L_D$ in $q$. The generalized forces generated by these elements are:

$$-J_D^*(q)\bar{R}_I(L_D(q))J_D(q)\dot{q} \in T_q^*\mathcal{Q}$$

where $\bar{R}_I$ is a diagonal matrix composed of elements representing the impedance form of each dissipative element. The tensor $J_D^*(q)\bar{R}_I(L_D(q))J_D(q)$ can be seen as a 2-covariant tensor resulting from the pullback of $\bar{R}_I(L_D(q))$.

### 3.13.5 The work-space and the interaction with the environment

In this subsection, we generalize the definition of work space for a generic mechanism. We define the *work space* as the set of relative positions corresponding to the edges of the interaction graph. This is the space in which, by definition, the robot exchanges energy with the environment.

**Definition 3.13.1 (Work space).** *If we index the edges of the interaction graph with an indexing function $i(\cdot)$, and as usual with $r(i(j))$ and $l(i(j))$ the vertices of the graph $G_I$ to which the $j$-th edge is pointing and from which it is departing, we define the interaction space $\mathcal{X}$ as:*

$$\mathcal{X} := SE_{l(i(1))}^{r(i(1))}(n) \times \cdots \times SE_{l(i(n_i))}^{r(i(n_i))}(n)$$

*where $n_i = e(G_I)$ denotes the number of edges in the graph $G_I$.*

It is possible to define a mapping having the following form:

$$L_I : \mathcal{Q} \to \mathcal{X}$$

which associates with each point in $\mathcal{Q}$ the relative configurations of all the bodies with which the environment can interact. We can then consider its Jacobian, which is defined as:

$$J_I(q) : T_q\mathcal{Q} \to T_{L_I(q)}\mathcal{X} \; ; \; \dot{q} \mapsto D_{L_I}(q)\dot{q}$$

where $D_{L_I}(q)$ is the differential of $L_I$ in $q$. The final form of the Lagrangian equations of the complete system, where we summed all the generalized forces, follows:

$$\begin{aligned} \frac{d}{dt}\left(\frac{\partial L}{\partial \dot{q}}\right) - \frac{\partial L}{\partial q} &+ J_D^*(q)\bar{R}_I(L_D(q))J_D(q)\dot{q} \\ &= J_K^*(q)P^*(q)\lambda + J_K^*(q)W_K + J_I^*(q)W_I \\ 0 &= P(q)J_K(q)\dot{q} \end{aligned} \tag{3.105}$$

where $W_K \in T^*_{L_K(q)}\bar{\mathcal{Q}}$ corresponds to the torque which can be applied to the kinematic pairs, $\lambda$ are constraint forces and where $W_I \in T^*_{L_I(q)}\mathcal{X}$ represents the forces that the environment can apply to the mechanism.

### 3.13.6 Elimination of the constraints.

It is now possible to calculate the generalized Poisson manifold, which defines a Hamiltonian formulation by means of the Legendre transformation of the Lagrangian function. This procedure and its analysis can be found in [54].

In [54] it is also shown that is only possible to get a proper Poisson bracket satisfying the Jacobi identity under certain integrability conditions.

In our case, the integrability condition is such that the distribution defined by

$$\Delta(q) := \ker(P(q)J_k(q))$$

should be integrable. If we have no nonholonomic pairs and no loop chains, $P(q)$ and $P^*(q)$ are null, the constraints are automatically satisfied and we indeed get an explicit expression of the dynamic equations of the mechanism. Intuitively, for a mechanism without kinematic loops, this procedure "works out" any holonomic pair at the beginning by choosing its configuration space as part of the total configuration space.

## 3.14 Conclusions

In this chapter we analysed the interconnection of mechanical, rigid, physical systems.

After defining all the composing elements, we have seen that the total interconnection can be described by a set of graphs.

We have also shown how the information of the interconnection is either stored in the Poisson bracket of the Hamiltonian system describing the complete system, or in a Dirac structure. Irreversible free energy dissipation results, which is represented by a negative definite symmetric bracket in the Hamiltonian system.

The power of Dirac structures is that it is a tool which can be used to express interconnection without the need of any causal specification and it nicely expresses network power continuity (Tellegen's theorem).

For a situation involving only holonomic pairs and no kinematic loops, we supplied a method that could directly give an explicit formulation of the equations of the mechanism. For a case where there were either nonholonomic pairs or kinematic loops present, on the other hand, we presented a way to find implicit equations for the system in a Lagrangian setting.

Methods which show how to go from a Lagrangian system with constraints to the corresponding Hamiltonian system are described in literature [54]

# 4. Intrinsically Passive Control

In most robotics applications, robots have to interact with an unknown environment in order to complete desired tasks. It is never possible to obtain an accurate model of the environment and this has consequences for the development of the controller which must preserve a proper interacting behavior during any situation.

A well-behaved system during interaction with unknown environments is a human being. In humans, it is known in physiology that there are delays in the communication between the brain and the muscles between 150 and 250 ms. Muscle spindles are "built in" to the muscles and have afferent nerves (nerves that send information back to the spinal cord and the central nervous system) that signal a weighted sum of muscle length and muscle velocity of shortening. In addition to sending information up the spinal column, those afferent nerves make excitatory connections through a single synapse (which is the fastest connection you can get) onto the afferent nerves that send motor commands out to the muscles so that stimulation of the muscle spindles evokes contraction of the corresponding muscle with a delay of 25 to 50 ms (depending on where the muscle is - arm muscles have shorter delays than leg muscles). This is the fastest feedback system in the body; it is responsible for the "knee-jerk" reflex that doctors use, and it looks a lot like a PD controller (though quite nonlinear and with adaptive gains). There is also a Golgi tendon organ that responds almost exclusively to actively generated muscle forces (and not to externally applied loads). This organ also makes a feedback connection via the spinal cord to the muscle it is in, but this time the action is inhibitory (acting to turn off the muscle). The combination of force, position and velocity feedback looks a lot like it acts (among other things) to regulate or manage the overall muscle impedance [72]. This short physiological deviation might justify the decomposition of a robotic control system in a way as the one reported in Figure 4.1.

Here the *Intrinsically Passive Control* (IPC) resembles the muscle behavior and it should preserve passivity under any contact situations with any passive environments, whereas the Supervisor takes care of the planning, scheduling and coordination of the task to be performed. In humans the local control is implemented by the *muscle spindles*: with the explained delays of communication with the brain, the system would simply be unstable without

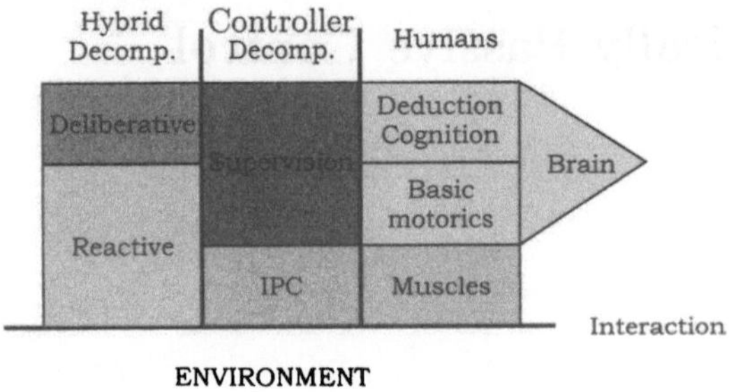


**Fig. 4.1.** A biological analogy of the control strategy

this local control of interaction. The IPC is characterized by the fact that it has two power ports and that it is passive. With one of the two ports it is connected to the system to be controlled, namely a robot, and with the other port it is connected to a supervisor module which is, hierarchically speaking, a higher lever of control. The major point is that when the supervisor does not supply any power, the controlled system will be passive. In this chapter, the structure of the IPC will be studied following the results which where first presented in [73].

The above techniques yield coupled stability under any model uncertainties and interaction of the plant with any other passive system. This is very important, for example for robotics applications where the robot can interact with unknown environments at any time. We can design the controller using knowledge of the plant in a model-based fashion. Non-exactness of the model of the plant could spoil performance but NEVER stability. Furthermore, as shown in [74], even with perfect knowledge of the plant, with a controller which is not passive by itself, instability can arise during unmodeled interaction through other possibly present power ports. A typical example is a controlled robot which unexpectedly interacts with the environment. This may happen because the environment cannot be merely considered a signal source; it is a dynamic system, as explained in [6].

A basic requirement for the strategies presented here is that we should be able to apply collocated control to back-drivable transmissions, as we will see later. Alternative methods can also be used in the case in which back-drivable systems are not present. These techniques do need force measurements.

## 4.1 Force versus Interactive control

During an interacting task of a robot with its environment, the forces which appear depend on both the robot, which we can influence, and the environment, which we cannot influence. This implies that strictly speaking there is no way to ensure that the contacting force reaches a certain value unless we have a *perfect* model of the environment, which is by definition never

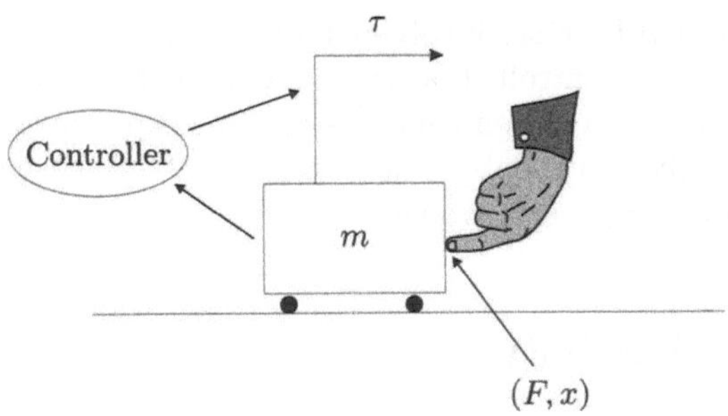


**Fig. 4.2.** Representation of the I/O MMP

the case. To formally explain the previous statement, consider the simple one-dimensional example of Figure 4.2. A control force $\tau$ acts on the mass $m$ at position $x$. The environment exerts a force $F$ represented in the figure by a hand. The power exchanged between the controlled mass and the environment is therefore $P = F\dot{x}$.

Using linear control, a certain robot dynamics is achieved, which can be expressed in the behavioral framework [41] in the following way:

$$R(s)\begin{pmatrix} F(s) \\ x(s) \end{pmatrix} = 0, \qquad R(s) \in \mathbb{R}^{1\times 2}[s]$$

where $R(s)$ is a polynomial matrix.

In the same way, the environment has its own dynamics that, assuming linearity for simplicity, can be represented in a similar way:

$$R_E(s)\begin{pmatrix} F(s) \\ x(s) \end{pmatrix} = 0, \qquad R_E(s) \in \mathbb{R}^{1\times 2}[s]$$

Consider two extreme cases:

1. The mass does not contact anything: $F(t) = 0\ \forall t$.
2. The mass position is fixed: $\dot{x}(t) = 0\ \forall t$.

In the first case, the value of $x(t)$ depends only on $R(s)$, which we can directly control. In this case we can effectively use what is called *position control.*

In the second case, the value of $F(t)$ depends only on $R(s)$, which we can directly control. In this case we can effectively use what is called *force control.*

In *all* other cases both $F(s)$ and $x(s)$ depend on both the controlled robot $R(s)$ and the environment $R_E(s)$:

$$\begin{pmatrix} R(s) \\ R_E(s) \end{pmatrix}\begin{pmatrix} F(s) \\ x(s) \end{pmatrix} = \begin{pmatrix} 0 \\ 0 \end{pmatrix}$$

The solution to the previous differential equation gives the unique solution for $F(t)$ and $x(t)$, which *also* depends on the imperfectly or poorly characterized environment $R_E(s)$.

We conclude that for tasks involving interaction it is questionable whether $F(t)$ and $x(t)$ can be controlled, as these partly depend on the variable, imperfectly characterized environment. Instead one can control the only entity that is intrinsically controllable and independent of the environment: the controlled behavior $R(s)$ as a whole.

## 4.2 Physical Passivity

Passivity is a very powerful concept used in the design of control strategies [75]. An overview of passivity concepts can be found in [61] and in [76].

The basic idea of passivity comes from physics and generalizes the concept of energy. In general, we consider a set of $G_i$ input-output mappings from a signal $u_i$ to $y_i$. In $L_2$ theory [61], $u_i$ and $y_i$ are supposed to belong to the same Hilbert space with inner product $\langle . \rangle_i$. We can then consider for each system $G_i$ the quantity $\langle u_i, y_i \rangle$ corresponding to the "generalized energy" of the system. The system is said to be passive if for each $u_i$, there always exists a constant finite $\beta$ such that $\langle u_i, G(u_i) \rangle \geq \beta$. This means that the system can supply only a finite amount of "generalized energy". If all the inputs and outputs $u_i, y_i$ belong to the same signal space, it is possible to consider interconnections satisfying a form of power continuity that is equivalent to Tellegen's theorem. These concepts have been largely used to prove statements about the stability of controlled systems.

The previous form of passivity is too general and actually not suitable for interacting systems. This is due to the fact that interacting physical systems are characterized by the exchange of real energy and if we want to consider passivity of the controlled system during interaction with an environment, we should consider physical power conjugates $u_i$ and $y_i$. This also implies that $u_i$ and $y_i$ do not actually have to belong to the same Hilbert space, but that they should just belong to dual vector spaces. This is once again a good argument to use bond graph techniques in the analysis of what we could call physical passivity.

It is important to realize [77] that, properly speaking, all physical systems in the universe are passive by the first law of thermodynamics, but they can be considered active depending on certain modeling assumptions like:

- Differences in time scale
- Differences in spatial bounds
- Modulated sources due to feedback

As an example of a difference in time scale we can consider a battery: it is clearly passive because it will eventually discharge, but for a short time it can be considered as an ideal voltage source. As an example of apparent active behavior due to differences in spatial bounds we could consider the gravitational force: we could model it as a constant source (constant gravitational force), but this is just a model approximation because the bound of our model is spatially restricted. Eventually, modulated sources can unexpectedly seem to generate energy. This is often a result of the application of feedback, which can, if not intrinsically passive, result in an infinite source of energy.

### 4.2.1 Collocated control

Now consider a mechanism which can interact with its actuators and with the environment. If we can measure positions and/or velocities at exactly the same point as where we actuate forces on the robot, we say that we can apply *collocated control*. In this case, it is possible to implement a controller which is attached to the mechanism in a power consistent way, such that it is physically passive itself. The result of interconnecting a robot with such a controller is in this case a physically passive system since the interconnection takes place through a power port due to the collocation of forces and their dual velocities. This means that if the robot interacts with a passive environment through another energetic port, this interaction will always be passive, even if discontinuous contacts, bouncing or other undesired effects take place. In [74] it is also noted that unless such an intrinsically passive controller is used, there are always environments destabilizing the interaction. Such a controller, which is passive by itself, is indeed an IPC.

### 4.2.2 Robustness and Performance

When designing an IPC, we can consider a model of the robot to be controlled. We can therefore use knowledge of the system in order to develop the controller in a model-based way.

Nevertheless, if the controller is intrinsically passive, ANY model error can influence the performance, but can NEVER influence the passivity of the controlled system, which will stay stable in ANY situation.

This is actually the only way we can approach the problem in order to ensure passive interaction with an unknown environment.

Strictly speaking, the previous discussion is only true in the case of continuous control, since physical energy is a continuous function of time. This implies that, in order to get as close as possible to the ideal case, high sampling rates will in general be necessary.

### 4.2.3 Supervision

In order to be able to perform certain tasks with the controlled robot, it is necessary to inject energy into the controlled system. This is unavoidable if we want to work with the robot, since by definition, work is energy. We will therefore have another subsystem in our control scheme that will be energetically connected to the intrinsically passive controller and which will supply energy to the system under supervision to implement the required

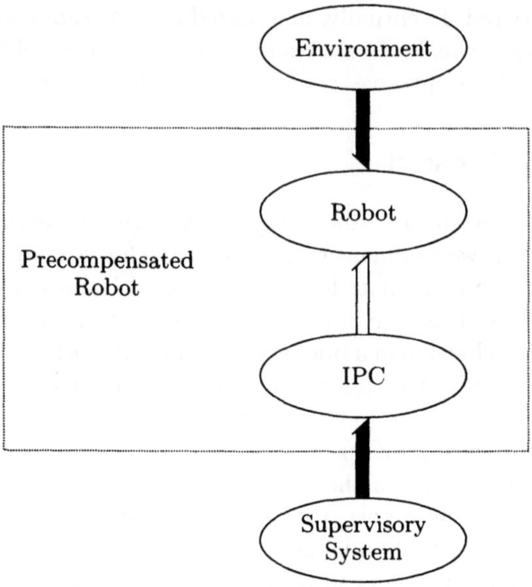


**Fig. 4.3.** The general form of an intrinsically passive control scheme.

tasks. The general proposed scheme can be found in Figure 4.3. As shown in the figure, we can consider the passive controller together with the robot as a pre-compensated system on which the supervisory system will act.

## 4.3 Interconnection of port-controlled Hamiltonian systems

We now develop an intrinsically passive controller having a Hamiltonian structure plus dissipation. We consider the interconnection between this controller and a Hamiltonian plant. As it was shown in Chapter 3, a very important feature of Generalized Port-Controlled Hamiltonian Systems (GPCHSs) is that their interconnection is still a GPCHS. To review this, with reference to Figure 4.4, consider two GPCHSs:

$$\dot{x}_i = (J_i - R_i)\frac{\partial H_i}{\partial x_i} + \begin{pmatrix} g_i^I & g_i^O \end{pmatrix}\begin{pmatrix} u_i^I \\ u_i^O \end{pmatrix} \tag{4.1}$$

$$\begin{pmatrix} y_i^I \\ y_i^O \end{pmatrix} = \begin{pmatrix} (g_i^I)^T \\ (g_i^O)^T \end{pmatrix}\frac{\partial H_i}{\partial x_i} \qquad i = 1, 2 \tag{4.2}$$

where state dependence has been omitted for notational clarity. The two systems can be interconnected through the interconnection ports $(u_i^I, y_i^I)$ by setting:

$$u_1^I = y_2^I \qquad \text{and} \qquad u_2^I = -y_1^I \tag{4.3}$$

Note that the minus sign in the previous equation is necessary for dealing consistently with power: $P_i = \langle u_i^I, y_i^I \rangle$ is the input power of system $i$ and, interconnecting the system through the "$I$" ports, we clearly need $P_1 = -P_2$. Figure 4.4 shows that this corresponds to the symplectic connection treated in Chapter 3 and shown in Figure 3.6. It is possible to see that the interconnected system results in:

$$\dot{x} =(J(x) - R(x))\frac{\partial H}{\partial x} + g(x)\begin{pmatrix} u_1^O \\ u_2^O \end{pmatrix} \tag{4.4}$$

$$\begin{pmatrix} y_1^O \\ y_2^O \end{pmatrix} =g^T(x)\frac{\partial H}{\partial x} \tag{4.5}$$

where $x = (x_1, x_2)^T$, $H(x) = H_1(x_1) + H_2(x_2)$ the sum of the two energies and

$$J(x) = \begin{pmatrix} J_1 & g_1^I (g_2^I)^T \\ -(g_2^I)(g_1^I)^T & J_2 \end{pmatrix}$$

$$R(x) = \begin{pmatrix} R_1 & 0 \\ 0 & R_2 \end{pmatrix}$$

and

$$g(x) = \begin{pmatrix} g_1^O \\ g_2^O \end{pmatrix}$$

where all dependencies of the matrices have been omitted for clarity. As expected, the interconnected system is therefore again a GPCHS with as ports the remaining ports. Furthermore, the total energy is the sum of the energies of the two systems. We can now specialize this result to the interconnection of a general controller to a plant with a canonical form. To summarize, we study the interconnection of a symplectic plant with a controller through a power port characterizing the collocated control. Such a plant can be represented using some local canonical coordinates as:

$$\begin{pmatrix} \dot{q} \\ \dot{p} \end{pmatrix} = \begin{pmatrix} 0 & I_n \\ -I_n & 0 \end{pmatrix} \begin{pmatrix} \frac{\partial H(q,p)}{\partial q} \\ \frac{\partial H(q,p)}{\partial p} \end{pmatrix} + \begin{pmatrix} 0 & 0 \\ g(q,p) & J_I^*(q) \end{pmatrix} \begin{pmatrix} u \\ W_I \end{pmatrix} \tag{4.6}$$

$$\begin{pmatrix} y \\ T \end{pmatrix} = \begin{pmatrix} 0 & g^T(q,p) \\ 0 & J_I(q) \end{pmatrix} \begin{pmatrix} \frac{\partial H(q,p)}{\partial q} \\ \frac{\partial H(q,p)}{\partial p} \end{pmatrix}.$$

where $W_I$ represents the wrenches that the environment can apply to the mechanism, $T$ the dual twists and $J_I(q)$ the interaction Jacobian (see Eq. (3.105)). An example of such a system is a fully actuated robotics system

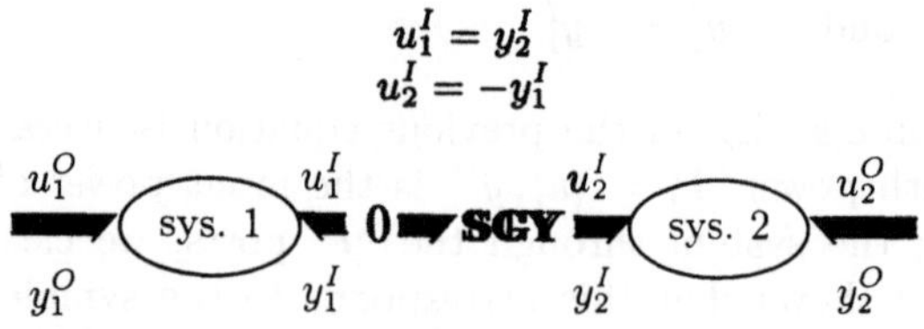


**Fig. 4.4.** Interconnection of two systems

with holonomic constraints, where the port $(u, y)$ is represented by the port of the actuator and the port $(T, W_I)$ as the interaction port.

Suppose that the second system, which corresponds to the IPC to be designed, corresponds to a generalized port-controlled Hamiltonian system for which we suppose that the Jacobi identity is satisfied for its Poisson bracket. When we choose some canonical coordinates $(q_c, p_c, r_c) = (q_{c1}, \ldots, q_{cm}, p_{c1}, \ldots, p_{cm}, r_{c1}, \ldots, r_{cp})$, this second system has the following form:

$$\begin{pmatrix} \dot{q}_c \\ \dot{p}_c \\ \dot{r}_c \end{pmatrix} = \begin{pmatrix} 0 & I_m & 0 \\ -I_m & 0 & 0 \\ 0 & 0 & 0_p \end{pmatrix} \begin{pmatrix} \frac{\partial H_c}{\partial q_c} \\ \frac{\partial H_c}{\partial p_c} \\ \frac{\partial H_c}{\partial r_c} \end{pmatrix} + \begin{pmatrix} g_{c1} & g_{s1} \\ g_{c2} & g_{s2} \\ g_{c3} & 0 \end{pmatrix} \begin{pmatrix} u_c \\ u_s \end{pmatrix} \tag{4.7}$$

$$\begin{pmatrix} y_c \\ y_s \end{pmatrix} = \begin{pmatrix} g_{c1}^T & g_{c2}^T & g_{c3}^T \\ g_{s1}^T & g_{s2}^T & 0 \end{pmatrix} \begin{pmatrix} \frac{\partial H_c}{\partial q_c} \\ \frac{\partial H_c}{\partial p_c} \\ \frac{\partial H_c}{\partial r_c} \end{pmatrix}.$$

The previous form of the input port is chosen because it still keeps the invariance of certain Casimir functions, which will be analyzed later.

If we couple the two systems (Eq. (4.6) with Eq. (4.7)), and suppose that the symplectic connection represented in Figure 4.4 is used, where in the figure $u_2^I = u_c$, $y_2^I = y_c$ and sys. 1 is the plant and sys. 2 is the controller, then the interconnected system may be written as:

$$\dot{x}_T = J_T(x_T)\frac{\partial H_T(x_T)}{\partial x_T} + \begin{pmatrix} 0 & 0 \\ J_I^*(q) & 0 \\ 0 & g_{s1} \\ 0 & g_{s2} \\ 0 & 0 \end{pmatrix} \begin{pmatrix} W_I \\ u_s \end{pmatrix} \tag{4.8}$$

$$\begin{pmatrix} T \\ y_s \end{pmatrix} = \begin{pmatrix} 0 & J_I(q) & 0 & 0 & 0 \\ 0 & 0 & g_{s1} & g_{s1} & 0 \end{pmatrix} \frac{\partial H_T(x_T)}{\partial x_T}$$

where $x_T^T = (q\ p\ q_c\ p_c\ r_c)$. This is a generalized Hamiltonian system with Hamiltonian function: $H_T(x_T) = H(q, p) + H_c(q_c, p_c, r_c)$, and it is defined with respect to the Poisson bracket associated with the structure matrix

$$J_T(x_T) = \begin{pmatrix} 0 & I_n & 0 & 0 & 0 \\ -I_n & 0 & gg_{c1}^T & gg_{c2}^T & gg_{c3}^T \\ 0 & -g_{c1}g^T & 0 & I_m & 0 \\ 0 & -g_{c2}g^T & -I_m & 0 & 0 \\ 0 & -g_{c3}g^T & 0 & 0 & 0_p \end{pmatrix} \tag{4.9}$$

where the dependence of the matrix component from the states has been omitted to simplify the notation.

The kernel of the structure matrix obviously has dimension $p$, hence if the Poisson bracket defined by $J_T(x_T)$ in Eq. (4.9) satisfies the Jacobi identity, then $p$ Casimir functions $c_i$ exist which stay constant during the dynamic evolution of the coupled system. When a Poisson framework is used, it is not easy to give conditions for the integrability of $J_T(x_T)$. With a Dirac structure approach, however, there are sufficient conditions that guarantee the integrability of the interconnected system under certain conditions imposed on the plant and the controller [78].

The Casimir functions can be calculated by integrating the one forms which belong to the kernel of the matrix $J_T(x_T)$. It is then easy to see that the Casimir functions $c_i$ with $i = 1, \dots, p$ should satisfy:

$$\frac{\partial c_i}{\partial q} = g(q,p)g_{c3}^T(q_c, p_c, r_c)\frac{\partial c_i}{\partial r_c} \tag{4.10}$$

and are functions only of $q$ and $r_c$ since we have $\frac{\partial c_i}{\partial p} = \frac{\partial c_i}{\partial q_c} = \frac{\partial c_i}{\partial p_c} = 0$. Supposing that the Poisson bracket corresponding to the connected system is integrable, then there are $p$ functions $c_i(q, r_c)$, $i = 1, \dots, p$ which are constant during the evolution of the system.

Furthermore, assume that the square matrix:

$$\frac{\partial c_i}{\partial r_{cj}} \quad i,j = 1, \dots, p \tag{4.11}$$

is nonsingular. Then we can define a change of coordinates from $(q, p, q_c, p_c, r_c)$ to $(q, p, q_c, p_c, c)$ for which:

$$\begin{pmatrix} \dot{q} \\ \dot{p} \\ \dot{q}_c \\ \dot{p}_c \\ \dot{c} \end{pmatrix} = \begin{pmatrix} 1 & 0 & 0 & 0 & 0 \\ 0 & 1 & 0 & 0 & 0 \\ 0 & 0 & 1 & 0 & 0 \\ 0 & 0 & 0 & 1 & 0 \\ \frac{\partial c}{\partial q} & 0 & 0 & 0 & \frac{\partial c}{\partial r_c} \end{pmatrix} \begin{pmatrix} \dot{q} \\ \dot{p} \\ \dot{q}_c \\ \dot{p}_c \\ \dot{r}_c \end{pmatrix}. \tag{4.12}$$

Using this change of coordinates, the structure matrix takes the form:

$$J'_T(x_T) = \begin{pmatrix} 0 & I_n & 0 & 0 & 0 \\ -I_n & 0 & g'{g'_{c1}}^T & g'{g'_{c2}}^T & 0 \\ 0 & -g'_{c1}{g'}^T & 0 & I_m & 0 \\ 0 & -g'_{c2}{g'}^T & -I_m & 0 & 0 \\ 0 & 0 & 0 & 0 & 0_p \end{pmatrix}. \tag{4.13}$$

Where $g'$ and $g'_{ci}$ are the result of the change of coordinates on $g$ and the $g_{ci}$.

The restricted system for each invariant leaf defined by $c_i(q, r_c) = c_i^0$, $i = 1, \ldots, p$ is:

$$\dot{\hat{x}}_T = \hat{J}_T(\hat{x}_T)\frac{\partial \hat{H}_T(\hat{x}_T)}{\partial \hat{x}_T} + \begin{pmatrix} 0 & 0 \\ J_I^*(q) & 0 \\ 0 & g_{s1} \\ 0 & g_{s2} \end{pmatrix}\begin{pmatrix} W_I \\ u_s \end{pmatrix} \tag{4.14}$$

$$\begin{pmatrix} T \\ y_s \end{pmatrix} = \begin{pmatrix} 0 & J_I(q) & 0 & 0 \\ 0 & 0 & g_{s1}^T & g_{s2}^T \end{pmatrix}$$

where $\hat{x}_T^T = (q\; p\; q_c\; p_c)$,

$$\hat{J}_T(\hat{x}_T) = \begin{pmatrix} 0 & I_n & 0 & 0 \\ -I_n & 0 & g' {g'_{c1}}^T & g' {g'_{c2}}^T \\ 0 & -g'_{c1} g'^T & 0 & I_m \\ 0 & -g'_{c2} g'^T & -I_m & 0 \end{pmatrix} \tag{4.15}$$

and $\hat{H}_T(q, p, q_c, p_c) = H(q,p) + H_c(q_c, p_c, f(q))$ where $f$ is such that $c_i(q, f(q)) = c_i^0$ and exists locally due to the assumption of Eq. (4.10).

This means that the system evolves on a leaf of the foliation defined by the Casimir functions with dimension $2(n+m)$. The restricted dynamics is then again a generalized Hamiltonian system given by Eq. (4.14) and defined by the Hamiltonian function being the total Hamiltonian function restricted to the leaf with respect to the Poisson bracket defined by $J_T(\hat{x}_T)$.

## 4.4 Intrinsically Passive Feedback

In this section, we will show that a port-controlled Hamiltonian system is a network model of the controller which is meaningful for design purposes. These methods are intrinsically passive for construction and useful for insight in the physics. In the previous section, we showed that the interconnection of two Hamiltonian systems through a power port is power consistent and therefore the energy of the coupled system will never increase. This means that if we consider a plant with which we can interact by means of collocated control[1], we may design a controller having a Hamiltonian structure and couple it to the plant in the same fashion as shown in the previous section. This was already noted in [58] for a simple case with no dynamic extension. Here the more general case with a dynamic extension will be treated and its advantages will be pointed out.

---

[1] We have collocated control when we can actuate and measure variables in the same physical position with no dynamics in between

### 4.4.1 Energy Shaping

Consider the two interconnected systems of Sect. 4.3, and suppose that the symplectic system represented by Eq. (4.6) corresponds to the system to be controlled, for example a robot manipulator, and that the system represented in Eq. (4.7) corresponds to the IPC.

The case in which the Poisson structure of the interconnected system is singular and fulfills the Jacobi identities has been analyzed in Sect. 4.3. Once the state of the controller is initialized, the interconnected system evolves on one of the leafs defined by the Casimir functions of the interconnected system. The controlled system may be interpreted as having modified the Hamiltonian function and having a dynamic extension of dimension $2m$. Such a controller generalizes the shaping of potential energy through Hamiltonian feedback of simple mechanical systems [58]. In that case the interaction map of the plant is $g(q) = \frac{\partial C(q)}{\partial q}$, where $C$ is a Casimir function; there is no symplectic part in the controller ($m = 0$) and the interaction map of the controller is $g_3 = Id$. The restriction of the dynamics of the controlled plant to an invariant leaf then corresponds to modifying the Hamiltonian function by adding a potential term. Although the procedure is very similar, there is important conceptual difference between this approach and other energy-shaping approaches based for instance on Euler-Lagrange systems [79]. A major difference is that in the latter case the two systems (the plant and the controller) are interconnected, not through the port variables (for mechanical systems: forces and velocities), but through forces and displacements, and hence the controller model may not easily be physically interpretable. Furthermore, in the presented approach it is possible to directly express the controller-plant interconnection, also if spatial geometry plays a role, as will be seen in the next chapter. This is not as straightforward in the Euler-Lagrange approach.

In the presented approach, the passivity of the coupled system can be directly deduced from the energy-consistent port interconnection of two passive systems; this interconnection is directly interpretable as passive with model uncertainty of any magnitude or with possible interaction of the plant with other systems through other power ports.

Furthermore, it might seem that we must measure $y$ for the actual implementation. This is in general not necessary if certain states of the plant are directly measurable. In the case of a robotic manipulator, it is not necessary to measure the port variables corresponding to the joints velocities, because it is sufficient to measure the positions of the joints.

This is possible only under certain conditions, namely when for the controller canonical coordinates of the form Eq. (4.7) exist for which $g_c^T = (0_{k\times m}\ 0_{k\times m}\ g_{c3}^T)$, and the conditions of Eq. (4.11) on the Casimir functions must be satisfied. In this case we can find a function $r_c = f(q)$ which is sufficient for integrating the controller's dynamics and therefore we can calculate the input for the plant. This is possible since $\dot{q}_c$ and $\dot{p}_c$ do not directly depend on $u_c$, which corresponds to the joint velocities. Note that the resulting total energy of the system is:

$$\hat{H}_T(q, p, q_c, p_c) = H(q, p) + H_c(q_c, p_c, f(q)) \ .$$

This means that the potential energy of the plant can be changed by creating an indirect dependency on $q$ in the controller's Hamiltonian function.

### 4.4.2 Damping Injection

So far we have considered lossless systems. This is clearly not very useful in practice because such a controller results in highly oscillating behaviors. It is therefore necessary to apply damping and this can be easily done if our controller possesses a real dynamic extension: $\dim q_c > 0$.

Before proceeding, the following simple result is relevant [73]:

**Theorem 4.4.1.** *If we consider instead of Eq. (4.8) the following system*

$$\dot{x}_T = (J_T(x_T) - R_T(x_T))\frac{\partial H_T(x_T)}{\partial x_T} \tag{4.16}$$

*where*

$$R_T(x_T) := \begin{pmatrix} 0_n & 0_n & 0 & 0 & 0 \\ 0_n & R_1 & R_{h1} & R_{h2} & 0 \\ 0 & R_{h1}^T & R_{c1} & R_{c12} & 0 \\ 0 & R_{h2}^T & R_{c12}^T & R_{c2} & 0 \\ 0 & 0 & 0 & 0 & 0_p \end{pmatrix}$$

*is a state-dependent, symmetric, positive semi-definite, two-contravariant tensor, then the Casimir functions of $J_T(x_T)$ are still conserved quantities.*

This means that the system keeps on moving on the leaf characterized by the initial value of the Casimir functions and this dissipation only influences the system's motion on the current leaf. Note that the previous result is a sufficient condition for what follows.

The dissipative elements $R_{hi}$ $i = 1, 2$ are hybrid in the sense that they are placed in the coupling part of the matrix and do not have much physical sense. From now on we consider them equal to zero. $R_1$ corresponds to viscous friction in the plant and $R_{c1}, R_{c2}, R_{c12}$ correspond to dissipative elements which we can implement within our controller. It is important to note that due to their form, the implementation of such a term only requires the state of the controller and is independent of the port variables. This means that in a robotic system, such a dissipative element can be implemented without measurements of joint velocities; only the controller states are sufficient.

Supposing in the worst case that $R_1 = 0$, the decrease in total energy will be:

$$\dot{H}_T(x_T) = -\begin{pmatrix} \frac{\partial H_c}{\partial q_c} & \frac{\partial H_c}{\partial p_c} \end{pmatrix} R_c(x_c) \begin{pmatrix} \frac{\partial H_c}{\partial q_c} \\ \frac{\partial H_c}{\partial p_c} \end{pmatrix} \tag{4.17}$$

where $R_c$ is in general semi-negative definite. If we choose $R_c$ negative definite, asymptotic stability can often easily be proven by La Salle's theorem.

It is therefore possible, as in [79] or [80], to inject damping in the system by means of a dynamic extension and without a direct measure of the port variables. The possible implementable damping seems more general than the damping presented in [79] or [80] where, compared with the presented one, the only term different from zero is $R_{c2}$. This method is valuable for practical robot control, where measurements of joints velocities are often NOT available.

### 4.4.3 The resulting controller

It is now possible to give the general form of the controller for the case treated in Sect. 4.3:

$$\begin{aligned} \begin{pmatrix} \dot{q}_c \\ \dot{p}_c \end{pmatrix} &= \begin{pmatrix} -R_{c1} & -R_{c12} + I_m \\ -R_{c12}^T - I_m & -R_{c2} \end{pmatrix} \begin{pmatrix} \frac{\partial H_c(q_c, p_c, f(q))}{\partial q_c} \\ \frac{\partial H_c(q_c, p_c, f(q))}{\partial p_c} \end{pmatrix} + g_s(q_c, p_c, f(q)) u_s \\ \begin{pmatrix} y_c \\ y_s \end{pmatrix} &= \begin{pmatrix} g_c^T(q_c, p_c, f(q)) \\ g_s^T(q_c, p_c, f(q)) \end{pmatrix} \begin{pmatrix} \frac{\partial H_c(q_c, p_c, f(q))}{\partial q_c} \\ \frac{\partial H_c(q_c, p_c, f(q))}{\partial p_c} \\ \frac{\partial H_c(q_c, p_c, f(q))}{\partial r_c} \end{pmatrix} \end{aligned} \tag{4.18}$$

where it is possible to see that the only external variable that must be measured is $q$ and that Eq. (4.17) holds.

## 4.5 A simple example

To illustrate the presented concepts, we use a simple one-dimensional example. In the next chapter a complex controller is developed; the techniques presented here will be used. Consider the control of a one-dimensional mass which can interact with an unknown environment that can apply an external force $F_{\text{ext}}$ to it. To control the interactive property of such a mass, we can think about developing an IPC which, when coupled to the robot's mass, gives a system equivalent to the one shown in Figure 4.5, as presented in [80]. The mass dynamics can be expressed by the following equations:

$$\begin{aligned} \begin{pmatrix} \dot{x} \\ \dot{p} \end{pmatrix} &= \begin{pmatrix} 0 & 1 \\ -1 & 0 \end{pmatrix} \begin{pmatrix} 0 \\ p/m \end{pmatrix} + \begin{pmatrix} 0 & 0 \\ 1 & 1 \end{pmatrix} \begin{pmatrix} F_{\text{ext}} \\ F_c \end{pmatrix} \\ \begin{pmatrix} \dot{x} \\ \dot{x} \end{pmatrix} &= \begin{pmatrix} 0 & 1 \\ 0 & 1 \end{pmatrix} \begin{pmatrix} 0 \\ p/m \end{pmatrix} \end{aligned}$$

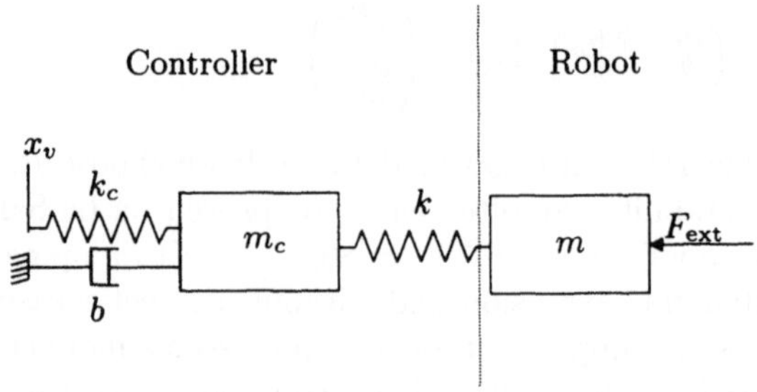


**Fig. 4.5.** The controlled system equivalent.

Where $x$ indicates the position of the robot's mass and $p$ its momentum. The controller can be represented by:

$$\begin{pmatrix} \dot{\Delta x_c} \\ \dot{p}_c \\ \dot{\Delta x} \end{pmatrix} = \begin{pmatrix} 0 & 1 & 0 \\ -1 & -b & 1 \\ 0 & -1 & 0 \end{pmatrix} \begin{pmatrix} k_c \Delta x_c \\ p_c/m_c \\ k\Delta x \end{pmatrix} + \begin{pmatrix} 0 & -1 \\ 0 & 0 \\ 1 & 0 \end{pmatrix} \begin{pmatrix} \dot{x} \\ \dot{x}_v \end{pmatrix} \tag{4.19}$$

$$\begin{pmatrix} -F_c \\ -F_v \end{pmatrix} = \begin{pmatrix} 0 & 0 & 1 \\ -1 & 0 & 0 \end{pmatrix} \begin{pmatrix} k_c \Delta x_c \\ p_c/m_c \\ k\Delta x \end{pmatrix}$$

where $\Delta x_c$ is the displacement of the spring $k_c$, $\Delta x$ the displacement of the spring $k$ and $p_c$ the momentum of the mass $m_c$. Furthermore, the Hamiltonian function of the robot is $H(p) = 1/(2m)p^2$ and the one of the controller $H(\Delta x, p, \Delta x_c) = 1/2(p_c^2/m_c + k\Delta x^2 + k_c \Delta x_c^2)$. Note that the variable $b$ indicates the symmetric dissipating part.

Because the total force $F_{tot}$ applied to the robot mass is equal to the one applied by the controller plus the one applied by the environment ($F_{ext}$), the coupled system will have the following equations:

$$\begin{pmatrix} \dot{x} \\ \dot{p} \\ \dot{\Delta x_c} \\ \dot{p}_c \\ \dot{\Delta x} \end{pmatrix} = \begin{pmatrix} 0 & 1 & 0 & 0 & 0 \\ -1 & 0 & 0 & 0 & -1 \\ 0 & 0 & 0 & 1 & 0 \\ 0 & 0 & -1 & -b & 1 \\ 0 & 1 & 0 & -1 & 0 \end{pmatrix} \begin{pmatrix} 0 \\ p/m \\ k_c \Delta x_c \\ p_c/m_c \\ k\Delta x \end{pmatrix} + \begin{pmatrix} 0 & 0 \\ 1 & 0 \\ 0 & -1 \\ 0 & 0 \\ 0 & 0 \end{pmatrix} \begin{pmatrix} F_{ext} \\ \dot{x}_v \end{pmatrix}$$

$$\begin{pmatrix} \dot{x} \\ -F_v \end{pmatrix} = \begin{pmatrix} 0 & 1 & 0 & 0 & 0 \\ 0 & 0 & -1 & 0 & 0 \end{pmatrix} \begin{pmatrix} 0 \\ p/m \\ k_c \Delta x_c \\ p_c/m_c \\ k\Delta x \end{pmatrix} .$$

The Casimir functions of the lossless part can be found by integrating the 1–forms belonging to the kernel of the structure matrix. Consider a general one form:

$$\omega = \bar{x}\, dx + \bar{p}\, dp + \bar{\Delta x}_c\, d(\Delta x_c) + \bar{p_c}\, dp_c + \bar{\Delta x}\, d(\Delta x).$$

To insure that this form is in the kernel of the Poisson tensor, we have:

$$\begin{cases} \bar{p} = \bar{p}_c = 0 \\ -\bar{x} - \bar{\Delta x} = 0 \\ -\bar{\Delta x}_c + \bar{\Delta x} = 0 \end{cases} \Rightarrow \left\{ \frac{\partial c}{\partial \Delta x_c} = \frac{\partial c}{\partial \Delta x} = -\frac{\partial c}{\partial x} \right.$$

which results in a Casimir function of the form:

$$c(x, \Delta x, \Delta x_c) = \Delta x_c - x + \Delta x$$

and indicating the value of $c$ at time $t = 0$ with $c_0$, we obtain:

$$\Delta x = (x - \Delta x_c) + c_0. \tag{4.20}$$

If we now choose the change of coordinates:

$$\begin{pmatrix} \dot{c} \\ \dot{p} \\ \dot{\Delta x}_c \\ \dot{p}_c \\ \dot{\Delta x} \end{pmatrix} = \begin{pmatrix} -1 & 0 & 1 & 0 & 1 \\ 0 & 1 & 0 & 0 & 0 \\ 0 & 0 & 1 & 0 & 0 \\ 0 & 0 & 0 & 1 & 0 \\ 0 & 0 & 0 & 0 & 1 \end{pmatrix} \begin{pmatrix} \dot{x} \\ \dot{p} \\ \dot{\Delta x}_c \\ \dot{p}_c \\ \dot{\Delta x} \end{pmatrix} \tag{4.21}$$

then we obtain the following system:

$$\begin{pmatrix} \dot{c} \\ \dot{p} \\ \dot{\Delta x}_c \\ \dot{p}_c \\ \dot{\Delta x} \end{pmatrix} = \begin{pmatrix} 0 & 0 & 0 & 0 & 0 \\ 0 & 0 & 0 & 0 & -1 \\ 0 & 0 & 0 & 1 & 0 \\ 0 & 0 & -1 & -b & 1 \\ 0 & 1 & 0 & -1 & 0 \end{pmatrix} \begin{pmatrix} 0 \\ p/m \\ k_c \Delta x_c \\ p_c/m_c \\ k \Delta x \end{pmatrix} + \begin{pmatrix} 0 \\ 1 \\ 0 \\ 0 \\ 0 \end{pmatrix} F_{\text{ext}} + \begin{pmatrix} 0 \\ 0 \\ -1 \\ 0 \\ 0 \end{pmatrix} \dot{x}_v \tag{4.22}$$

from which we can ignore the first line representing the Casimir function.

It is possible to see with La Salle's theorem that with $F_{\text{ext}} = 0$ and $\dot{x}_v = 0$, the system is asymptotically stable, implementing effectively damping injection. This can be seen by observing that $\dot{H} = -b(p_c/m_c)^2$, and therefore $\dot{H} = 0$ iff $p_c = 0$. It is then possible to see that only the state $(x, p, \Delta x, p_c, \Delta x_c) = 0$ is the largest invariant set contained in the set of states with $p_c = 0$, which also implies that $c_0 = 0$.

Note that in order to implement the controller of Eq. (4.19), we do NOT need to measure $\dot{x}$ directly. This is due to the fact that with Eq. (4.20), we can integrate the controller's dynamics with only knowledge about $x$ and $\Delta x_c$:

$$\begin{aligned} \begin{pmatrix} \dot{\Delta x}_c \\ \dot{p}_c \end{pmatrix} &= \begin{pmatrix} 0 & 1 \\ -1 & -b \end{pmatrix} \begin{pmatrix} k_c \Delta x_c \\ p_c/m_c \end{pmatrix} + \begin{pmatrix} 0 \\ k(x - \Delta x_c + c_0) \end{pmatrix} + \begin{pmatrix} -1 \\ 0 \end{pmatrix} \dot{x}_v \\ -F_c &= k(x - \Delta x_c + c_0). \end{aligned} \tag{4.23}$$

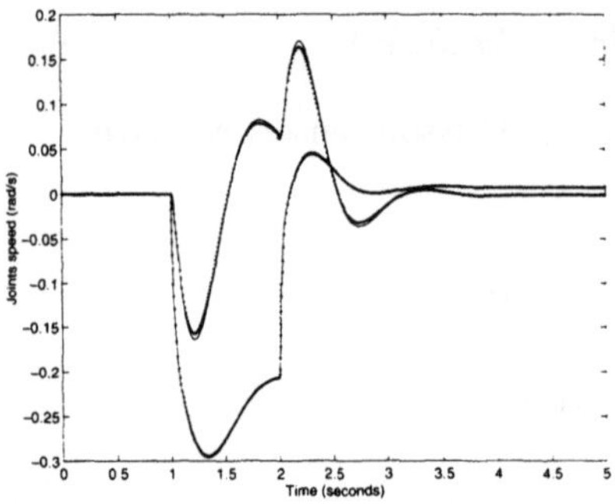


Fig. 4.6. Velocity response to $x$ pulse force.

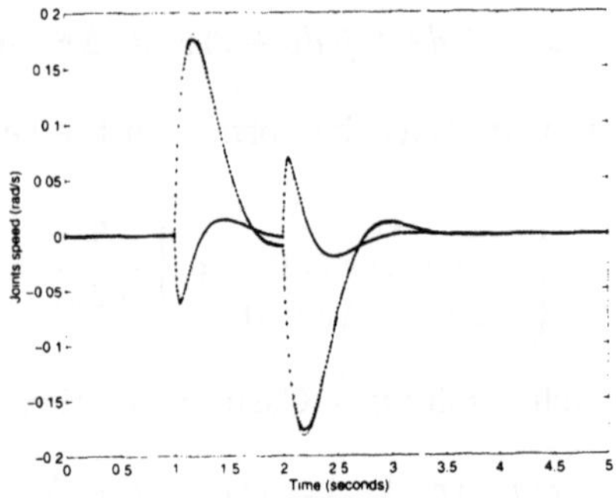


Fig. 4.7. Velocity response to $y$ pulse force.

That it is not necessary to measure the velocity of the robot is also evident in a bond graph of the controlled system. In a bond graph, the necessity to measure velocity in order to implement damping is represented by a causal path from the robot inertia element to the dissipative element; sketching a bond graph of the controlled system shows that with the presented scheme we do not get such a causal path.

### 4.5.1 Actuator's Saturations

With the proposed controller, it is also possible to passively handle force actuators saturations, which are always present [80]. This cannot easily be done with other techniques. If we know that an actuator has a saturation $F_{\text{sat}}$ and the force we apply to the robot exceeds this value, we can choose a non-linear relation for the spring $k$, and implement it by using a saturated block. This is still a perfectly passive non-linear spring. The result will be that both the robot and the controller's mass will never receive a force higher than $F_{\text{sat}}$ from spring $k$. Even if non-linear, the spring $k$ is still passive, which ensures passivity of the total controlled system.

### 4.5.2 The multidimensional case

The multidimensional case of the idea presented in Figure 4.5 is reported in detail in [80], where a method is given to calculate the various parameters. After choosing a desired effective interactive stiffness in Cartesian coordinates by means of a power-continuous transformation represented by the robot Jacobian, we can connect this Cartesian spring to $n$ controllers' masses ($m_c$ in Figure 4.5) where $n$ corresponds to the number of actuators. We therefore have for each actuator a mass $m_c$, a spring $k$ and a damping $b$, but the interactive stiffness corresponding to $k_c$ is expressed in the work space and connected to the rest by means of the robot Jacobian (see [80]).

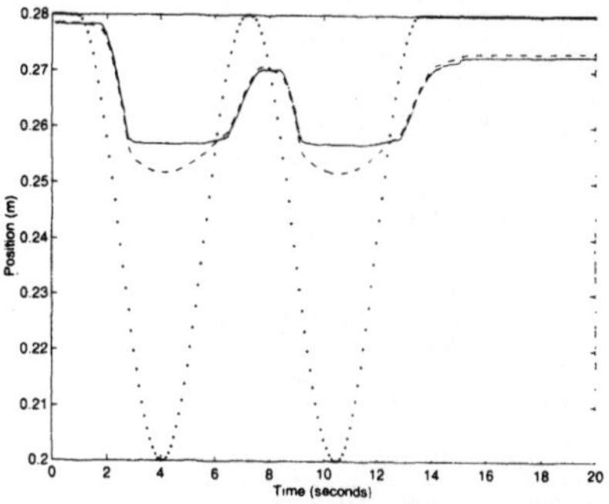


**Fig. 4.8.** Experiment: $x_v$, $x_c$, $x_{\text{robot}}$ vertical motion with table collision.

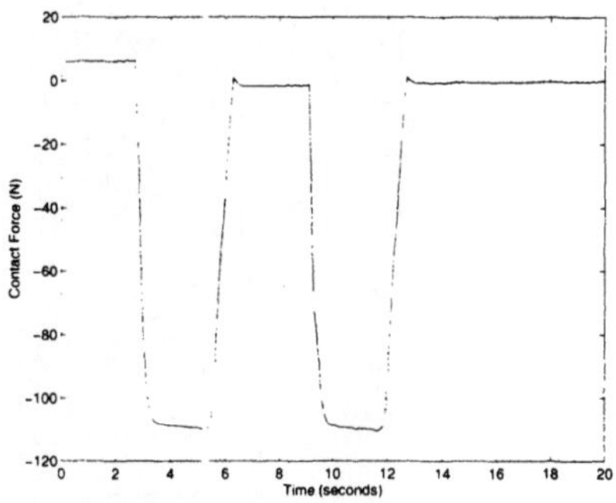


**Fig. 4.9.** Experiment: contact force during vertical motion with table collision.

**Simulations.** In some simulations the ideal case with velocity feedback was compared with the presented idea. A robot with two degrees of freedom has been used for this purpose. The robot is considered at a stable posture such that the end effector position corresponds to the virtual position. Two external force pulses of 100 $N$ of the duration of 2 $sec$ and starting at 2 $sec$ have been simulated for both the direct velocity feedback and the injected damping. The parameters were $b_x = b_y = 500\ Ns/m$, the desired Cartesian stiffness $k_x = 10\ N/m$ and $k_y = 1000\ N/m$. Applying the analyses described in [80], we can calculate that $k_{max} = 10^5\ N/m$. Figure 4.6 plots the joint velocities of the pulse force in the $x$ direction and Figure 4.7 the joint velocities of the pulse force in the $y$ direction. The dotted lines correspond to the cases with velocity measurement. A good match can be seen.

**Experimental results.** In order to test the applicability of the concept, a first one-dimensional experiment has been performed with the vertical prismatic axis of a SCARA Robot shown in Figure 4.10. The virtual position was supplied as a sinusoid on the motion of this axis in such a way that during the cycle it would collide with the table.

### 4.5.3 The Bosch Turbo SCARA

In Figure 4.8, the dotted line corresponds to the virtual position $x_v$, the continuous line corresponds to the position of the robot $x$ and the dashed line to the position of the controller's mass $x_c$. The phase lag between $x_v$ and $x$, when not in contact, is due to the low stiffness chosen to prevent high contact forces.

The corresponding interaction force is reported in Figure 4.9. Some bias and nonrepeatability in the force measurement is clearly noticeable and due to the nonoptimal quality of the sensor.

**Fig. 4.10.** The BOSCH TurboSCARA.

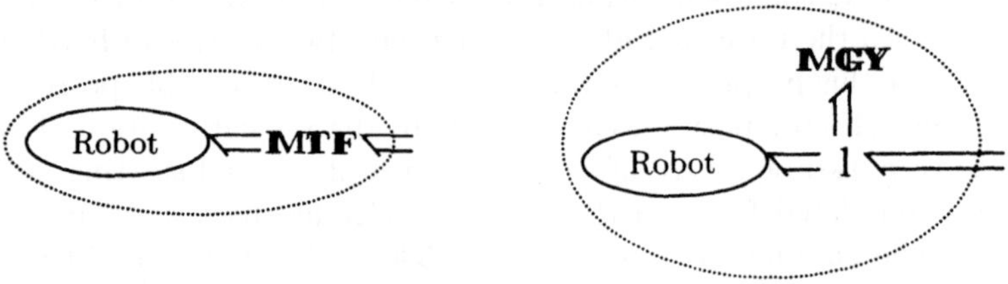


**Fig. 4.11.** Power-continuous precompensation.

It is possible to see that the energy is properly dissipated and free motion and collision of the system are stable.

## 4.6 Considerations on Power-Continuous Compensations

So far we only modified the robot dynamics by means of an intrinsically passive Hamiltonian controller which can store energy itself. It is also possible to consider a precompensation which is power continuous (it neither stores nor dissipates energy) and would change the system dynamics. To be as explicit as possible, we will consider a robot that is completely actuated and from which we can measure all the generalized velocities. Such a plant can be represented in canonical coordinates, as in Eq. (4.6), where $g(q, p) = I_n$. We could achieve power-continuous compensations by means of bond-graph transformers or gyrator-like elements at the control port. For a transformer, a picture of such a precompensation is given on the left side of Figure 4.11. Such a compensation does not have any effect on the dynamics of the system if there are zero inputs.

Another power-continuous compensation that could be used is the one represented on the right side of Figure 4.11. Such a compensation always has an effect on the dynamics evolution of the robot, even if the new input is zero.

### 4.6.1 Transformer-like compensations

With reference to the Lagrangian formulation treated in Sect. 3.3, we could characterize a robot in Riemannian geometric terms by means of its configuration manifold $\mathcal{Q}$ and a metric $m_{ij}$ on $\mathcal{Q}$ characterizing its kinetic energy. The inertia matrix represents the metric $m_{ij}$ in joint coordinates.

From basic arguments of Riemannian geometry, which have also been described in [81] and then in [82], if the curvature (see Sect. A.7.7) of the Riemann manifold $(\mathcal{Q}, m_{ij})$ is zero everywhere, there exist coordinates for the manifold $\mathcal{Q}$ in which the metric is represented by an identity matrix. Such coordinates are called Euclidean coordinates [20]. This means that for robots with inertia matrix of zero curvature, there exist coordinates in which the so-called Coriolis and centrifugal forces disappear (the Christoffel's symbols are identically equal to zero). The Jacobian of the coordinate change from the coordinates used to express the robot port in Figure 4.11 and these new special coordinates then characterize a transformer. This transformer is such that if we were to look from the new port, the newly compensated robot would look linear.

In such a way we can effectively implement feedback linearization of a robot in a completely intrinsically passive way. The problem is that the class of robots with a zero curvature inertia matrix is very restricted. However, for the design of new robots, such a feature could be considered.

### 4.6.2 Antisymmetric compensations

Consider a robot of which the dynamics equations have the form Eq. (4.6) where $g(q,p) = I_n$. For such a robot, a compensation like the right one of Figure 4.11 corresponds to:

$$u = S(q,p)\dot{q} + u^* \ , \ \text{with } S(q,p) = -S^T(q,p) \tag{4.24}$$

where $(u, y)$ are the power conjugate variables of the original robot port, and $(u^*, y^*)$ are the power conjugate variables of the compensated robot, and the **MGY** is characterized by the antisymmetric element $S(q,p)$. The equations of the new-compensated Hamiltonian system are:

$$\begin{pmatrix} \dot{q} \\ \dot{p} \end{pmatrix} = \begin{pmatrix} 0 & I_n \\ -I_n & S(q,p) \end{pmatrix} \begin{pmatrix} \frac{\partial H(q,p)}{\partial q} \\ \frac{\partial H(q,p)}{\partial p} \end{pmatrix} + \begin{pmatrix} 0 \\ I_n \end{pmatrix} u^*$$

$$y^* = (0 \; I_n) \begin{pmatrix} \frac{\partial H(q,p)}{\partial q} \\ \frac{\partial H(q,p)}{\partial p} \end{pmatrix} . \tag{4.25}$$

Such a power-continuous compensation therefore changes the Poisson structure of the original system. This implies that such a compensation can drastically change the system dynamics even if the compensation is power continuous. When such an element is used, the redistribution of the energy in the various directions of $\mathcal{Q}$ is changed and also the duality of positions and momenta. Furthermore, if there is an additional condition, namely the integrability of the structural matrix of Eq. (4.25), new canonical coordinates $Q(q,p), P(q,p)$ exist.

A first question that presents itself is whether it is possible to change the Riemannian curvature of the Lagrangian system that expresses the robot dynamics in order to bring it to zero, and then to search for a coordinate change like the one discussed in Sect. 4.6.1.

In Lagrangian terms, $S(q,p)$ corresponds to a 2-form $S_{ij}$ on $\mathcal{Q}$ whose antisymmetric structure justifies power continuity. We express Eq. (3.10) with the Christoffel symbols (see Sect. A.7.8) and use index tensor notation, and as a result of this precompensation, the Lagrangian equations expressing the motion of the robot become:

$$\ddot{q}^l + \Gamma^l_{jk}\dot{q}^j\dot{q}^k = M^{li}S_{ij}\dot{q}^j + M^{li}\tau^*_i . \tag{4.26}$$

Since the term in which the Christoffel symbols appear is quadratic in the velocities, we consider a 3 covariant tensor $S_{kij}$, antisymmetric in the last two indices. Therefore, we can consider $S_{kij}\dot{q}^k = S_{ij}$ as our power-continuous element. This operation changes our connection from $\Gamma^l_{jk}$ to $\Gamma^l_{jk} + M^{li}S_{kij}$. Unfortunately, in order to be uniquely associated to a metric, the connection should be torsion free [20], which is equivalent to saying that the Christoffel symbols should be symmetric in the lower indexes. This implies that $S_{kij} = S_{jik}$. This is incompatible with the asymmetry of the first two indexes since a 3 covariant tensor having two antisymmetric indices and the others symmetric is identically zero. In this case :

$$S_{kij} = S_{jik} = S_{ijk} = -S_{ikj} = -S_{kij} \equiv 0$$

where we used the symmetry of the first two indices, which can easily be verified from the other hypothesis. This implies that there is no power-continuous element which can change the symmetric part of a connection that is related to the curvature. This unfortunately proves the following:

**Theorem 4.6.1.** *It is not possible to change the curvature of a Lagrangian system by means of a passive feedback of the presented form. Only the torsion can be changed.*

Nevertheless, there could be ways to exploit such a power-continuous compensation, for example, one could passively compensate the Coriolis and centrifugal effects partially or totally. This remains a topic of research worthwhile to be pursued.

## 4.7 Conclusions

In this chapter we made a step towards systematic way to develop intrinsically passive controllers with a Hamiltonian structure plus dissipation. The utility of a dynamic extension in the controller as a generalization of the example in [58] has been shown. Even if the final implementation result, Eq. (4.18), is similar to the result of [79], it has been deducted from a more physical perspective: the power-continuous interconnection of systems with a physical structure.

Furthermore, within this framework, a physical way to implement damping in a controlled system was presented. The process follows physical reasoning, which helps intuition and ensures stability.

Simulations and experiments have been presented which indicated the proper functioning of the idea. This method is NOT intended for the estimation of velocity, but for the direct implementation of damping.

The advantage of the presented damping method compared to those of others is its ability to ensure passivity, also in cases where the actuators are saturated and when there are uncertainties in the model, and its capacity of supplying physical intuition. Stability of the system is easily understood and proven from passivity concepts and it does not depend on system bandwidths. A detailed model of the robot is not necessary and the proposed scheme has therefore intrinsic robust properties.

# 5. A Novel Impedance Grasping Strategy

This chapter presents an intrinsically passive control strategy for robotics grasping tasks. This can be seen as a nontrivial, geometric application of the techniques presented in the previous chapter. It is shown that robot control design synthesized as a spatial interconnection between the robot and the controller can be easily handled. The idea is based on what is called the *Virtual Object Concept* [83] and can be used both for tips grasp and full grasp. One of the major advantages of the presented strategy is the passive nature of the algorithm and the physical intuition it supplies due to the description of the controller as a spatial interconnection of physical elements. Most of the grasping strategies known in literature deal with the control of the tip contact forces [10]. Since grasping is concerned with the interaction of a robotic hand with the environment, strategies which consider the control of interaction explicitly seem more appropriate. One of the most problematic phenomena of some force control strategies is that stability cannot be ensured if very restrictive features of the object to be grasped are not assumed like its stiffness and friction. Furthermore, a force control strategy is not suitable to control the change between no-contact and contact as it has been explained in Sect. 4.1. This is due to the fact that force control is only meaningful in contact since it is not possible to apply a force different than zero in free space.

For these reasons, a grasping technique based on more physical reasoning and passivity seems worth pursuing. The presented technique is based on a strategy which shapes the potential energy of the system in order to achieve a desired compliance and injects some damping to ensure asymptotic stability and a proper behavior. These techniques have been already used in the past [79, 80], but here a nontrivial geometry plays a role.

Such an impedance strategy does not have the shortcomings of other grasping techniques: it is strictly passive in steady-state situations for any passive environment and the supplied energy in moving tasks is directly controllable.

The compliance control of each finger allows for rolling, slipping, and whole-hand grasping in a natural way.

## 5.1 Controllable Springs

For the control strategy which will be presented, it is necessary to consider springs with two hinge points for which the minimum potential energy relative position can be varied. This is analog to a spring the natural length of which can be varied. Such an action changes the spring-stored energy and this implies that, in order to properly describe this process in an energetical consistent way, we need an additional power port through which we can control this action.

If we consider as a simple example a linear, one-dimensional spring with finite length $x_l$, and stiffness $k$, its energy function can be expressed as:

$$E(x) = \frac{1}{2}k(x - x_l)^2.$$

The energetic port of the spring is then characterized by the effort-flow pair $(k(x - x_l), \dot{x})$. If we consider the possibility of varying the final length $x_l$, we need to consider an additional port. The new energy function should be considered, then, as a function of $(x - x_l)$ and not only as a function of $x$ and the new energetic port, which is used to control the finite length, should be described by the effort-flow pair $(k(x_l - x), \dot{x}_l)$: the spring's state change is $(\dot{x} - \dot{x}_l)$.

For a geometric spring, as treated in Sect. 3.6.1, a "finite length" corresponds to the relative configuration $r_i^j$ for which the energy function has its minimum. By following the line of what done in Sect. 3.6.1, we could consider as an energy function for our elastic element, a function of $(h_j^i \ o \ r_i^j)$ which has its minimum at the identity of $SE_i(n)$ or equivalently a function of $(h_i^j \ o \ r_j^i)$ which has its minimum at the identity of $SE_j(n)$. These two choices would then correspond respectively to the energy functions of Eq. (3.34) and Eq. (3.35). The *effective state* of this variable length spring would therefore be $(h_i^j \ o \ r_j^i)$.

In order to explain better what will be done in Sect. 5.4, we choose a slightly different but equivalent approach to describe variable springs. If we want to describe a variable spring connecting body $B_b \in \mathcal{E}_b$ to body $B_i \in \mathcal{E}_i$, we consider an additional space $\mathcal{E}_{v(i)}$ which will be called the *supporting space* for the variable spring. We then describe an energy function on the relative position of $v(i)$ with respect to $\mathcal{E}_i$, which we indicate as $h_{v(i)}^i \in SE_{v(i)}^i(3)$ as usual. From a composition of isometries, we obtain:

$$h_{v(i)}^i = h_b^i \ o \ h_{v(i)}^b \tag{5.1}$$

We now control the effective minimal potential energy-relative position of $b$ with respect to $i$ by varying $h_{v(i)}^b$. We can then analyse how the state $h_{v(i)}^i$ of the variable spring changes as a function of time. The following result gives the searched relation.

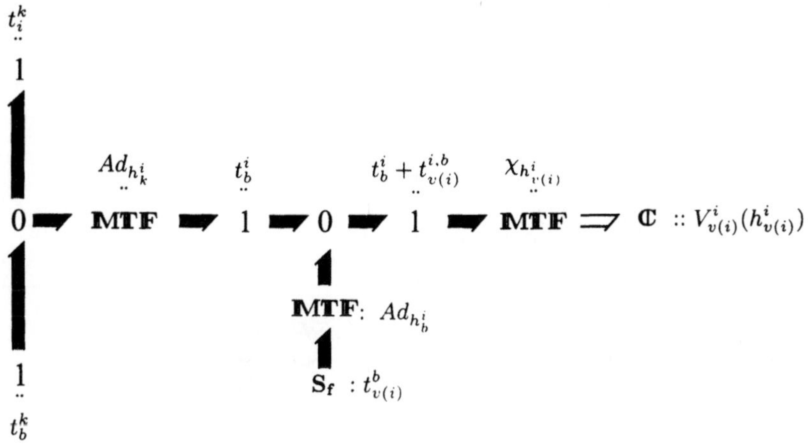


**Fig. 5.1.** The bond graph of a variable length spatial spring.

**Theorem 5.1.1 (Variable length spring state).** *Given a spring with the positive definite energy function* $V^i_{v(i)}(\cdot)$ *of* $h^i_{v(i)} \in SE^i_{v(i)}(3)$ *such that* $h^i_{v(i)} = h^i_b \; o \; h^b_{v(i)}$, *the following identity holds:*

$$t^i_{v(i)} = t^i_b + Ad_{h^i_b} t^b_{v(i)}. \tag{5.2}$$

**Proof.** This is a straightforward application of Theorem 1.3.8. ∎

A representation in coordinates-free bond graphs of such a variable spring is reported in Figure 5.1, where $t^k_b$ and $t^k_i$ represent the hinge points where the two bodies attached to the spring are connected and $t^b_{v(i)}$ is the twist which is used to change the equilibrium point of the spring. The mapping $\chi_{h^i_{v(i)}}$ represents the inverse of the right intrinsic map of Figure 1.9; it is defined in Eq. (3.47). Note that this varying spring is not symmetric, but symmetry can be easily built in. This feature is not explained here for the sake of clarity.

## 5.2 Physical Controller Structure

In this section we describe the structure of the IPC from a conceptual point of view. Following the philosophy of Hogan's Physical equivalence principle [2], we are going to create an IPC which has a directly interpretable physical equivalent system: it is described as a spatial interconnection of physical elements.

In Figure 5.2, two fingertips of a robotic hand are represented as ellipsoids. These two tips are the extremities of robotic chains corresponding to the fingers of a robotic hand to be controlled. We always talk about a robotic hand, but the techniques here presented can be equivalently used for the control of coordinated robots.

**Fig. 5.2.** The intuitive idea of the grasping

The goal of the proposed controller is to create the equivalent effect as the drawn springs and the mass corresponding to the sphere in Figure 5.2. This sphere is called the *virtual object*. This means that the controlled system should dynamically behave as the spatial interconnected system presented in Figure 5.2 in which for clarity only part of the robot mechanism is shown.

One of the springs, called the *hand configuration spring*, is connected at one side to a $V$ indicating what is known in impedance control [2] as the *virtual position* of the hand. The position of the $V$ can be changed by the supervision system (see Sect. 4.2.3) and its result is a global motion of the hand.

In Figure 5.2, only two fingertips are drawn to explain the concepts, but we can have $n$ fingers. The minimum of the potential energy which is function of the relative position of these springs is controllable by the supervisory system using the techniques presented in Sect. 5.1. When the hand is free to move, the robotic system configuration tends to the configuration of minimum potential energy of the system allowed by the kinematic constraints.

If it is ensured that each hand motion also implies motion of the virtual object, we can ensure asymptotically stable behavior by the creation of a damping force on the virtual object which dissipates free energy[1].

In free space, we can therefore control the global position of the hand leaving the minimum energy-relative positions of the fingers' springs unchanged and changing the hand's virtual position, and we can change the configura-

[1] Energy as such cannot be dissipated because of the first principle of thermodynamics. When it is said that energy is dissipated, we actually mean dissipation of free energy, which can be defined for an isothermal system as the Legendre transformation of the energy with respect to the entropy [45].

tion of the fingers with respect to each other by changing the equilibrium position of their springs.

In case we are grasping an object, the springs of the fingers will not longer have the minimum potential energy configuration anymore and their stored energy can be used to quantify physically a *grasp energy* representative of the hardness of the grasp. A maximum grasp energy can then be related to the material of the object to be grasped.

It is important to realize that the springs we will define in our controller, are spatial springs [21, 67, 68] and not trivial translational springs. This implies that we can cleverly choose the center of stiffnesses [21, 67, 68] of the springs in such a way that we can specify the way the grasp reacts to disturbing external forces. By means of proper choices, we can easily control rolling contacts around a desired nominal grasp.

### 5.2.1 The Virtual Object Dynamics

Within the controller, the dynamics of the virtual object is simulated. With reference to Eq. (3.46), we can write the equation of the virtual body as follows:

$$\begin{pmatrix} \dot{h}_b^0 \\ \dot{m}^b \end{pmatrix} = \begin{pmatrix} 0 & \chi_{h_b^0} \; o \; Ad_{h_b^0} \\ -Ad_{h_b^0} \; o \; \chi^*_{h_b^0} & m^b \wedge \end{pmatrix} \begin{pmatrix} \frac{\partial H_B}{\partial h_b^0} \\ \frac{\partial H_B}{\partial m^b} \end{pmatrix} + \begin{pmatrix} 0 \\ Ad^*_{h_b^0} \end{pmatrix} w_{\text{tot}}^0 \tag{5.3}$$

$$t_b^0 = \begin{pmatrix} 0 & Ad_{h_b^0} \end{pmatrix} \begin{pmatrix} \frac{\partial H_B}{\partial h_b^0} \\ \frac{\partial H_B}{\partial m^b} \end{pmatrix}$$

where $b$ indices the virtual body, $w_{\text{tot}}^0$ the total wrench applied to the virtual body expressed in the inertial space, $t_b^0$ the twist of the virtual body with respect to the inertial space, $m^b$ the momentum of body $b$ in its own space and $H_B(h_b^0, m^b)$ the total energy of the virtual body.

Analogously to what done in Sect. 3.11.1 we can rewrite the previous equations in a more compact form:

$$\begin{aligned} \dot{x}_B &= J_B(x_B)\frac{\partial H_B(x_B)}{\partial x_B} + \phi_B(x_B) w_{\text{tot}}^0 \\ t_b^0 &= \phi_B^*(x_B)\frac{\partial H_B(x_B)}{\partial x_B} \end{aligned} \tag{5.4}$$

where $x_B := (h_b^0, m^b)$ and the other assignments should be obvious from Eq. (5.3).

### 5.2.2 The Springs

We can now consider the equations of the $n+1$ springs represented in Figure 5.2, where $n$ is the number of fingers of the considered robotics hand and the $(n+1)$-th spring is the spring connecting the virtual object to the hand's virtual position (V in Figure 5.2). This last spring does not have variable length and is a function of $h_b^{v(b)}$ where $v(b)$ indicates the space corresponding to the $V$ of Figure 5.2.

If we index with $b$ the virtual body space and with $v(i)$ the additional supporting space needed to create a variable spring, the effective twist of the $i$-th spring $t^i_{v(i)} = t^i_b + Ad_{h^i_b} t^b_{v(i)}$ (see Sect. 5.1) can be expressed as:

$$t^i_{v(i)} = \begin{pmatrix} Id & Ad_{h^i_b} \end{pmatrix} \begin{pmatrix} t^i_b \\ t^b_{v(i)} \end{pmatrix}. \tag{5.5}$$

This implies that the general Hamiltonian equations for the used springs with variable length are:

$$\begin{aligned} \dot{h}^i_{v(i)} &= \begin{pmatrix} \chi_{h^i_{v(i)}} & \chi_{h^i_{v(i)}} \; o \; Ad_{h^i_b} \end{pmatrix} \begin{pmatrix} t^i_b \\ t^b_{v(i)} \end{pmatrix} \\ \begin{pmatrix} w^i_b \\ w^b_{v(i)} \end{pmatrix} &= \begin{pmatrix} \chi^*_{h^i_{v(i)}} \\ Ad^*_{h^i_b} \; o \; \chi^*_{h^i_{v(i)}} \end{pmatrix} \frac{\partial V^i_{v(i)}(h^i_{v(i)})}{\partial h^i_{v(i)}} \end{aligned} \tag{5.6}$$

where $\chi_{h^j_i}$ and $\chi^*_{h^j_i}$ are defined in Eq. (3.47). and the pair $(t^i_b, w^i_b)$ corresponds to the power port where are attached the bodies connected to the two end points $b$ and $i$ of the spring, whereas the pair $(t^b_{v(i)}, w^b_{v(i)})$ represents the port which is used to change the effective minimal potential relative position (see Figure 5.1). In the previous equation, $w^i_b$ is the wrench that the body attached to the extreme $b$ of the elastic element applies to the spring, expressed in the space $\mathcal{E}_i$ of the other body.

It is convenient to express the motion of each of the fingertips and of the virtual object in a common space. We use for it the inertial space, which is indexed with 0, as usual.

Applying Eq. (1.30), we obtain the following identities:

$$t^i_b = \begin{pmatrix} Ad_{h^i_0} & -Ad_{h^i_0} \end{pmatrix} \begin{pmatrix} t^0_b \\ t^0_i \end{pmatrix} \Rightarrow \begin{pmatrix} w^0_b \\ w^0_i \end{pmatrix} = \begin{pmatrix} Ad^*_{h^i_0} \\ -Ad^*_{h^i_0} \end{pmatrix} w^i_b \tag{5.7}$$

where $w^0_b$ is the wrench that the virtual object applies to the $i$-th spring and $w^0_i$ is the wrench that tip $i$ applies to the $i$-th spring, both expressed in space 0.

If furthermore we consider $h_b^{v(b)}$ the state of the $(n+1)$-th spring, it is then possible to collect all the springs equations and give a complete expression which results in:

$$\dot{x}_S = \phi_S(x_S) \begin{pmatrix} t_b^0 \\ t_{v(b)}^0 \\ t_{\text{tips}}^0 \\ t_{\text{var}}^b \end{pmatrix}$$

$$\begin{pmatrix} w_b^0 \\ w_{v(b)}^0 \\ w_{\text{tips}}^0 \\ w_{\text{var}}^b \end{pmatrix} = \phi_S^*(x_S)\frac{\partial H_S(x_S)}{\partial x_S} \tag{5.8}$$

where:

$$x_S = (h_{v(1)}^1, \ldots, h_{v(n)}^n, h_b^{v(b)}), \qquad H_S(x_S) = V_b^{v(b)}(h_b^{v(b)}) + \sum_{i=1}^{n} V_{v(i)}^i(h_{v(i)}^i),$$

$$\phi_S = \begin{pmatrix} \phi_b & \phi_{v(b)} & \phi_{\text{tips}} & \phi_{\text{var}} \end{pmatrix},$$

$t_{v(b)}^0$ is the twist of the virtual hand position represented by a $V$ in Figure 5.2 with respect to the inertial frame and

$$t_{\text{tips}}^0 = \begin{pmatrix} t_1^0 \\ \vdots \\ t_n^0 \end{pmatrix} \qquad t_{\text{var}}^b = \begin{pmatrix} t_{v(1)}^b \\ \vdots \\ t_{v(n)}^b \end{pmatrix}$$

$$\phi_b = \begin{pmatrix} \chi h_{v(1)}^1 \; o \; Ad_{h_0^1} \\ \vdots \\ \chi h_{v(n)}^n \; o \; Ad_{h_0^n} \\ \chi_{h_b^{v(b)}} \; o \; Ad_{h_0^{v(b)}} \end{pmatrix} \phi_{v(b)} = \begin{pmatrix} 0 \\ \vdots \\ 0 \\ -\chi_{h_b^{v(b)}} \; o \; Ad_{h_0^{v(b)}} \end{pmatrix}$$

$$\phi_{\text{tips}} = \begin{pmatrix} -\chi h_{v(1)}^1 \; o \; Ad_{h_0^1} & \cdots & 0 \\ \vdots & \ddots & \vdots \\ 0 & \cdots & -\chi h_{v(n)}^n \; o \; Ad_{h_0^n} \\ 0 & \cdots & 0 \end{pmatrix}$$

$$\phi_{\text{var}} = \begin{pmatrix} \chi h_{v(1)}^1 \; o \; Ad_{h_b^1} & \cdots & 0 \\ \vdots & \ddots & \vdots \\ 0 & \cdots & \chi h_{v(n)}^n \; o \; Ad_{h_b^n} \\ 0 & \cdots & 0 \end{pmatrix}.$$

In the previous notation, some dependencies have been omitted for the sake of notational clarity and the definitions of Eq. (3.47) have been used.

### 5.2.3 Interconnection object-springs

By combining Eq. (5.4) and Eq. (5.8), and realizing that $w^0_{\text{tot}} = -w^0_b$ we get:

$$\begin{pmatrix} \dot{x}_B \\ \dot{x}_S \end{pmatrix} = \begin{pmatrix} J_B & -\phi_B \phi_v^* \\ \phi_v \phi_B^* & 0 \end{pmatrix} \begin{pmatrix} \frac{\partial H_C}{\partial x_B} \\ \frac{\partial H_C}{\partial x_S} \end{pmatrix} + \begin{pmatrix} 0 & 0 & 0 \\ \phi_{\text{tips}} & \phi_{v(b)} & \phi_{\text{var}} \end{pmatrix} \begin{pmatrix} t^0_{\text{tips}} \\ t^0_{v(b)} \\ t^b_{\text{var}} \end{pmatrix}$$

$$\begin{pmatrix} w^0_{\text{tips}} \\ w^0_{v(b)} \\ w^b_{\text{var}} \end{pmatrix} = \begin{pmatrix} 0 & \phi^*_{\text{tips}} \\ 0 & \phi^*_{v(b)} \\ 0 & \phi^*_{\text{var}} \end{pmatrix} \begin{pmatrix} \frac{\partial H_C}{\partial x_B} \\ \frac{\partial H_C}{\partial x_S} \end{pmatrix} . \tag{5.9}$$

Where $H_C(x_B, x_S) = H_B(x_B) + H_S(x_S)$. The previous system is a generalized port-controlled Hamiltonian system of the form presented in Sect. 3.4.1 and it is therefore lossless. In order to achieve an asymptotically stable behavior, we must add damping. Consider the inputs of Eq. (5.9) equal to zero; then we get a holonomic Hamiltonian system with six degrees of freedom corresponding to the configuration of the virtual object in space. This implies that we could find canonical coordinates with six generalized positions, six generalized dual momenta and the other coordinates would be Casimir functions for the Hamiltonian system.

### 5.2.4 Creating Damping

We can inject damping in the physical equivalent system presented in Figure 5.2: by means of control we can create a viscous friction force applied to the virtual object. Intuitively, we could think of the virtual object as moving in a fluid with high viscosity which would extract energy from its motion irreversibly.

We can add such a viscous effect by considering an antisymmetric two contravariant tensor $R_t$, as was done in Sect. 4.4.2. We can therefore subtract a symmetric, semi-positive definite tensor $R_B$ from the antisymmetric tensor $J_B$ of Eq. (5.9). $R_B$ has the following form:

$$R_B = \begin{pmatrix} R_h & 0 \\ 0 & R_m \end{pmatrix} \tag{5.10}$$

where the off-diagonal terms have been chosen equal to zero because otherwise they are not physically interpretable. The term $R_m$ is the element representing the usual viscous force since its effect on the virtual object dynamics would be to apply an additional wrench to it, equal to:

$$w^b_d = R_m \frac{\partial H_C}{\partial m^v} = R_m t_b^{b,0} \tag{5.11}$$

where the last identity is a consequence of what was said right after Remark. 3.7.3, p.100 and since $R_m$ is positive definite, it represents a dissipative wrench. Implementing such a term, the derivative of the energy of the controller is:

$$\dot{H} = -\langle \partial H_C / \partial h_b^0, R_h \, \partial H_C / \partial h_b^0 \rangle - \langle t_b^{b,0}, R_m \, t_b^{b,0} \rangle + \langle w_{\text{tips}}^0, t_{\text{tips}}^0 \rangle + \langle w_{v(b)}^0, t_{v(b)}^0 \rangle + \langle w_{\text{var}}^b, t_{\text{var}}^b \rangle \quad (5.12)$$

where $\langle , \rangle$ is the natural dual product of a co-vector on a vector. Due to the positive semidefinite hypothesis of $R_B$, the elements of the first line of Eq. (5.12) will never increase the energy of the controller and if either $t_b^{0,b}$ or $\partial H_C / \partial h_b^0$ are different than zero, these elements will decrease the energy irreversibly if we suppose $R_h, R_m$ to be positive definite.

For practical reasons, we consider only $R_m \neq 0$ and $R_h = 0$, as will be seen in Sect. 5.4.4.

## 5.3 The Controlled Hand

Consider now the robotic hand to be controlled as a fully actuated holonomic robotic mechanism with configuration manifold $\mathcal{Q}$. In this specific case, we mean with a fully actuated robotics system a system for which any generalized force $\tau \in T_q^* \mathcal{Q}$ can be applied at any configuration $q \in \mathcal{Q}$.

The *tips Jacobian* for this system is a linear, configuration-dependent mapping of the following form:

$$J_{\text{tips}}(q) : T_q \mathcal{Q} \rightarrow \underbrace{se_0(3) \times \ldots \times se_0(3)}_{n-\text{times}} \; ; \; \dot{q} \mapsto J_{\text{tips}}(q) \dot{q} \quad (5.13)$$

which maps a configuration velocity $\dot{q}$ to the twists of the tips of the hand:

$$t_{\text{tips}}^0 = J_{\text{tips}}(q) \dot{q} \quad (5.14)$$

We can consider a second Jacobian called the 'interaction Jacobian'. This Jacobian was treated in Sect. 3.13.5. It maps motions of the robot to motions of the links with which the environment can interact. In the general case, for whole-hand manipulation, these could be all the links of the hand. We indicate the interaction Jacobian with $J_I(q)$.

For notational convenience, we define $\hat{\phi}_I(q) := J_I^*(q)$ and $\hat{\phi}_{\text{tips}} := J_{\text{tips}}^*(q)$, which are respectively the adjoints of the interaction Jacobian and the tips Jacobian.

Due to the hypothesis of holonomicity and full actuation of the robotic system, we can write its dynamic equation in the following form:

$$
\begin{pmatrix}\dot{q}\\ \dot{p}\end{pmatrix}=\begin{pmatrix}0 & I\\ -I & 0\end{pmatrix}\begin{pmatrix}\frac{\partial H_R(q,p)}{\partial q}\\ \frac{\partial H_R(q,p)}{\partial p}\end{pmatrix}+\begin{pmatrix}0 & 0\\ \hat{\phi}_I(q) & I\end{pmatrix}\begin{pmatrix}W\\ \tau\end{pmatrix}
$$

$$
\begin{pmatrix}T\\ \dot{q}\end{pmatrix}=\begin{pmatrix}0 & \hat{\phi}_I^*(q)\\ 0 & I\end{pmatrix}\begin{pmatrix}\frac{\partial H_R(q,p)}{\partial q}\\ \frac{\partial H_R(q,p)}{\partial p}\end{pmatrix} \tag{5.15}
$$

where $H_R(q,p)=E_k(q,p)+E_p(q)$ is the total energy of the robotic hand.

The dual pair $(T,W)$ corresponds to the power port through which the robot can exchange energy with the environment. The pair $((q,\dot{q}),(q,\tau))$ corresponds to the energy port of the actuators. It is a dual pair, but it is configuration dependent and not yet suitable to be interconnected with the controller.

In order to solve the problem and to create the desired interconnection as shown in Figure 5.2, we can consider the dual relation of Eq. (5.14), represented by the adjoint of $J_{\text{tips}}(q)$:

$$
\tau=J_{\text{tips}}^*(q)\begin{pmatrix}-w_1^0\\ \vdots\\ -w_n^0\end{pmatrix}=-J_{\text{tips}}^*(q)w_{\text{tips}}^0=-\hat{\phi}_{\text{tips}}(q)w_{\text{tips}}^0 \tag{5.16}
$$

where $-w_i^0$ is the wrench applied by the finger springs $i$ to tip $i$, which is equal and opposite to the wrench $w_i^0$ that tip $i$ applies to the spring $i$ once it is connected to it.

After few calculations, it is possible to get the final equations of the interconnected system:

$$
\dot{x}_T=(J_T(x_T)-R_T(x_T))\frac{\partial H_T(x_T)}{\partial x_T}+\phi_T(x_T)\begin{pmatrix}W\\ t_{v(b)}^0\\ t_{\text{var}}^b\end{pmatrix}
$$

$$
\begin{pmatrix}T\\ w_{v(b)}^0\\ w_{\text{var}}^b\end{pmatrix}=\phi_T^*(x_T)\frac{\partial H_T(x_T)}{\partial x_T} \tag{5.17}
$$

where $x_T:=(q,p,x_B,x_S)$, $H_T(x_T)=H_R(q,p)+H_C(x_B,x_S)$,

$$
J_T(x_T):=\begin{pmatrix}0 & I & 0 & 0\\ -I & 0 & 0 & -\hat{\phi}_{\text{tips}}\phi_{\text{tips}}^*\\ 0 & 0 & J_B & -\phi_B\phi_v^*\\ 0 & \phi_{\text{tips}}\hat{\phi}_{\text{tips}}^* & \phi_v\phi_B^* & 0\end{pmatrix},\quad R_T(x_T):=\begin{pmatrix}0&0&0&0\\ 0&0&0&0\\ 0&0&R_B&0\\ 0&0&0&0\end{pmatrix},
$$

$$
\phi_T(x_T):=\begin{pmatrix}0 & 0 & 0\\ \phi_I(q) & 0 & 0\\ 0 & 0 & 0\\ 0 & \phi_{v(b)} & \phi_{\text{var}}\end{pmatrix}.
$$

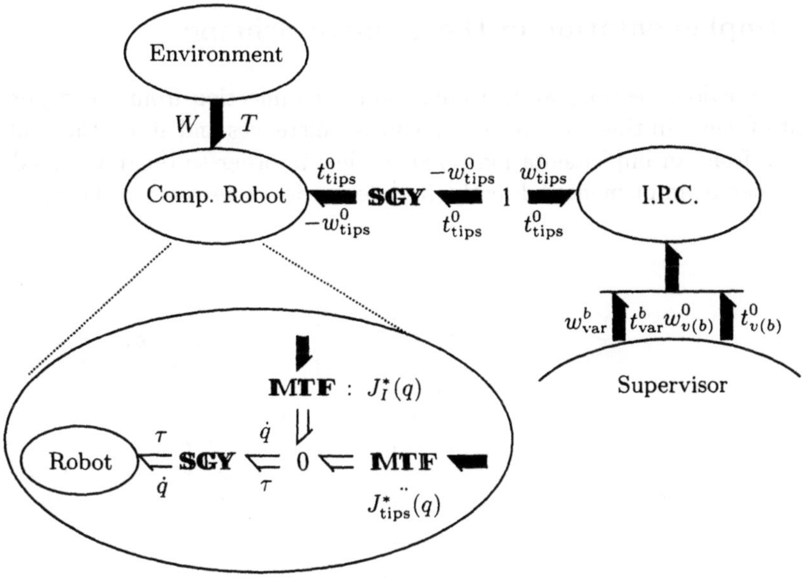


**Fig. 5.3.** The interconnection between the robot and the intrinsically passive controller.

In Figure 5.3 a bond graph representation of the power-continuous interconnection is shown. It is possible to see that the symplectic interconnection achieved corresponds to the one reported in Figure 3.7.

The supervisor can change the virtual position of the hand by supplying a $t^0_{v(b)}$ different than zero and it can change the position of minimal potential energy of the springs by suppling a $t^b_{\text{var}}$ different than zero. This is also used to compensate for the kinematic constraints of the robot as it will be explained in Sect. 5.5. At the same time it can monitor the energy supplied to the controlled robot. When the twists of the supervisor are zero, only the environment can supply energy to the controlled system.

The dissipation implemented in the IPC and the friction present in the real robot ensures an asymptotically stable behavior.

*Remark 5.3.1.* Note that the designed controller only uses kinematic information of the robot to be controlled, namely the Jacobian of Eq. (5.14) which in Figure 5.3 has actually been considered as part of the Compensated Robot. This Jacobian is used as a power-continuous transformation and therefore kinematic model mismatching will never create instability.

## 5.4 Implementation of the Control Scheme

In the previous section, we analyzed the interconnection from a conceptual point of view. In this section we will choose references and study the control system from an implementation point of view in order to build a procedure which can be implemented directly. In order to do so, we need to choose some

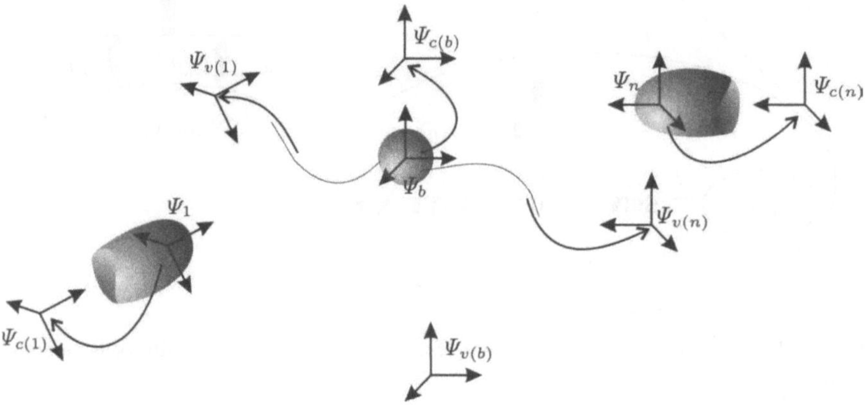


**Fig. 5.4.** The basic frames.

references for the various bodies of the system:

- **Tips' frames**
  With reference to Figure 5.4, assign to each tip $i$ a Cartesian coordinate system $\Psi_i$, which can be visualized as a frame rigidly connected to the tip. It is useful to introduce a second coordinate system for the $i$-th tip whose origin is called *center of stiffness*[2]; this second coordinate system is indicated with $\Psi_{c(i)}$.
- **The Virtual Object frame**
  With reference to Figure 5.4, consider the virtual object to be of spheric shape and uniform density. Consider for it a coordinate frame $\Psi_b$. Consider its inertial properties described in this frame as a mass $m$ and inertia tensor $jI$, where $I$ indicates an identity matrix. It is useful to introduce a second coordinate system for the virtual object. This is indicated with $\Psi_{c(b)}$.
- **The Support Spaces frames**
  As shown in Sect. 5.1, to describe variable springs, we need supporting spaces for each spring $i$ which we indicated with $v(i)$. We will have as many of these spaces as the tips we are considering.

[2] In a work of Fasse [64] this frame is called center of compliance. As shown by Lončarić [21], the center of stiffness and the center of compliance coincide only in very restrictive cases.

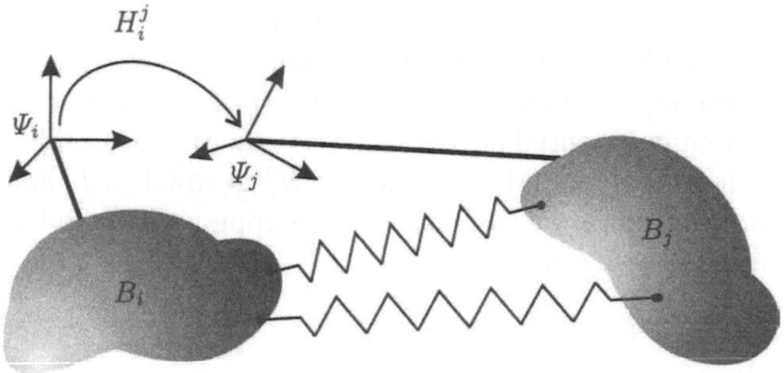


**Fig. 5.5.** Spatial spring between two bodies.

- **The Hand Virtual frame**
  In order to control the interaction between the grasped object and the environment, an extra space, which we indicated with $v(b)$, is needed. This space is connected with the $V$ of Figure 5.2. We choose one coordinate frame for this space, which we indicate with $\Psi_{v(b)}$.

### 5.4.1 Suitable Springs energy functions

We will now give suitable energy functions which can be used to implement the equations of the 3D springs. These energy functions have been introduced in the excellent work of Fasse [64] and are here reported in a different way and with many more additional details. Some recent results can be also found in [70]. As already said, a potential energy function is a function of the relative position of two bodies. These two bodies are for the $n$ tips springs the tips $i$ and the corresponding supporting spaces $v(i)$ and for the hand spring, the body space $b$ and the space $v(b)$.

To express these functions analytically, we have to choose coordinates for the various spaces. To describe the energies of the fingers springs, we choose for the tip the coordinates $\Psi_{c(i)}$ and for the supporting space $v(i)$ the Cartesian coordinates $\Psi_{v(i)}$. The relative position $h_i^{v(i)}$ in these coordinates will therefore be:

$$H_{c(i)}^{v(i)} := \Psi_{c(i)} \, o \, h_{v(i)}^{i} \, o \, \Psi_{v(i)}^{-1}$$

The energy function we describe is such that it has a minimum in $H_{c(i)}^{v(i)} = I$, which corresponds to a relative position of $\mathcal{E}_{v(i)}$ and $\mathcal{E}_i$ such that the two frames $\Psi_{c(i)}$ and $\Psi_{v(i)}$ coincide.

The common origin of these two frames at equilibrium is called the *center of stiffness* because we choose the stiffness expressed in this frame as having a special form.

For generality of exposition, with reference to Figure 5.5, we index as $i$ and $j$ the two bodies between which we want to consider a spring, and two coordinates frames $\Psi_i$ and $\Psi_j$ attached respectively to the body $i$ and $j$ such that in a minimum potential energy corresponding to an equilibrium position we can conclude that the relative position $H_i^j$ is equal to $I$. Following what was done in Sect. 3.6.1, we can consider a mapping of the following form as the generalized elastic force:

$$dV : SE(3) \rightarrow T^*SE(3); H_i^j \mapsto (H_i^j, F_i^j)$$

As already stated in Sect. 3.6.1, the center of stiffness is the linearization of the previous map at the identity ($H_i^j = I$). In this point, by definition $\Psi_i = \Psi_j$.

A differential of the previous map is a linear map of the following form:

$$K : se(3) \rightarrow se^*(3) \; ; \; \delta T \mapsto K\delta T \tag{5.18}$$

where $K$ is a matrix that can be represented in the coordinates in which $\delta T \in se(3)$ is expressed. In our case this is expressed using $\Psi_i = \Psi_j$.

The element $\delta T$ of Eq. (5.18) is an infinitesimal twist represented in vector form. To be more specific, consider the representation $H_i^j$ of a relative position. At a certain instant, the corresponding numerical representation of the twist $t_i^j$ will be:

$$\tilde{T}_i^j = \begin{bmatrix} \Omega_i^j & v_i^j \\ 0 & 0 \end{bmatrix} = \dot{H}_i^j H_j^i = \begin{bmatrix} \dot{R}_i^j & \dot{p}_i^j \\ 0 & 0 \end{bmatrix} \begin{bmatrix} R_j^i & p_j^i \\ 0 & 1 \end{bmatrix} \tag{5.19}$$

where $\Omega_i^j$ is an antisymmetric matrix. We can associate to $\Omega_i^j$ a unique vector $\omega_i^j \in \mathbb{R}^3$ such that for all $x \in \mathbb{R}^3$ we have $\Omega_i^j x = \omega_i^j \wedge x$ where $\wedge$ is the usual vector product of two vectors. In general we indicate such an operation with the operator tilde so that in the case just explained we would indicate:

$$\Omega_i^j = \tilde{\omega}_i^j = \begin{bmatrix} 0 & -\omega_3 & \omega_2 \\ \omega_3 & 0 & -\omega_1 \\ -\omega_2 & \omega_1 & 0 \end{bmatrix} \text{ with } \omega_i^j = \begin{bmatrix} \omega_1 \\ \omega_2 \\ \omega_3 \end{bmatrix}.$$

This implies that we can consider a vector representation of the matrix $\tilde{T}_i^j$, which we indicate with $T_i^j$ and define as:

$$T_i^j = \begin{bmatrix} \omega_i^j \\ v_i^j \end{bmatrix} \Leftrightarrow \tilde{T}_i^j = \begin{bmatrix} \tilde{\omega}_i^j & v_i^j \\ 0 & 0 \end{bmatrix}.$$

An analogous expression can be given for wrenches, for which we have:

$$W_i^j = \begin{bmatrix} m_i^j \\ f_i^j \end{bmatrix} \Leftrightarrow \tilde{W}_i^j = \begin{bmatrix} \tilde{f}_i^j & m_i^j \\ 0 & 0 \end{bmatrix}$$

with $m_i^j$ corresponding to the angular torque. We can then partition Eq. (5.18) in order to show the rotational and translational components in the following way:

$$\begin{bmatrix} m_i^j \\ f_i^j \end{bmatrix} = \begin{bmatrix} K_o & K_c \\ K_c^T & K_t \end{bmatrix} \begin{bmatrix} \delta\theta_i^j \\ \delta p_i^j \end{bmatrix} \tag{5.20}$$

where we indicate with $\delta T = \left[(\delta\theta_i^j)^T \ (\delta p_i^j)^T\right]^T$. The matrix $K$ as defined here is always symmetric and therefore $K_o$ and $K_t$, which are respectively called *rotational stiffness* and *translational stiffness* are also symmetric. Lončarić [21] has shown that $K_c = K_c^T$ corresponds to a maximum decoupling between rotation and translation. In this case, the point corresponding to the coinciding origins of the coordinate systems $\Psi_i$ and $\Psi_j$ at equilibrium is called *center of stiffness*. In our search for proper energy functions, we consider a symmetric $K_c$, which implies that the origin of our reference frames are implicitly chosen at the center of stiffness at equilibrium. By means of the matrix identity:

$$\tilde{v} = A\tilde{w} + \tilde{w}A^T \Leftrightarrow v = (\mathrm{tr}(A)I - A^T)w \tag{5.21}$$

and the hypothesis that $K_o$, $K_t$ and $K_c$ are all symmetric, we can express Eq. (5.20) in an equivalent form, namely:

$$\begin{aligned} \tilde{m}_i^j &= 2\,\mathrm{as}(G_o\delta\tilde{\theta}_i^j) + 2\,\mathrm{as}(G_c\delta\tilde{p}_i^j) \\ \tilde{f}_i^j &= 2\,\mathrm{as}(G_c\delta\tilde{\theta}_i^j) + 2\,\mathrm{as}(G_t\delta\tilde{p}_i^j) \end{aligned} \tag{5.22}$$

where $K_x = (\mathrm{tr}(G_x)I - G_x)$ for $x = t, o, c$ and the $G_x$ are called *co-stiffnesses.* The operator as() returns the antisymmetric part of the matrix given as an argument. Since $K_x$ is symmetric, there exist a conformal transformation corresponding to a rotation of the coordinate system, such that we have:

$$K_x = R_x\Gamma_x R_x^T \quad \text{where} \quad R_x^{-1} = R_x^T \tag{5.23}$$

and $\Gamma_x$ is a diagonal matrix of principal stiffnesses in the directions corresponding to the columns of the orthonormal matrix $R_x$ which are expressed in the coordinates $\Psi_i = \Psi_j$ at equilibrium. It is easy to see that to any diagonal matrix $\Gamma_x$, we can associate a unique diagonal matrix $\Lambda_x$ for which:

$$\Gamma_x = \mathrm{tr}(\Lambda_x)I - \Lambda_x \quad \text{and} \quad \Lambda_x = \frac{1}{2}\mathrm{tr}(\Gamma_x)I - \Gamma_x. \tag{5.24}$$

The $\Lambda_x$ corresponding to the $\Gamma_x$ of Eq. (5.23) are called *principal co-stiffnesses.* It is then possible to see that for each $K_x$, we have a corresponding $G_x$, given by:

$$G_x = \frac{1}{2}\mathrm{tr}(K_x)I - K_x \quad \text{and} \quad K_x = \mathrm{tr}(G_x)I - G_x. \tag{5.25}$$

*Remark 5.4.1.* It is therefore possible, from a practical point of view, to choose a center of stiffness and $K_o, K_t, K_c$ by means of choosing principal directions and corresponding stiffness values. From these $K_x$ we can then compute the corresponding co-stiffnesses $G_x$. We should then find an energy function $V$ parameterized by the $G_x$ such that the linearization of $dV$ around the origin would result in the relations of Eq. (5.22).

We can decompose the total energy in three energies: translational, rotational and couple energies:

$$V(R_i^j, p_i^j) = V_t(R_i^j, p_i^j) + V_o(R_i^j) + V_c(R_i^j, p_i^j). \tag{5.26}$$

For the previous energies, $V_t(\cdot)$ should just depend on $K_t$ and $G_t$, $V_o(\cdot)$ only on $K_o$ and $G_o$ and $V_c(\cdot)$ only on $K_c$ and $G_c$.

**Orientational Energy.** Let us start with the orientational energy. If we take the differential of the rotational energy to be chosen, we get:

$$dV_o(R_i^j) = V_o(R_i^j + dR_i^j) - V_o(R_i^j). \tag{5.27}$$

Since we have for any two vectors $v, w \in \mathbb{R}^3$ that:

$$v^T w = \frac{1}{2}\,\mathrm{tr}(\tilde{v}\tilde{w}),$$

because this energy function is only a function of a rotation, we would like to be able to express this differential as

$$dV_o(R_i^j) = m_o^T \delta\theta_j^i = \frac{1}{2}\,\mathrm{tr}(\tilde{m}_o \tilde{\delta\theta}_j^i) \tag{5.28}$$

where $m_o$ would correspond to the rotational part of the wrench $W_j^i$ that body $j$ would apply to the spring expressed in frame $\Psi_i$. Nevertheless, for what was explained in Remark. 3.6.1, p.90, $W_j^i = -W_i^{i,j}$, and therefore $m_o$ is also equal to the torque that the spring applies to body $i$ expressed in $\Psi_i$.

This implies that $\tilde{m}_o$ can be anything expressed by:

$$\tilde{m}_o(R_i^j) = 2\,\mathrm{as}(x(R_i^j))$$

where $x$ is any matrix dependent on $R_i^j$ which should be found. Furthermore, we want the differential of this torque to be such that in the neighborhood of the identity it satisfies the chosen local behavior expressed by Eq. (5.22). This implies that:

$$\tilde{m}_o(I + \tilde{\delta\theta}_j^i) = 2\,\mathrm{as}(G_o \tilde{\delta\theta}_j^i).$$

A straightforward function realizing this latter equation is

$$\tilde{m}_o(R_i^j) = -2\,\mathrm{as}(G_o R_i^j) \tag{5.29}$$

where we need a minus sign so that we create not a maximum but a minimum of the potential energy at the identity. This implies with Eq. (5.28) that:

$$dV_o(R_i^j) = -\operatorname{tr}(\operatorname{as}(G_o R_i^j)\tilde{\delta\theta}_j^i) = -\operatorname{tr}(G_o R_i^j \tilde{\delta\theta}_j^i) \tag{5.30}$$

where the last equality results from the following matrix identity[3]:

$$\operatorname{tr}(AB) = \operatorname{tr}(\operatorname{sy}(A)\operatorname{sy}(B)) + \operatorname{tr}(\operatorname{as}(A)\operatorname{as}(B)). \tag{5.31}$$

Eventually, since:

$$R_i^j \tilde{\delta\theta}_j^i = dR_i^j = (R_i^j + dR_i^j) - R_i^j$$

we can infer that a $V_o(R_i^j)$ satisfying Eq. (5.27) and Eq. (5.30) is:

$$V_o(R_i^j) = -\operatorname{tr}(G_o R_i^j). \tag{5.32}$$

*Remark 5.4.2.* Note that the previous energy function exactly corresponds to the one presented in [64], but here we deduced it from the local behavior around the identity which can be specified by a proper choice of $K_o$. Note also that the element $(k,l)$ of the matrix $R_i^j$ can be interpreted as the scalar product of the $k$-th axis of $\Psi_j$ with the $l$-th axis of $\Psi_i$. This means that when $\Psi_i$ and $\Psi_j$ coincide, $R_i^j$ is the identity which corresponds to collineation of the coordinates frames. The element $(k,l)$ of $G_o$ can therefore be seen as a weight for the collineation of the axis $k$ of $\Psi_i$ and the axis $l$ of $\Psi_j$. This has been the starting point for getting this energy function in [69]. With this interpretation, the necessity of the minus sign of Eq. (5.29) should become clear: the scalar product of two vectors is maximum when the vectors are colinear; to let it become a minimum we need to invert its sign.

We have therefore proven the following:

**Theorem 5.4.1 (Orientational Elastic Wrenches).** *A spring with elastic energy that is given by Eq. (5.32) and connecting body $i$ and $j$ would apply in a relative position $(p_i^j, R_i^j)$ a wrench $W^i = \left[m_o^T\ f_o^T\right]^T$ to body $i$ and expressed in frame $\Psi_i$ such that:*

$$\tilde{m}_o = -2\operatorname{as}(G_o R_i^j) \tag{5.33}$$

$$\tilde{f}_o = 0 \tag{5.34}$$

**Translational Energy.** Strangely enough, the translational energy is more involved than the orientational one. This is due to the fact that it is not physically meaningful to describe a purely translational anisotropic spring. This means that either we define a spring which generates a force proportional to the distance of the origins of $\Psi_i$ and $\Psi_j$ using a scalar constant which is independent from the direction, or any directional dependence also implies the generation of a torque between the two bodies.

[3] The operator sy(·) indicates the symmetric part of a matrix and as(·) the anti-symmetric one.

**Theorem 5.4.2 (Anisotropic translational springs).** *Purely translational anisotropic springs do not exist: either a spring is isotropic*[4] *or it also generates a torque.*

**Proof.** First consider the following relation treated in Remark. 3.6.1, p.90:

$$w_j^i = -Ad^*_{h_i^j} w_i^j. \tag{5.35}$$

After the choice of coordinates $\Psi_i$ and $\Psi_j$ we can express this latter relation as:

$$W_j^i = -Ad^T_{H_i^j} W_i^j \tag{5.36}$$

where we have

$$Ad_{H_i^j} = \begin{bmatrix} R_i^j & 0 \\ \tilde{p}_i^j R_i^j & R_i^j \end{bmatrix} \Rightarrow Ad^T_{H_i^j} = \begin{bmatrix} R_j^i & \tilde{p}_j^i R_j^i \\ 0 & R_j^i \end{bmatrix} \tag{5.37}$$

In order to obtain a purely translational force, the following should go:

$$\begin{bmatrix} 0 \\ f_j^i \end{bmatrix} = -\begin{bmatrix} R_j^i & \tilde{p}_j^i R_j^i \\ 0 & R_j^i \end{bmatrix} \begin{bmatrix} 0 \\ f_i^j \end{bmatrix}$$

which can be true for any force $f_i^j$ if and only if:

$$\tilde{p}_j^i f_i^{i,j} = 0 \Leftrightarrow \tilde{p}_j^i f_j^i = 0 \Leftrightarrow p_j^i \wedge f_j^i = 0 \Leftrightarrow \exists k \in \mathbb{R}\ ;\ f_j^i = k\, p_j^i$$

for a scalar $k$. The scalar $k$ corresponds to the constant describing the stiffness of the isotropical translational spring. ■

Theorem 5.4.2 implies that in the formulation of translational energy we should also consider the relative orientation of $i$ and $j$ so that it is defined correctly for the general anisotropic case. The usual potential energy we would use for a translational spring, and that would result in the proper contribution for Eq. (5.20), would be a quadratic form of the position difference of the origins of $\Psi_i$ and $\Psi_j$, namely:

$$V_t(p_i^j) = \frac{1}{2}(p_i^j)^T K_t (p_i^j) \tag{5.38}$$

where $K_t$ can be interpreted as being described with the coordinates of $\Psi_j$ at all times. As it is also shown in [67], the choice of translational potential energy of Eq. (5.38) creates the problems described above since there is no dependence from a relative rotation. The problems are due to an asymmetry in the energy function which describes the stiffness only in reference $\Psi_j$. To solve these difficulties, we can "symmetrize" this energy, as [67], and choose:

[4] Isotropic comes from the Greek *isos* (equal) and *tropos* (rotation) which means direction independent.

$$V_t(p_i^j, p_j^i) = \frac{1}{4}(p_i^j)^T K_t(p_i^j) + \frac{1}{4}(p_j^i)^T K_t(p_j^i) \quad (5.39)$$

where the equality of the $K_t$ in the two frames creates the desired symmetry. Note that at equilibrium, $\Psi_i = \Psi_j$ and we have an energy equivalent to Eq. (5.38). The energy of Eq. (5.39) does depend on the relative orientation since it is equal to:

$$V_t(p_i^j, R_i^j) = \frac{1}{4}(p_i^j)^T K_t(p_i^j) + \frac{1}{4}((R_i^j)^T p_i^j)^T K_t(R_i^j)^T p_i^j \quad (5.40)$$

where the identity $p_j^i = -R_j^i p_i^j$ can be shown when considering that $H_j^i = (H_i^j)^{-1}$. It can be proven that $p^T K_t p = -\operatorname{tr}(\tilde{p} G_t \tilde{p})$ for any $p \in \mathbb{R}^3$ and therefore Eq. (5.40) is equal to:

$$V_t(p_i^j, R_i^j) = -\frac{1}{4}\operatorname{tr}(\tilde{p}_i^j G_t \tilde{p}_i^j) - \frac{1}{4}\operatorname{tr}(\tilde{p}_i^j R_i^j G_t R_j^i \tilde{p}_i^j) \quad (5.41)$$

where we used the matrix identity $\operatorname{tr}(R^T A R) = \operatorname{tr}(A)$. The following result gives an expression for the wrenches generated by an elastic energy with such an energy function.

**Theorem 5.4.3 (Translational Elastic Wrenches).** *A spring with elastic energy given by Eq. (5.41) and connecting body i and j would apply in a relative position* $(p_i^j, R_i^j)$ *a wrench* $W^i = \begin{bmatrix} m_t^T & f_t^T \end{bmatrix}^T$ *to body i, and such that this wrench expressed in frame* $\Psi_i$ *is:*

$$\tilde{m}_t = -\operatorname{as}(G_t R_j^i \tilde{p}_i^j \tilde{p}_i^j R_i^j) \quad (5.42)$$

$$\tilde{f}_t = -R_j^i \operatorname{as}(G_t \tilde{p}_i^j) R_i^j - \operatorname{as}(G_t R_j^i \tilde{p}_i^j R_i^j) \quad (5.43)$$

**Proof.** First of all, we should consider that in a certain configuration $(p_i^j, R_i^j)$, the differential of the energy function should be such that:

$$dV_t(p_i^j, R_i^j) = (m_j^i)^T \delta\theta_j^i + (f_j^i)^T \delta p_j^i = \frac{1}{2}\operatorname{tr}(\tilde{m}_j^i \tilde{\delta\theta}_j^i) + \frac{1}{2}\operatorname{tr}(\tilde{f}_j^i \tilde{\delta p}_j^i) \quad (5.44)$$

where $\delta T_j^i = \begin{bmatrix} \delta\theta_j^{i\,T} & \delta p_j^{i\,T} \end{bmatrix}^T$ is an infinitesimal twist of body $j$ with respect to $i$ and expressed in $\Psi_i$. Furthermore, for the reasons explained in Remark. 3.6.1, p.90, we have $m_t := -m_i^{i,j} = m_j^i$ and $f_t := -f_i^{i,j} = f_j^i$. This implies that if we calculate the differential of the energy function and we shape it to a form like Eq. (5.44), we immediately obtain $m_t$ and $f_t$. We will therefore calculate this differential. We get:

$$dV_t(p_i^j, R_i^j) = V_t(p_i^j + dp_i^j, R_i^j + dR_i^j) - V_t(p_i^j, R_i^j) \quad (5.45)$$

where

$$dH_i^j = \begin{bmatrix} dR_i^j & dp_i^j \\ 0 & 0 \end{bmatrix} = H_i^j \delta \tilde{T}_j^i = \begin{bmatrix} R_i^j & p_i^j \\ 0 & 1 \end{bmatrix} \begin{bmatrix} \tilde{\delta\theta}_j^i & \delta p_j^i \\ 0 & 0 \end{bmatrix} = \begin{bmatrix} R_i^j \tilde{\delta\theta}_j^i & R_i^j \delta p_j^i \\ 0 & 0 \end{bmatrix}. \tag{5.46}$$

We can split the energy in the two parts reported in Eq. (5.41) such that:

$$dV_t(p_i^j, R_i^j) = dV_t^{(1)}(p_i^j, R_i^j) + dV_t^{(2)}(p_i^j, R_i^j) \tag{5.47}$$

where

$$\begin{aligned} dV_t^{(1)}(p_i^j, R_i^j) &= V_t^{(1)}(p_i^j + R_i^j \delta p_j^i, R_i^j + R_i^j \tilde{\delta\theta}_j^i) - V_t^{(1)}(p_i^j, R_i^j) = \\ &-\frac{1}{4}\operatorname{tr}((p_i^j + R_i^j \delta p_j^i)^{\tilde{}}\, G_t (p_i^j + R_i^j \delta p_j^i)^{\tilde{}}\,) - \frac{1}{4}\operatorname{tr}(\tilde{p}_i^j G_t \tilde{p}_i^j) = \\ &-\frac{1}{4}\operatorname{tr}(\tilde{\delta p}_j^i R_j^i G_t R_i^j \tilde{\delta p}_j^i) - \frac{1}{4}\operatorname{tr}(\tilde{p}_i^j G_t R_i^j \tilde{\delta p}_j^i R_j^i) - \frac{1}{4}\operatorname{tr}(R_i^j \tilde{\delta p}_j^i R_j^i G_t \tilde{p}_i^j). \end{aligned} \tag{5.48}$$

Above we again used the identity $(Rp)^{\tilde{}} = R\tilde{p}R^T$. The first of the previous three terms is a second-order term one which we can therefore discard. By applying the identity $\operatorname{tr}(AB) = \operatorname{tr}(BA)$ we get:

$$\begin{aligned} dV_t^{(1)}(p_i^j, R_i^j) &= -\frac{1}{2}\operatorname{tr}\left(\frac{R_j^i \tilde{p}_i^j G_t R_i^j - (R_j^i \tilde{p}_i^j G_t R_i^j)^T}{2}\, \tilde{\delta p}_j^i\right) = \\ &\frac{1}{2}\operatorname{tr}(-\operatorname{as}(R_j^i \tilde{p}_i^j G_t R_i^j) \tilde{\delta p}_j^i) \end{aligned} \tag{5.49}$$

and since

$$-\operatorname{as}(R_j^i \tilde{p}_i^j G_t R_i^j) = -R_j^i \operatorname{as}(\tilde{p}_i^j G_t) R_i^j = -R_j^i \operatorname{as}(G_t \tilde{p}_i^j) R_i^j$$

this clearly gives the first term of Eq. (5.43). We can now analyse the differential of the second term $V_t^{(2)}(\cdot)$ of Eq. (5.41). We get:

$$\begin{aligned} dV_t^{(2)}(p_i^j, R_i^j) &= \\ &-\frac{1}{4}\operatorname{tr}((p_i^j + R_i^j \delta p_j^i)^{\tilde{}}\, (R_i^j + R_i^j \tilde{\delta\theta}_j^i) G_t (R_j^i - \tilde{\delta\theta}_j^i R_j^i)(p_i^j + R_i^j \delta p_j^i)^{\tilde{}}\,) \\ &+\frac{1}{4}\operatorname{tr}(\tilde{p}_i^j R_i^j G_t R_j^i \tilde{p}_i^j). \end{aligned} \tag{5.50}$$

Discarding second-order terms from the previous equation, we obtain:

$$
\begin{aligned}
&dV_t^{(2)}(p_i^j, R_i^j) = \\
&-\frac{1}{4}\operatorname{tr}(\tilde{p}_i^j R_i^j G_t R_j^i R_i^j \tilde{\delta p}_j^i R_j^i - \tilde{p}_i^j R_i^j G_t \tilde{\delta\theta}_j^i R_j^i \tilde{p}_i^j + \\
&\tilde{p}_i^j R_i^j \tilde{\delta\theta}_j^i G_t R_j^i \tilde{p}_i^j + R_i^j \tilde{\delta p}_j^i R_j^i R_i^j G_t R_j^i \tilde{p}_i^j) = \\
&-\frac{1}{4}\operatorname{tr}(R_j^i \tilde{p}_i^j R_i^j G_t \tilde{\delta p}_j^i + \tilde{\delta p}_j^i G_t R_j^i \tilde{p}_i^j R_i^j) - \frac{1}{4}\operatorname{tr}(\tilde{p}_i^j R_i^j (\tilde{\delta\theta}_j^i G_t - G_t \tilde{\delta\theta}_j^i) R_j^i \tilde{p}_i^j) = \\
&-\frac{1}{2}\operatorname{tr}\left(\frac{R_j^i \tilde{p}_i^j R_i^j G_t - (R_j^i \tilde{p}_i^j R_i^j G_t)^T}{2}\tilde{\delta p}_j^i\right) - \\
&\qquad \frac{1}{4}\operatorname{tr}(\tilde{p}_i^j \tilde{p}_i^j R_i^j (\tilde{\delta\theta}_j^i G_t - G_t \tilde{\delta\theta}_j^i) R_j^i)
\end{aligned}
\tag{5.51}
$$

and since we have:

$$
\begin{aligned}
&-\frac{1}{4}\operatorname{tr}(\tilde{p}_i^j \tilde{p}_i^j R_i^j (\tilde{\delta\theta}_j^i G_t - G_t \tilde{\delta\theta}_j^i) R_j^i) = \\
&-\frac{1}{4}\operatorname{tr}(\tilde{p}_i^j \tilde{p}_i^j R_i^j \tilde{\delta\theta}_j^i G_t R_j^i) + \frac{1}{4}\operatorname{tr}(\tilde{p}_i^j \tilde{p}_i^j R_i^j G_t \tilde{\delta\theta}_j^i R_j^i) = \\
&-\frac{1}{4}\operatorname{tr}(G_t R_j^i \tilde{p}_i^j \tilde{p}_i^j R_i^j \tilde{\delta\theta}_j^i) + \frac{1}{4}\operatorname{tr}(R_j^i \tilde{p}_i^j \tilde{p}_i^j R_i^j G_t \tilde{\delta\theta}_j^i) = \\
&-\frac{1}{2}\operatorname{tr}\left(\frac{G_t R_j^i \tilde{p}_i^j \tilde{p}_i^j R_i^j - R_j^i \tilde{p}_i^j \tilde{p}_i^j R_i^j G_t}{2}\tilde{\delta\theta}_j^i\right) = \\
&\qquad -\frac{1}{2}\operatorname{tr}(\operatorname{as}(G_t R_j^i \tilde{p}_i^j \tilde{p}_i^j R_i^j)\tilde{\delta\theta}_j^i)
\end{aligned}
\tag{5.52}
$$

we finally obtain:

$$
dV_t^{(2)}(p_i^j, R_i^j) = \frac{1}{2}\operatorname{tr}(-\operatorname{as}(R_j^i \tilde{p}_i^j R_i^j G_t)\tilde{\delta p}_j^i) + \frac{1}{2}\operatorname{tr}(-\operatorname{as}(G_t R_j^i \tilde{p}_i^j \tilde{p}_i^j R_i^j)\tilde{\delta\theta}_j^i))
\tag{5.53}
$$

and considering that

$$
-\operatorname{as}(R_j^i \tilde{p}_i^j R_i^j G_t) = -\operatorname{as}(G_t R_j^i \tilde{p}_i^j R_i^j)
$$

this gives the other terms of Eq. (5.42) and Eq. (5.43). ∎

Clearly, the wrench which the spring applies to the body $j$ is opposite to the one applied to $i$ (see Remark. 3.6.1, p.90). Observe also that around the identity the computed elastic wrench gives the desired behavior specified by $K_t$ and expressed in Eq. (5.22).

**Coupling Energy.** From Eq. (5.22), the coupling energy should be such that the corresponding torque and force linearization at the identity should satisfy:

$$
m_c(0 + R_i^j \delta p_j^i, I + I\tilde{\delta\theta}_j^i) = -2\operatorname{as}(G_c \tilde{\delta\theta}_j^i) \tag{5.54}
$$

$$
f_c(0 + R_i^j \delta p_j^i, I + I\tilde{\delta\theta}_j^i) = -2\operatorname{as}(G_c \tilde{\delta p}_j^i). \tag{5.55}
$$

It is possible to see that a simple energy function satisfying the previous relations at the identity and reported in [64] is

$$V_t(p_i^j, R_i^j) = \mathrm{tr}(G_c R_j^i \tilde{p}_i^j). \tag{5.56}$$

It is then possible to give the following result:

**Theorem 5.4.4 (Coupling Elastic Wrenches).** *A spring with elastic energy given by Eq. (5.56) and connecting body $i$ and $j$ would apply in a relative position $(p_i^j, R_i^j)$ a wrench $W^i = \left[m_c^T \; f_c^T\right]^T$ to body $i$ and expressed in frame $\Psi_i$ such that:*

$$\tilde{m}_c = -2\,\mathrm{as}(G_c \tilde{p}_i^j R_i^j) \tag{5.57}$$

$$\tilde{f}_c = -2\,\mathrm{as}(G_c R_i^j) \tag{5.58}$$

**Proof.** Reasoned along the same line as for the translational energy we get:

$$\begin{aligned}
dV_c(p_i^j, R_i^j) &= V_c(p_i^j + dp_i^j, R_i^j + dR_i^j) - V_c(p_i^j, R_i^j) = \\
&\mathrm{tr}(G_c(R_j^i - \tilde{\delta\theta}_j^i R_j^i)(p_i^j + R_i^j \delta p_j^i)^{\sim}) - \mathrm{tr}(G_c R_j^i \tilde{p}_j^i) = \\
&= \mathrm{tr}(G_c R_j^i R_i^j \tilde{\delta p}_j^i R_j^i - G_c \tilde{\delta\theta}_j^i R_j^i \tilde{p}_i^j) = \\
&\mathrm{tr}(R_j^i G_c \tilde{\delta p}_j^i) - \mathrm{tr}(R_j^i \tilde{p}_i^j G_c \tilde{\delta\theta}_j^i) = \\
&\qquad \frac{1}{2}\mathrm{tr}(2\,\mathrm{as}(R_j^i G_c)\tilde{\delta p}_j^i) + \frac{1}{2}\mathrm{tr}(-2\,\mathrm{as}(R_j^i \tilde{p}_i^j G_c)\tilde{\delta\theta}_j^i)
\end{aligned} \tag{5.59}$$

which after applying some properties of the antisymmetric part of the product of two matrices gives the result to be proven. ■

### 5.4.2 Storing Positions

The position of each of the frames $\Psi_x$ in Figure 5.4 can be associated to the necessary changes of coordinates and relative positions from $\Psi_x$ to a fixed frame $\Psi_0$ that is attached to the inertial space. We indicate such a matrix with $H_x$. More precisely we have that:

$$\begin{aligned}
H_i &:= H_i^0 = \Psi_0 \; o \; h_i^0 \; o \; \Psi_i^{-1} \\
H_{c(i)} &:= H_i H_{c(i)}^i \\
H_b &:= H_b^0 = \Psi_0 \; o \; h_b^0 \; o \; \Psi_b^{-1} \\
H_{c(b)} &:= H_b H_{c(b)}^b \\
H_{v(i)} &:= H_{v(i)}^0 = \Psi_0 \; o \; h_{v(i)}^0 \; o \; \Psi_{v(i)}^{-1} \\
H_{v(b)} &:= H_{v(b)}^0 = \Psi_0 \; o \; h_{v(b)}^0 \; o \; \Psi_{v(b)}^{-1}
\end{aligned}$$

where $H^i_{c(i)}$, $H^b_{c(b)}$ are fixed changes of coordinates within the same space.

It is then easy to calculate the matrices which are needed in order to compute the elastic wrenches, namely:

$$H^{v(i)}_{c(i)} = (H_{v(i)})^{-1} H_{c(i)} \text{ and } H^{v(b)}_{c(b)} = (H_{v(b)})^{-1} H_{c(b)}. \tag{5.60}$$

The direct kinematics of the robotic hand should then be available as a map of the following form, which can be calculated with the techniques presented in Chapter 3:

$$L(q) : \mathcal{Q} \to \underbrace{SE(3) \times \ldots \times SE(3)}_{n \text{ times}} ; \; q \mapsto (H_1(q), \ldots, H_n(q)) \tag{5.61}$$

and the body position $H_b$ is a state of the controller. The coordinates changes represented by $H^i_{c(i)}$, $H^b_{c(b)}$ are fixed, known, and chosen at the beginning, as will be shown later.

### 5.4.3 The Wrenches of the system

If we use the proposed energy functions and we want to create an elastic force between two bodies $i$ and $j$ by means of control, we should proceed as follows.

**Algorithm 5.4.5 (Calculation of elastic wrenches)**

1. *Choose a relative position $r^j_i$ of minimal potential energy.*
2. *In this relative position, choose a common point which will be the center of stiffness.*
3. *Choose two coordinate systems $\Psi_i$ and $\Psi_j$ for $i$ and $j$ respectively, which have their origin in the center of stiffness and coincide at the equilibrium relative position $r^j_i$.*
4. *Choose the desired $K_t$, $K_o$, $K_c$ which are expressed at equilibrium in the frames $\Psi_i = \Psi_j$.*
5. *Calculate the corresponding $G_t$, $G_o$, $G_c$ with Eq. (5.25).*
6. *With $H^j_i$, the total wrench generated by the spring on body $i$ and expressed in $\Psi_i$ is the sum of the wrenches of the orientational, translational and coupling energies $W^i = \left[(m^i)^T \; (f^i)^T\right]^T$ with:*

$$\tilde{m}^i = -2\,\mathrm{as}(G_o R^j_i) - \mathrm{as}(G_t R^i_j \tilde{p}^j_i \tilde{p}^j_i R^j_i) - 2\,\mathrm{as}(G_c \tilde{p}^j_i R^j_i) \tag{5.62}$$

$$\tilde{f}^i = -R^i_j\,\mathrm{as}(G_t \tilde{p}^j_i) R^j_i - \mathrm{as}(G_t R^i_j \tilde{p}^j_i R^j_i) - 2\,\mathrm{as}(G_c R^j_i) \tag{5.63}$$

7. *The wrench $W^j$ that the spring applies to body $j$ will be $W^j = -Ad^T_{H^i_j} W^i$, which implies that:*

$$m^j = -R^j_i m^i - \tilde{p}^j_i R^j_i f^i \tag{5.64}$$

$$f^j = -R^j_i f^i \tag{5.65}$$

If we apply Alg. 5.4.5 to the pair of bodies $(i, v(i))$ for the springs connecting the tips to the virtual body, and to the pair $(b, v(b))$ for the hand spring, we can compute the elastic wrenches in the system. If we choose the $\Psi_{c(i)}$ as coordinates for the finger tips, $\Psi_{v(i)}$ as coordinates for the supporting bodies $v(i)$, $\Psi_{c(b)}$ for the virtual body and $\Psi_{v(b)}$ for the hand virtual position, we have as representative matrices $H_{c(i)}^{v(i)}$ for the tips springs and $H_{c(b)}^{v(b)}$ for the hand.

We can then choose the desired stiffnesses for each of these springs, which we indicate as[5]:

$$K^{(i)} = \begin{bmatrix} K_o^{(i)} & K_c^{(i)} \\ K_c^{(i)} & K_t^{(i)} \end{bmatrix} \quad i = 1 \dots n \quad \text{and} \quad K^b = \begin{bmatrix} K_o^b & K_c^b \\ K_c^b & K_t^b \end{bmatrix} . \tag{5.66}$$

With Alg. 5.4.5 we can calculate therefore for each $H_{c(i)}^{v(i)}$ and $H_{c(b)}^{v(b)}$ the elastic wrenches applied to the tips $W^{c(i)}$, the elastic wrenches applied to the supporting spaces $W^{v(i)}$ and the direct elastic wrenches applied to the virtual body $W^{c(b)}$.

**Wrenches on the tips and control torques.** Once we obtain the wrenches $W^{c(i)}$ for each time step, we can directly calculate the torques we have to apply to the robot in order to get the same effect that these virtual springs would have.

Use a numerical representation of the Jacobian of Eq. (5.14) where the twists $t_i^0$ are expressed in a common fixed frame $\Psi_0$, and indicate it with $\bar{J}_{\text{tips}}(q)$.

Since the wrenches $W^{c(i)}$ of each fingertip are expressed in the frame $\Psi_{c(i)}$, we need to transform them first to the fixed space in the frame $\Psi_0$ using:

$$W^{0,c(i)} = Ad^T_{H_0^{c(i)}} W^{c(i)} \tag{5.67}$$

the torques to set to the actuators will therefore be:

$$\tau = \bar{J}^T_{\text{tips}}(q) \begin{bmatrix} Ad^T_{H_0^{c(1)}} & \cdots & 0 \\ \vdots & \ddots & \vdots \\ 0 & \cdots & Ad^T_{H_0^{c(n)}} \end{bmatrix} \begin{bmatrix} W^{c(1)} \\ \vdots \\ W^{c(n)} \end{bmatrix} . \tag{5.68}$$

**Elastic Wrench on the virtual object.** The total elastic wrench applied to the virtual object is the sum of all the wrenches generated by the springs attached to it.

We considered each tip spring energy as a function of the relative position of each tip $i$ and the supporting space $v(i)$. As shown in Eq. (5.6), this formulation is actually used to describe a spring which connects each tip to

[5] Note that at the center of stiffness, the coupling stiffness matrices $K_c^{(i)}$ and $K_c^b$ are symmetric.

the virtual object. When we look at the dual equation of Eq. (5.5), we see that the wrench applied to the body and the wrench applied to the supporting space $v(i)$ are the same. This can be thought of intuitively by considering the space $v(i)$ rigidly connected to $b$ at a certain instant. We therefore only need to express these wrenches in the right coordinates. If we express all the wrenches in the coordinate $\Psi_b$, we get:

$$W^b_{\text{elastic}} = \sum_{i=1}^{n} Ad^T_{H_b^{v(i)}} W^{v(i)} + Ad^T_{H_b^{c(b)}} W^{c(b)} \tag{5.69}$$

where $W^{v(i)}$ and $W^{c(b)}$ are calculated using Alg. 5.4.5.

### 5.4.4 Simulation of the Virtual Object dynamics

So far we have completely analyzed how to calculate the torques to supply to the robot as a function of the positions of the finger tips and of $H_b$. The matrix $H_b$ is a time-varying matrix describing the dynamic evolution of the virtual object which is the dynamic extension of the controller.

We have chosen the virtual object with a very simple inertial structure on purpose and its dynamics can be easily simulated in real time within the controller. Considering Eq. (5.3) and the coordinates systems used, we can give a coordinate expression for the virtual object dynamics.

Assuming the virtual body in a gravitational-less environment, it does not have potential energy and its Hamiltonian is:

$$H_B(H_b, M^b) = \frac{1}{2}(M^b)^T \begin{bmatrix} \frac{1}{j}I_3 & 0 \\ 0 & \frac{1}{m}I_3 \end{bmatrix} (M^b) \quad m, j \in \mathbb{R}. \tag{5.70}$$

The map $h_b^0$ of Eq. (5.3) is represented by the matrix $H_b^0 = H_b$. and therefore, the first Hamiltonian equation is:

$$\dot{H}_b = H_b \tilde{T}_b^{b,0} \quad \text{with} \quad T_b^{b,0} = \frac{\partial H_B}{\partial M^b} = \begin{bmatrix} \frac{1}{j}I_3 & 0 \\ 0 & \frac{1}{m}I_3 \end{bmatrix} (M^b) \tag{5.71}$$

and the second equation is:

$$\dot{M}^b = M^b \wedge T_b^{b,0} + W^b_{\text{tot}} \tag{5.72}$$

where the equation corresponds to Eq. (3.42) and $W^b_{\text{tot}}$ is the total wrench applied to the body. If we assume the $R_h$ of Sect. 5.2.4 equal to zero, the dissipating wrench is $W^b_{\text{diss}} = R_m T_b^{b,0}$ and if we use Eq. (5.69), the total wrench applied to the virtual body is:

$$W^b_{\text{tot}} = W^b_{\text{elastic}} + W^b_{\text{diss}} \tag{5.73}$$

With the previous equations it is therefore possible to calculate $\dot{H}_b$ and $\dot{M}^b$,

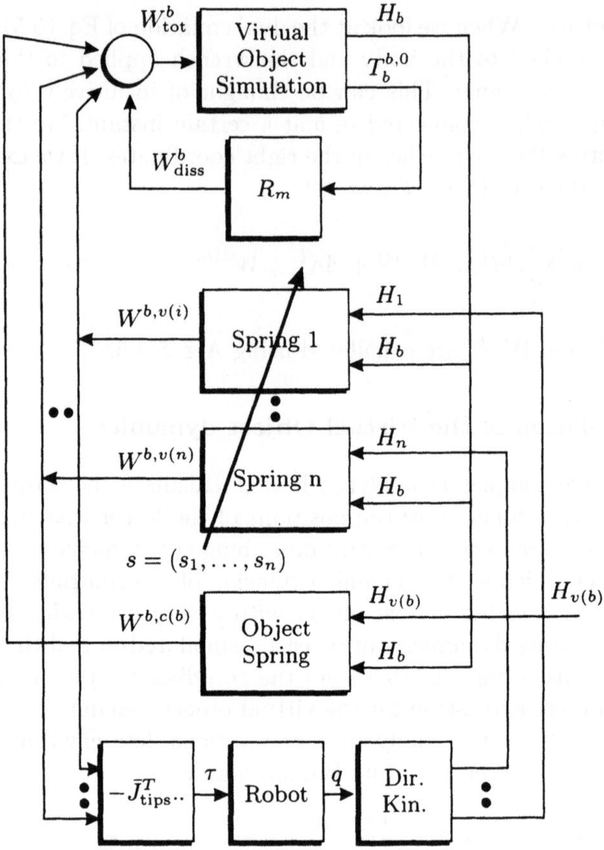


**Fig. 5.6.** The simplified grasping control scheme.

which are the state rates of the controller. The integration of $\dot{H}_b$ should be done with care since the matrix $H_b$ should remain in $SE(3)$ during all steps. In order to do that, the orthonormal matrix $R_b^0$ within $H_b^0$ should be normalized after each step in order to keep it within $SO(3)$ or even better, it should be converted to a unit quaternion and then integrated. With the integration of $H_b$, the position of the body is known at each step and the elastic wrenches can be calculated again closing the control loop. A simplified representation of the control scheme is reported in Figure 5.6

### 5.4.5 Varying the length of springs

The variation of the length of a spring, can be easily taken into account by considering the supervisory control twist $t^b_{v(i)}$ reported in Figure 5.1. If we indicate the expression of this twist in the base $\Psi_b$ with $T^b_{v(i)}$, we can express the effective state for the variable springs as:

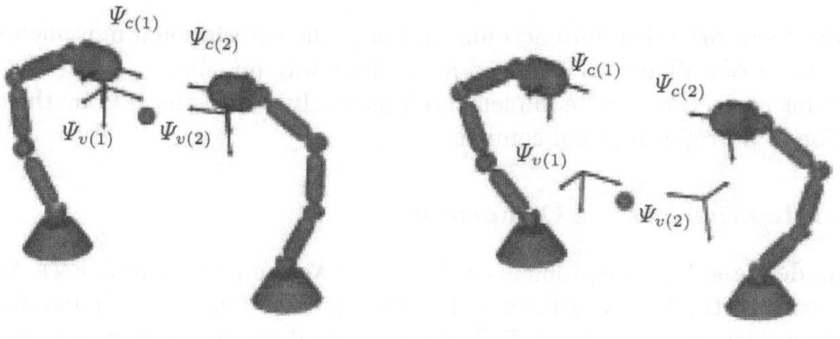


**Fig. 5.7.** The kinematical problem.

$$H_{c(i)}^{v(i)} = H_b^{v(i)} H_{c(i)}^b \tag{5.74}$$

The effective state will be therefore changed either by a change of $H_b^{v(i)}$ due to a supervisory control or by a change of $H_{c(i)}^b = (H_b)^{-1} H_{c(i)}$ corresponding to a change of the relative position of the virtual body with respect to the tip $i$. This energetic port can be used to consider the kinematic structure of the robot as it will be explained in the next section.

## 5.5 Taking the Robot Kinematics into account

One of the problems of the procedure previously explained, is that the kinematic constraints of the robot to be controlled are not considered anywhere. As a consequence, in order for the system to behave properly, the supervisor should vary the rest length of the springs in a suitable way.

To understand the nature of the problem, just consider a two finger system as the one reported in Figure 5.7. Suppose that, in order to perform a certain task, we want to lower the tips of the fingers toward the base. To do so, it would seem obvious to pull the virtual object down so that all the tips would follow as it is shown on Figure 5.7 from left to right.

Unfortunately, if we perform this operation without changing the rest length of the tip's springs, we could reach a configuration like the one shown on the right of Figure 5.7 where we can see that the tips do NOT follow the virtual object downward.

This is due to the fact that, since the system has limited degrees of freedom, the tip simply cannot follow both the translation and orientation of the $\Psi_{v(i)}$ frames. The result is that the system will reach a minimum of the potential energy which is a trend between the translational and orientational energy.

What we instead would like the system to do, is that the rest length of the finger's springs is changed in such a way that the orientation constraints of the robot are taken into account and only the translational movements of the center of stiffness are followed in the best way possible.

This correction can be implemented passively within the IPC by the definition of a proper internal compensation.

### 5.5.1 Internal Passive Compensation

To understand how the problem can be solved we can proceed as follows. With reference to the frames defined in Figure 5.4, a motion of the virtual object is characterized by the twist $T_b^{0,0}$ where as usual the frame $\Psi_0$ represents an inertial frame. We can analyse at a certain instant a virtual point $\bar{c}(i)$ rigidly connected to the virtual body and coincident with the center of compliance of the $i$-th tip $c(i)$. We can then calculate its linear speed due to a motion of the virtual object:

$$P^0_{\bar{c}(i)} = H^0_b P^b_{\bar{c}(i)} \tag{5.75}$$

where $P^k_{\bar{c}(i)}$ is the four dimensional homogeneous coordinate of the point $\bar{c}(i)$ in the frame $\Psi_k$. To calculate its speed we have:

$$v^{0,0}_{\bar{c}(i)} := \dot{P}^0_{\bar{c}(i)} = \dot{H}^0_b P^b_{\bar{c}(i)} = \dot{H}^0_b H^b_0 H^0_b P^b_{\bar{c}(i)} = T^{0,0}_b P^0_{\bar{c}(i)}$$

It is easy to see that:

$$v^{0,0}_{\bar{c}(i)} = Q_{\bar{c}(i)} T^{0,0}_b$$

where

$$Q_{\bar{c}(i)} := \begin{pmatrix} -\tilde{p}^0_{\bar{c}(i)} & I_3 \end{pmatrix},$$

and $\tilde{p}^0_{\bar{c}(i)}$ is the skew symmetric matrix corresponding to the three dimensional position vector $p^0_{\bar{c}(i)}$. An expression of the differential kinematics of the tip is given by a properly defined geometric Jacobian which is such that:

$$\begin{pmatrix} \omega^{c(i),0}_{c(i)} \\ v^{c(i),0}_{c(i)} \end{pmatrix} = T^{c(i),0}_{c(i)} = J_i(q)\dot{q} = \begin{pmatrix} J^\omega_i(q) \\ J^v_i(q) \end{pmatrix} \dot{q}$$

where $v^{c(i),0}_{c(i)}$ is the linear velocity with respect to the inertial frame of a virtual point rigidly connected to the tip and passing through the origin of $\Psi_{c(i)}$.

Using the previous partition we have therefore:

$$v^{c(i),0}_{c(i)} = J^v_i(q)\dot{q} \Rightarrow v^0_{c(i)} = R^0_{c(i)}(q) J^v_i(q)\dot{q}$$

where now $v^{0,0}_{c(i)}$ is the velocity of the center of compliance with respect to the inertial frame expressed in the inertial frame and due to a motion of the robot and $R^0_{c(i)}(q)$ is the rotation matrix from $\Psi_{c(i)}$ to $\Psi_0$.

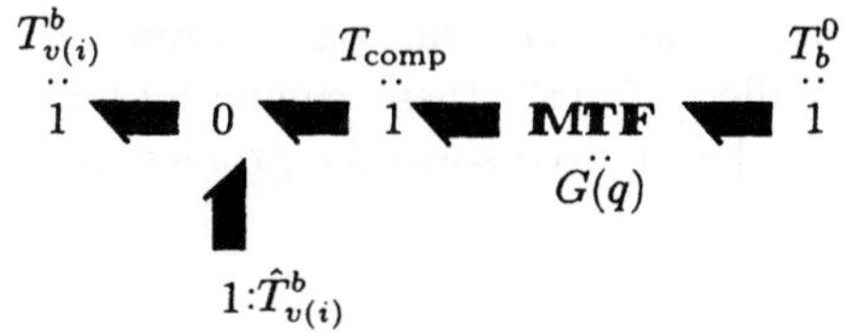


**Fig. 5.8.** The bond graph of the passive internal compensation.

It is now possible to calculate the joint velocities which better follow a certain velocity of the center of compliance. This can be done using a pseudoinversion since metrics are available in $T_q\mathcal{Q}$ and $\mathcal{E}_*$. A metric for $\mathcal{E}_*$ is clearly the Euclidean metric and as a metric for $T_q\mathcal{Q}$ the inertia matrix can be used. Doing so, it is possible to define:

$$K^+(q) := (R^0_{c(i)} J^v_i(q))^+$$

which will give the $\dot{q}$ which better follows a velocity of the center of compliance. We can now calculate the joints speed which better follow $v^{0,0}_{\bar{c}(i)}$:

$$\dot{q} = K^+(q) v^0_{\bar{c}(q)} \tag{5.76}$$

and the corresponding total motion of the tip using the geometric Jacobian:

$$T^{c(i),0}_{c(i)} = J_i(q) K^+(q) Q_{\bar{c}(i)} T^0_b. \tag{5.77}$$

On the other hand, we have the identity:

$$T^{c(i),0}_{v(i)} = T^{c(i),0}_b + T^{c(i),b}_{v(i)} \tag{5.78}$$

which implies that:

$$T^b_{v(i)} = Ad_{H^b_{c(i)}} T^{c(i),0}_{v(i)} - Ad_{H^b_0} T^0_b. \tag{5.79}$$

To achieve the described goal, it is then sufficient to set $T^{c(i),0}_{v(i)} = T^{c(i),0}_{c(i)}$ to finally obtain:

$$T^{b,b}_{v(i)} = \underbrace{\left(Ad_{H^b_{c(i)}} J_i(q) K^+(q) Q_{\bar{c}(i)} - Ad_{H^b_0}\right)}_{G(q)} T^{0,0}_b. \tag{5.80}$$

We can therefore define a modulated transformer based on $G(q)$ which automatically will compensate the shortcoming of the algorithm. A bond graph representation of the compensation of one tip spring is reported in Figure 5.8 where $\hat{T}^b_{v(i)}$ denotes the new twist that the supervisor can use to modify the rest lengths of that spring as desired and $T_{\text{comp}}$ is the compensation twist just explained and calculated in Eq. (5.80).

A lot of simulations have been run on a screw bondgraph multibody simulation package called `20sim`[6]. these simulations have shown that the internal compensation does indeed solve the problem as desired.

## 5.6 Grasping strategies

Suppose we have an object to be grasped and we want to plan a tip grasp configuration. We can describe this grasp configuration by defining the desired positions of the tips with respect to a "grasp focus" positioned somewhere between the fingers [84]. If we represent the focus by a frame $H_f$, the desired grasp configuration is described by $n$ relative configurations $H_f^1, \ldots, H_f^n$.

If we position the virtual body at the focus ($H_b = H_f$) and consider it as a representative position for the whole grasp, to achieve the desired configuration it should be $H_b^i = H_f^i \; \forall i$.

If the hand would just touch the object to be grasped without exerting any force, the energy of the controller's springs should be zero, which implies that in such a situation $H_{c(i)} = H_{v(i)}$.

It is than possible to chose a representative of the supervisory twist $\hat{T}_{v(i)}^b$ as:

$$\hat{T}_{v(i)}^b = \dot{s}_i \begin{pmatrix} 0_3 \\ p_i^f \end{pmatrix}$$

where $p_i^f$ is the position of the tip $i$ with respect to the focus expressed in $b$. The twist $\hat{T}_{v(i)}^b$ is nothing else than a pure translational twist in the direction of $p_i^f$. In order to perform a tip grasp, we have to proceed as follows:

**Algorithm 5.6.1 (Tip grasping algorithm)**

1. *Choose the location of the center of stiffness for the object by choosing* $H_b^{c(b)}$.
2. *Choose a* $H_{v(b)}$ *far from the object to be grasped and in the neighborhood of the hand.*
3. *For each* $i$, *increase* $s_i$ *sufficiently in order to open the hand (Opening Hand).*
4. *Move the virtual position of the object,* $H_{v(b)}$, *along a proper trajectory to* $H_f H_{c(b)}^b = H_f (H_b^{c(b)})^{-1}$.
5. *Decrease the* $s_i$ *to a proper value which represents a sufficiently high grasping internal energy (Closing Hand).*
6. *Change* $H_{v(b)}$ *to move the object and do what is needed.*

[6] `20sim` is an excellent simulation package which allows to use bond graphs and screw theory. It has been developed by Control Product B.V., Enschede, The Netherlands.

Proper choices of the center of stiffness for each tip would then specify how the robot reacts to small unexpected motions of the object due to external forces. In this way, it would also be possible to handle rolling contacts by exploiting the contact friction, but this requires further research.

The presented control system can not only be used to implement tips grasps but in the same way as for tips grasp, it is possible to open the hand and move the virtual position $H_{v(b)}$ over an object to be grasped rather than at its center. At the same time we can close the hand by decreasing the $s_i$ and the result is that the hand folds around the object to be grasped implementing a full grasp.

## 5.7 Conclusions

In this chapter, an intrinsically passive algorithm for controlling multi-limbed robotic systems based on physical concepts has been presented. A major advantage is the ensured stability in any interactive situation for any environment. This because Liapunov stability is ensured choosing as Liapunov function the equivalent energy of the robot-controller coupled system and their power-continuous interconnection. This equivalent energy is composed of the potential energy of the controller springs, the kinetic energy of the robot and of the virtual object.

# 6. IPC in Telemanipulation

In this chapter we study the interconnection of two port controlled Hamiltonian systems through a transmission line with delay. The contributions of the chapter are firstly a geometrical, multi-dimensional, power consistent exposition of telemanipulation of Intrinsically Passive Controlled physical systems as introduced in Chapter 4 [1, 85], with a clarification on impedance matching, and secondly a system theoretic condition for the adaptation of a general port controlled Hamiltonian system with dissipation (PCHD system) to a transmission line. These results have been presented in [9]. Experimental results on an Internet implementation are also presented.

A lot of work has been done in the field of telemanipulation. Some of the contributions specifically address the problem of time delays due to the actual transmission through a line of non-neglectable length. The problem was firstly addressed in [86] for a one dimensional case, and then extended in [87], where important considerations on the line causality and extensions with adaptation techniques are treated.

In [9] of which this chapter is an extension, a geometrical multi-dimensional case is presented, which uses digital transmission of data in order to create a perfectly bilateral telemanipulation system on a transmission line with varying, non-neglectable delays: the Internet.

## 6.1 Geometric scattering

Scattering variables are well known in network theory. To the best of the author's knowledge, the first works which present scattering variables from a geometrical point of view are [88, 89, 90, 9]. This allows to implement a telemanipulating system from an intrinsically geometric point of view.

The main idea is as follows. Given any vector space $\mathcal{V}$, we can consider the vector space:

$$\mathcal{D} := \mathcal{V} \times \mathcal{V}^*$$

On $\mathcal{D}$ there exist a canonical, symmetric, two covariant tensor called *+pairing*. This symmetric, non-degenerate 2_form, is defined by the bilinear operation:

$$\langle (f_1, e_1), (f_2, e_2) \rangle_+ := \langle e_1, f_2 \rangle + \langle e_2, f_1 \rangle$$

where $(f_i, e_i) \in \mathcal{D}$ and $\langle e_i, f_j \rangle$ denotes the intrinsic dual pairing. Using this tensor it is also possible to give a geometric definition of a Dirac structure as a 'self-orthogonal' subspace for the +pairing [88, 90, 91]. We can define a base for $\mathcal{V}$ as the columns of a matrix $B$:

$$B := \begin{pmatrix} e_1 \dots e_n \end{pmatrix}$$

and the dual base as the columns of a matrix $B_*$:

$$B_* := \begin{pmatrix} e_1^* \dots e_n^* \end{pmatrix}$$

such that $B_*^T B = B^T B_* = I$. We can then define the corresponding base matrix for $\mathcal{D}$ as the columns of a matrix $\bar{B}$:

$$\bar{B} := \begin{pmatrix} B & 0 \\ 0 & B_* \end{pmatrix}$$

and eventually, the adjoint matrix $\bar{B}_*$:

$$\bar{B}_* := \begin{pmatrix} B_* & 0 \\ 0 & B \end{pmatrix}$$

A representation of the plus pairing is then:

$$T_{ij} = \bar{B}_* \begin{pmatrix} 0 & I \\ I & 0 \end{pmatrix} \bar{B}_*^T \tag{6.1}$$

In order to define subspaces in a coordinate free way, we need a metric on $\mathcal{V}$ which corresponds to a characteristic impedance $B_* Z B_*^T$. We can define a 2 contravariant tensor on $\mathcal{D}$ based on $Z$ as:

$$Y^{li} = \bar{B} \begin{pmatrix} Z^{-1} & 0 \\ 0 & Z \end{pmatrix} \bar{B}^T, \tag{6.2}$$

and then consider the eigenvalues of

$$L_j^l := Y^{li} T_{ij} = \bar{B} \begin{pmatrix} 0 & Z^{-1} \\ Z & 0 \end{pmatrix} \bar{B}_*^T. \tag{6.3}$$

$\lambda$ is an eigenvalue of $L_j^l$ if

$$\lambda \begin{pmatrix} f \\ e \end{pmatrix} = \begin{pmatrix} 0 & Z^{-1} \\ Z & 0 \end{pmatrix} \begin{pmatrix} f \\ e \end{pmatrix} \tag{6.4}$$

which implies that it should be

$$\begin{cases} \lambda f = Z^{-1} e \\ \lambda e = Z f \end{cases} \Rightarrow \lambda^2 f = Z^{-1} Z f$$

This implies that if $f \neq 0$ we obtain that the eigenvalues should be $\lambda = \pm 1$. In order to have meaningful eigenvalues, if $f = 0$, it should be $e \neq 0$ and we would get again in a similar way that in this case again $\lambda = \pm 1$. We can therefore define two eigensubspaces associated to the eigenvalue $+1$ and $-1$ respectively. These two subspaces which turn out to be of the same dimension as $\mathcal{V}$ are clearly dependent on $Z$. We denote this as follows:

$$\mathcal{D} = \mathcal{S}_Z^+ \oplus \mathcal{S}_Z^- \tag{6.5}$$

which implies that for each $Z$ there is a unique way[1] to express a power pair $(f, e) \in \mathcal{D}$ as the sum of two elements $s^+ \in \mathcal{S}_Z^+$ and $s^- \in \mathcal{S}_Z^-$. Furthermore, it is possible to see that:

$$\mathcal{S}_Z^+ = \ker \left(I \; -Z^{-1}\right) \bar{B}_*^T = \operatorname{Im} \bar{B} \begin{pmatrix} Z^{-1} \\ I \end{pmatrix} \frac{N}{\sqrt{2}} \tag{6.6}$$

and

$$\mathcal{S}_Z^- = \ker \left(Z \; I\right) \bar{B}_*^T = \operatorname{Im} \bar{B} \begin{pmatrix} -I \\ Z \end{pmatrix} \frac{N^{-1}}{\sqrt{2}}. \tag{6.7}$$

where $N$ is the symmetric square root of $Z$ ($Z = NN$) which always exists under the hypothesis that $Z$ is symmetric and positive semi-definite (see Theorem 6.3.1). The last terms of Eq. (6.6) and Eq. (6.7) are used for normalization such that the columns of the matrices whose image is considered are orthonormal in the induced norms.

It is now possible to check whether $\mathcal{S}_Z^-$ and $\mathcal{S}_Z^+$ are orthogonal for the plus-pairing. Using the image representations, it is possible to see that this is the case iff the following matrix is identically zero:

$$\frac{1}{2} N^{-T} \left(-I \; Z^T\right) \begin{pmatrix} 0 & I \\ I & 0 \end{pmatrix} \begin{pmatrix} Z^{-1} \\ I \end{pmatrix} N = \frac{1}{2} N^{-T} (Z^T Z^{-1} - I) N \tag{6.8}$$

which is the case iff the tensor $Z$ is symmetric. We will shortly see that this condition is also essential to achieve the power decomposition which is fundamental for the scattering representation.

By restricting the +pairing to $\mathcal{S}_Z^+$ we obtain an inner product on $\mathcal{S}_Z^+$, and by restricting it to $\mathcal{S}_Z^-$ we obtain an inner product on $\mathcal{S}_Z^-$. Once again, using the image representation of $\mathcal{S}_Z^+$, it is possible to see that the induced inner product on $\mathcal{S}_Z^+$ using as base of $\mathcal{S}_Z^+$ the columns of

$$S_Z^+ := \bar{B} \begin{pmatrix} Z^{-1} \\ I \end{pmatrix} \frac{N}{\sqrt{2}} \tag{6.9}$$

[1] The author would like to thank Johan Hamberg of the Swedish Defence Research Agency for pointing out a problem concerning the unicity of the scattering subspaces.

is:

$$\frac{1}{2}N^T \begin{pmatrix} Z^{-T} & I \end{pmatrix} \begin{pmatrix} 0 & I \\ I & 0 \end{pmatrix} \begin{pmatrix} Z^{-1} \\ I \end{pmatrix} N$$
$$= N(Z^{-T} + Z^{-1})N = N\,\mathrm{sym}(Z^{-1})N = I \quad (6.10)$$

This implies that the chosen base $S_Z^+$ is orthonormal in the induced inner product. Similarly, using the image representation of $\mathcal{S}_Z^-$, it is possible to see that the induced inner product on $\mathcal{S}_Z^-$ using as base for $\mathcal{S}_Z^-$ the columns of

$$S_Z^- := \bar{B}\begin{pmatrix} -I \\ Z \end{pmatrix}\frac{N^{-1}}{\sqrt{2}} \quad (6.11)$$

is:

$$\frac{1}{2}N^{-T} \begin{pmatrix} -I & Z^T \end{pmatrix} \begin{pmatrix} 0 & I \\ I & 0 \end{pmatrix} \begin{pmatrix} -I \\ Z \end{pmatrix} N^{-1} =$$
$$-N^{-1}(Z + Z^T)N^{-1} = -N^{-1}\,\mathrm{sym}(Z)N^{-1} = -I \quad (6.12)$$

Since the previous matrix is $-I$, we will use as induced product, minus the result of Eq. (6.12). Once again, this shows that the chosen base is orthonormal for the induced inner product. It is now possible to present the most important theorem regarding geometric scattering which was also presented in [88, 90] under less general conditions.

**Theorem 6.1.1 (Scattering Power Decomposition).** *Given any $(f, e) \in \mathcal{D}$ and any positive definite, symmetric, two covariant tensor $Z$, the following relation holds:*

$$\langle e, f\rangle = \frac{1}{2}||s_Z^+||_+^2 - \frac{1}{2}||s_Z^-||_-^2$$

*where $s_Z^+ \in \mathcal{S}_Z^+$, $s_Z^- \in \mathcal{S}_Z^-$, $(f, e) = s_Z^+ + s_Z^-$ and $||\ ||_+$ and $||\ ||_-$ are respectively the induced inner products on $\mathcal{S}_Z^+$ and $\mathcal{S}_Z^-$.*

**Proof.** Due to Eq. (6.5), we are not restricting ourself if we consider:

$$\begin{pmatrix} f \\ e \end{pmatrix} = \frac{\bar{B}}{\sqrt{2}}\left(\begin{pmatrix} Z^{-1} \\ I \end{pmatrix} N s_+ + \begin{pmatrix} -I \\ Z \end{pmatrix} N^{-1} s_-\right) \quad (6.13)$$

In this case, we have that:

$$\langle e, f\rangle = \frac{1}{2}(s_+^T s_+ - s_- s_- - s_+^T s_- + s_-^T s_+). \quad (6.14)$$

Which directly proves the result using Eq. (6.10) and the negative of Eq. (6.12). ∎

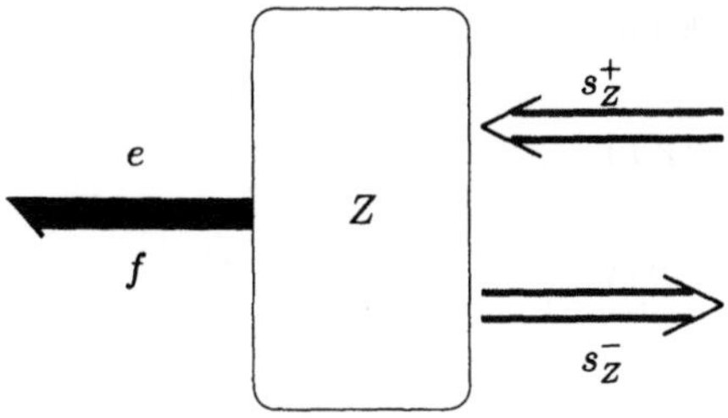


**Fig. 6.1.** *Scattering transformation.*

*Remark 6.1.1.* It is now evident why the scattering subspaces should be orthogonal using the +pairing: only under this condition we obtain the previous decomposition which is fundamental because it shows that we can algebraically write the power flow as the sum of a positive and negative power depending only on the two scattering variables. This can be interpreted as power going in opposite directions as shown in Figure 6.1 where it is shown in bond graph notation that the power bond has indeed the same direction as the variables $s_Z^+$ due to the accordance of sign. Intuitively $s_Z^+$ can be thought of as a wave transporting power in the direction of the bond and $s_Z^-$ in the opposite one.

### 6.1.1 Dimension of the space of scattering decompositions

Due to the fact that $Z$ should be symmetric, and due to their representations, it is possible to see that the dimension of the possible scattering decomposition is parameterized by the symmetric tensors $Z$ and it is therefore equal to $n(n+1)/2$.

### 6.1.2 Plus-product invariance

The set of all possible scattering sub-spaces could be also defined using the possible changes of coordinates which would leave the +pairing invariant, or equally the possible motions of subspaces which would leave the +pairing invariant.

### 6.1.3 Scattering decomposition

Using Eq. (6.13), in the chosen base, the mapping relating efforts and flows to scattering variables is:

$$\begin{cases} f = \dfrac{N^{-1}}{\sqrt{2}}(s_+ - s_-) \\ e = \dfrac{N}{\sqrt{2}}(s_+ + s_-) \end{cases} \tag{6.15}$$

and inverting the relations:

$$\begin{cases} s_+ = \dfrac{N^{-1}}{\sqrt{2}}(e + Zf) \\ s_- = \dfrac{N^{-1}}{\sqrt{2}}(e - Zf) \end{cases} \tag{6.16}$$

## 6.2 Causality and sign issues

It has been shown in [86] that in order to preserve passivity with a transmission line connecting two systems, the power port connected to the transmission line can be "coded and decoded" in scattering variables. The coded signal $s_Z^-$ can be sent on the line and it will be used by the other side as the incoming signal $s_Z^+$. The total energy stored on the line is therefore the integral of the traveling signal. Since the variable $s_Z^+$ is always an input for the two systems attached to the line, we have two causal possibilities:

- Computing $e$ and $s_Z^-$ as a function of $f$ and the incoming wave $s_Z^+$ .
- Computing $f$ and $s_Z^-$ as a function of $e$ and the incoming wave variable $s_Z^+$.

It has been shown in [87] that there are multiple reasons for choosing the last of the two options. Furthermore, if we want to have a perfectly symmetric system, the causalities at both sides should be the same. A first possibility, from a purely causal point of view, would be to let the line behave as a gyrative action in such a way that for the line length and delay tending to zero, the effort supplied by one side would become the input flow of the other and vice versa. Unfortunately such a system cannot work because in steady state, when the master and slave do not move, the velocities should be zero ($f = 0$), but at the same time, we want that a reflection of a force different from zero could take place if necessary (that is, $e \neq 0$). Since with a symplectic connection the two variables $f$ and $e$ are equal, this is not possible. This implies that the systems at both sides should have an *impedance* causality, and that the line in the limit of its length tending to zero should *not* behave as a gyrative action.

From this we conclude we have to choose exactly the same scattering mapping on both sides, and connect the departing wave of one side to the incoming wave of the other side. This has an important consequence: if the line length and its delay are tending to zero, then we get a causal inconsistency since 'the line' should supply the same power variable at both sides, namely the flow $f$. This would correspond to an algebraic loop with no delays and implies that the energy storage of a finite length line "fixes" the causal problem exactly as a mass would do to connect two physical systems through springs.

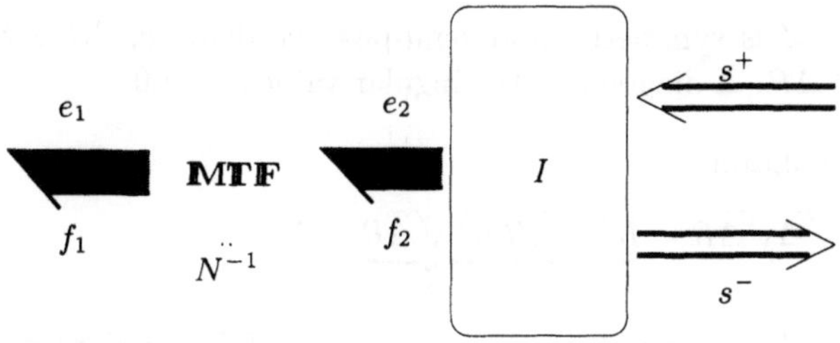


**Fig. 6.2.** *Scaling of impedance.*

## 6.3 Line Impedance adaptation

The case in which $Z = I$ corresponds to an impedance which maps the base $B$ to its intrinsic dual $B_*$. In this case and for the chosen coordinates, it is possible to see that $I$ is indeed the impedance felt by the system attached to the power port when we suppose no power coming from the line ($s^+ = 0$): looking at another causal form of Eq. (6.16) for $Z = I$

$$e = \sqrt{2}s^+ - f \rightarrow e = -f$$

which corresponds to an identity impedance in the chosen coordinates. The general impedance decomposition reported in Eq. (6.16) can also be interpreted in a different way using what is called in bond graphs a *transformer.* The resulting scheme is given in Figure 6.2. The equation characterizing a transformer with matrix transformation $N^{-1}$ are

$$\begin{aligned} f_1 &= N^{-1} f_2 \\ e_2 &= N^{-T} e_1 \end{aligned} \tag{6.17}$$

where in our case $N$ is a square, in general time-varying, non-singular matrix. The impedance seen at the $(e_1, f_1)$ port is the matrix $Z_1$ such that $e_1 = Z_1 f_1$ and substituting the transformer equations:

$$e_1 = N^T e_2 = N^T f_2 = (N^T N) f_1$$

which implies that $Z_1 = N^T N$. This is a trivial, well known result in network theory. $Z_1$ is a positive definite, symmetric, 2-contravariant tensor, and therefore:

$$P = e_1^T f_1 = f_1^T Z_1^T f_1 \geq 0 \tag{6.18}$$

A question arises: is it possible to find an **MTF** of Figure 6.2 such that the impedance seen from $(e_1, f_1)$ can get any symmetric desired value $Z$?
The answer is given by the following trivial linear algebra result:

**Theorem 6.3.1.** *Given any symmetric, positive semidefinite matrix $Z$ there exists always a symmetric matrix $N$ such that*

$$Z = N^T N = N^2$$

**Proof.** Since $Z$ is symmetric and semi-positive definite, $\exists R \in SO(n)$ such that $Z = R^T \Delta R$, $\Delta$ diagonal with singular values $\lambda_i \geq 0$. ■

Therefore we obtain

$$Z = R^T \sqrt{\Delta}\sqrt{\Delta} R = \underbrace{R^T \sqrt{\Delta} R}_{N} \underbrace{R^T \sqrt{\Delta} R}_{N} = N^2$$

With this result we can state that *all* meaningful impedances (symmetric and positive definite) $Z$ can be generated by a proper choice of a transformer $N^{-1}$.

This implies that since the identity scattering transformation can be expressed as found previously by

$$\begin{cases} s^+ = \dfrac{1}{\sqrt{2}}(e_2 + f_2) \\ s^- = \dfrac{1}{\sqrt{2}}(e_2 - f_2) \end{cases}$$

and using Eq. (6.17)

$$\begin{cases} s^+ = \dfrac{1}{\sqrt{2}}(N^{-1}e_1 + N f_1) \\ s^- = \dfrac{1}{\sqrt{2}}(N^{-1}e_1 - N f_1) \end{cases}$$

Eventually, we obtain the scattering transformation for a generic, multidimensional impedance $Z$ which corresponds indeed to Eq. (6.16):

$$\begin{cases} s^+ = \dfrac{N^{-1}}{\sqrt{2}}(e_1 + Z f_1) \\ s^- = \dfrac{N^{-1}}{\sqrt{2}}(e_1 - Z f_1) \end{cases} \tag{6.19}$$

As already seen, $Z$ is a fundamental parameter for the line, which characterizes the wave variables $s^+$, $s^-$, and directly effects the system behavior.

It is important to know that, in a real analog transmission line like a coaxial cable or a twisted pair, the impedance is obviously a physical characteristic of the line which we cannot influence. At the contrary, in a digital transmission line like the one considered in this chapter, only data are sent and the scattering mapping of Figure 6.2 corresponds to an algorithmic implementation which codes and decodes the sent and received data. Future work will formally analyse the correctness of this analogy.

## 6.4 Impedance Matching

Impedance matching is a well-known problem in transmission lines. The energy received from the line has to be absorbed by master and slave systems.

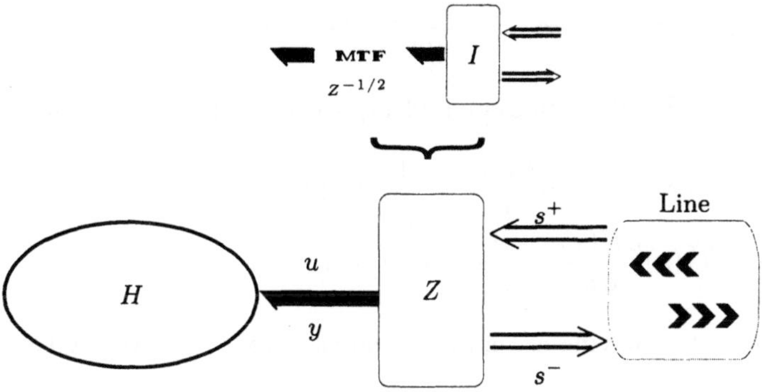


**Fig. 6.3.** *Power interconnection with an Hamiltonian system.*

Once the impedance $Z$ seen at the power port of Figure 6.3 is chosen (model of the line), a system with the "same impedance" needs to be connected at the end of the line to avoid waves reflections. This guarantees continuity of impedance with respect to the line.
A general system theoretic condition for matching of a general physical system connected to a line as in Figure 6.3 can now be stated as follows:

**Principle 6.4.1 (Matching Condition)** *The system seen at the scattering side of the transformation of Figure 6.3 and having $s^+$ as input and $s^-$ as output has to be of relative degree $\geq 1$ (that is, the system should have no direct feedthrough).*

This implies that there should not be an algebraic relation between the waves $s^+$ and $s^-$, which is exactly equivalent to the idea of undiscriminated reflection of power. In intuitive terms, the power should be first somehow "processed" by the master (resp. slave) before some information is sent back to the slave (resp. master).

Now, we want to investigate what conditions Principle 6.4.1 imposes on a generic PCHD system, as treated in Sect. 3.4.1, connected at the end of the line as in Figure 6.3. Since we consider port controlled generalized Hamiltonian systems (both master and slave sides), we have:

$$\begin{cases} \dot{x} = (J(x) - R(x))\dfrac{\partial H(x)}{\partial x} + G(x)f \\ e = G^T(x)\dfrac{\partial H(x)}{\partial x} \end{cases} \tag{6.20}$$

From the scattering transformation of Eq. (6.19), we can obtain the port variables as function of the wave variables:

$$s^+ + s^- = \sqrt{2}N^{-1}e \ \Rightarrow \ e = \frac{1}{\sqrt{2}}N(s^+ + s^-)$$

$$s^+ - s^- = \sqrt{2}Nf \;\Rightarrow\; f = \frac{1}{\sqrt{2}}N^{-1}(s^+ - s^-)$$

and thus the Hamiltonian system of Eq. (6.20) is transformed to:

$$\begin{cases} \dot{x} = (J(x) - R(x))\dfrac{\partial H(x)}{\partial x} + \dfrac{1}{\sqrt{2}}G(x)N^{-1}(s^+ - s^-) \\ e = G^T(x)\dfrac{\partial H(x)}{\partial x} = \dfrac{1}{\sqrt{2}}N(s^+ + s^-) \end{cases}$$

The new system having as input $s^+$ and as output $s^-$ is thus given as:

$$\begin{cases} \dot{x} = (J(x) - R(x) - G(x)N^{-1}N^{-1}G^T(x))\frac{\partial H(x)}{\partial x} + \\ \qquad\qquad \sqrt{2}N^{-1}G(x)s^+ \\ s^- = \sqrt{2}N^{-1}G^T(x)\frac{\partial H(x)}{\partial x} - s^+ \end{cases}$$

Hence we conclude that the input $s^+$ is directly fed through to the output $s^-$. This implies that any power arriving from the line is sent back independently of the state of the system connected to the line. Thus the Hamiltonian system of Eq. (6.20) does not satisfy Principle 6.4.1, and is not general enough for impedance matching.

Hence, in order to meet Principle 6.4.1, we have to enlarge the class of PCHD systems. We do this by considering PCHD systems of the extended form

$$\begin{cases} \dot{x} = (J(x) - R(x))\dfrac{\partial H(x)}{\partial x} + G(x)f \\ e = G^T(x)\dfrac{\partial H(x)}{\partial x} + B(x)f, \end{cases} \tag{6.21}$$

with $B(x) \geq 0$ a newly added dissipation matrix.
In this case we obtain, using Eq. (6.19) and Eq. (6.21), the new output equation:

$$s^- = \left(\frac{B(x)}{\sqrt{2}}N^{-1} + \frac{1}{\sqrt{2}}N\right)^{-1} G^T(x)\frac{\partial H(x)}{\partial x} + Fs^+ \tag{6.22}$$

where

$$F = \left(B(x)N^{-1} + N\right)^{-1}\left(B(x)N^{-1} - N\right)$$

which implies, using Principle 6.4.1, that to have impedance matching we must have:

$$B = NN = Z$$

since this implies $F = 0$.

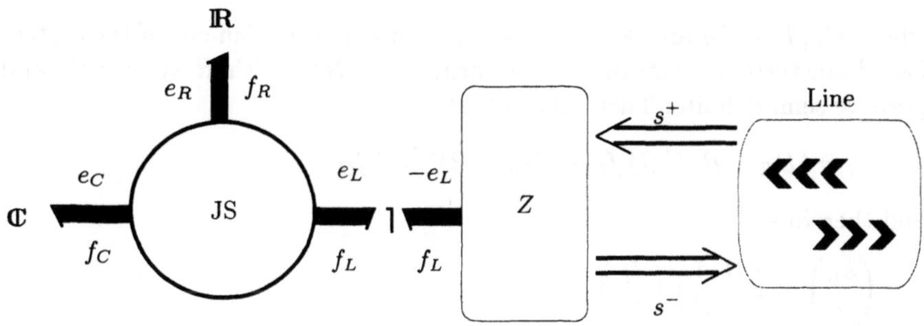


**Fig. 6.4.** The interconnection description

Thus, a system of the extended PCHD form of Eq. (6.21), for which $B(x)$ is equal to the impedance $Z$ of the line to which it is connected, guarantees the matching condition expressed in Principle 6.4.1, and eliminates any indiscriminate reflection of power.

Using the previous setting, it is even possible to give a measure for the level of matching if this is not perfect. This can be easily done considering the induced norm of $F$. From Eq. (6.22) it is possible to see that $F$ is a mapping of the following form:

$$F : \mathcal{S}^+ \rightarrow \mathcal{S}^- \; ; \; s^+ \mapsto Fs^+$$

and since $\mathcal{S}^+$ and $\mathcal{S}^-$ are normed spaces, we can define the following induced norm for $F$:

$$||F|| := \sup_{s \neq 0} \frac{||Fs||_-}{||s||_+}$$

If this norm is zero, perfect matching is obtained, if this norm is 1 perfect reflection takes place.

### 6.4.1 An interconnection approach

The previous conclusion can be obtained similarly using the framework of Dirac structures. For simplicity, we consider a network structure with reference to Figure 6.4 relating three ports: the power port connected to the line $(f_L, e_L)$, the power port connected to a dissipating component $(f_R, e_R)$ and a power port connected to a storage element $(f_C, e_C)$. The network structure can be expressed by a linear mapping of the form:

$$\begin{pmatrix} e_L \\ e_C \\ e_R \end{pmatrix} = \begin{pmatrix} D_L & G_1 & G_2 \\ -G_1^T & D_C & G_3 \\ -G_2^T & -G_3^T & D_R \end{pmatrix} \begin{pmatrix} f_L \\ f_C \\ f_R \end{pmatrix}$$

where $D_L, D_C, D_R$ are skew-symmetric. A dissipating element of the system has characteristic equations of the form $e_R = Rf_R$ with $R$ symmetric and positive semi-definite. This implies that

$$f_R = (D_R - R)^{-1}G_2^T f_L + (D_R - R)^{-1}G_3^T f_C$$

and therefore

$$\begin{pmatrix} e_L \\ e_C \end{pmatrix} = \begin{pmatrix} B & A \\ C & D \end{pmatrix} \begin{pmatrix} f_L \\ f_C \end{pmatrix}$$

where:

$$B := D_L + G_2(D_R - R)^{-1}G_2^T \tag{6.23}$$

$$A := G_1 + G_2(D_R - R)^{-1}G_3^T \tag{6.24}$$

$$C := -G_1^T + G_3(D_R - R)^{-1}G_2^T \tag{6.25}$$

$$D := D_C + G_3(D_r - R)^{-1}G_3^T \tag{6.26}$$

Applying the scattering transformation to the power port $(f_L, -e_L)$ it is possible to obtain:

$$s^- = -(N + N^{-1}B)(N - N^{-1}B)^{-1}s^+ + Kf_c$$

where

$$K = -\frac{1}{\sqrt{2}}(N + N^{-1}B)(N - N^{-1}B)^{-1}N^{-1}A + N^{-1}A.$$

This implies for Principle 6.4.1 that for adaptation we need to have $(N + N^{-1}B) = 0$ and therefore $B = -Z$ which implies:

$$D_L + G_2(D_R - R)^{-1}G_2^T = -Z$$

with $Z$ symmetric. This implies that necessarily $D_L = 0$ and furthermore, if we suppose $G_2$ to be square and non-singular, that also $D_R = 0$, implying that:

$$R^{-1} = G_2^{-1}ZG_2^{-T}.$$

From the previous analysis, we can conclude that the adaptation is independent of the state of the system and only depending on the system interconnection and its dissipative term.

## 6.5 Spatial telemanipulation

The presented theory can be used to passively implement spatial telemanipulation. With this is meant that the developed theory is well posed in a coordinate free setting and therefore it is possible to choose, for example

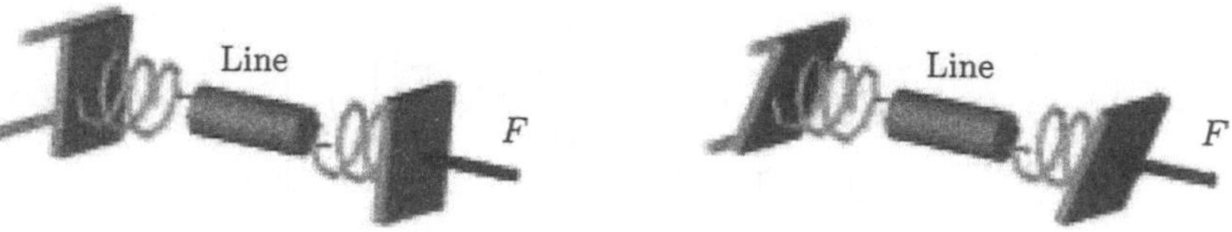


**Fig. 6.5.** A spatial telemanipulation example

$$V = se(3) \times \ldots \times se(3).$$

In this case the transmitted power variables will be a set of twists and their dual wrenches. To keep decoupled twists during transmission, the chosen line impedance should be of the form:

$$Z = \begin{pmatrix} Z_1 & 0 & \ldots & 0 \\ 0 & \ddots & \ddots & \vdots \\ \vdots & \ddots & \ddots & 0 \\ 0 & \ldots & 0 & Z_n \end{pmatrix}.$$

This can be used in complex telemanipulation systems using IPC techniques like the ones presented in Chapter 5 in which variations of geometric spring's length is controlled by a twist, an element in $se(3)$. The importance of the presented theory becomes relevant in these kind of applications.

### 6.5.1 A three dimensional simulation

To test the geometric scattering, a first simple three dimensional simulation has been implemented and reported in Figure 6.5. The two sides are composed of only a massive block, a spatial spring and some additional damping. The spring is connected on one side to the block and on the other side to the line through geometric scatterisation of $se(3)$. On one side a linear constant force $F$ in the middle of the block has been applied. As a consequence of this, both blocks started moving until the block on the other side touched some obstacles represented by the two bars on the left side. Due to this contact, the block on the left side tilted and this motion is passed to the side where the force is applied resulting in the configuration on the right side of Figure 6.5.

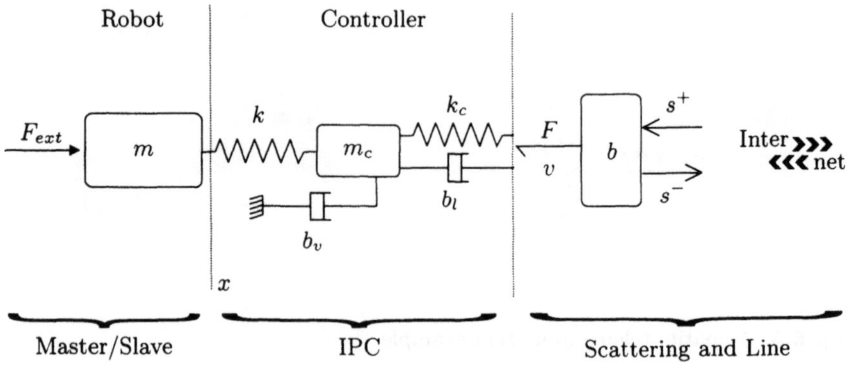


**Fig. 6.6.** One dimensional teleoperator.

## 6.6 Damping injection in telemanipulation

It is now possible to use the developed theory for an illustrative case. Let's consider a one dimensional "robot" composed of a single mass. It was shown in Sect. 4.5 that to control its interactive behavior by using only position measurements, it is possible to use the IPC reported in Figure 6.6 where the supervision power port $(F, v)$ was there connected to a supervisory controller. For a telemanipulation set-up, instead of connecting that port to a supervision controller, we can consider two identical systems and connect their supervision ports to the two extreme of a communication line as schematically shown in Figure 6.7.

Let's refer now to the one-dimensional teleoperator scheme shown in Figure 6.6. In the scalar case we have $Z = b \in \mathbb{R}^+$ and from Eq. (6.19) we obtain the usual scattering transformation with *force* and *velocity* as dual variables for a mechanical Hamiltonian system:

$$\begin{cases} s^+ = \dfrac{1}{\sqrt{2b}}(F + bv) \\ s^- = \dfrac{1}{\sqrt{2b}}(F - bv) \end{cases} \tag{6.27}$$

Considering Figure 6.6, the Hamiltonian system of Figure 6.3 can be now represented by the robot

$$\begin{aligned} \begin{pmatrix} \dot{x} \\ \dot{p} \end{pmatrix} &= \begin{pmatrix} 0 & 1 \\ -1 & 0 \end{pmatrix} \begin{pmatrix} 0 \\ p/m \end{pmatrix} + \begin{pmatrix} 0 & 0 \\ 1 & 1 \end{pmatrix} \begin{pmatrix} F_{ext} \\ F_c \end{pmatrix} \\ \begin{pmatrix} v_{ext} \\ v_c \end{pmatrix} &= \begin{pmatrix} 0 & 1 \\ 0 & 1 \end{pmatrix} \begin{pmatrix} 0 \\ p/m \end{pmatrix} \end{aligned} \tag{6.28}$$

and the controller

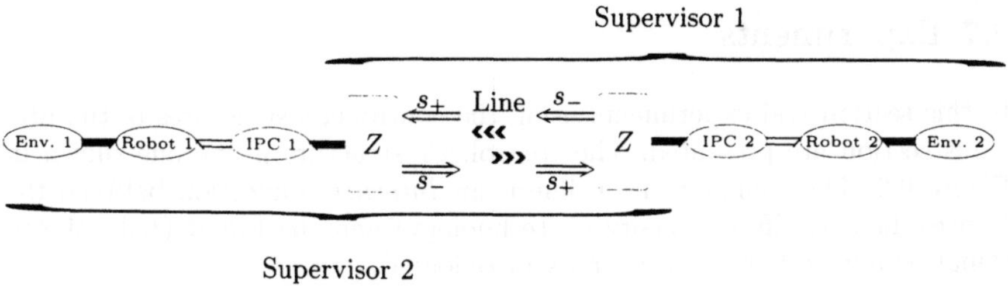


**Fig. 6.7.** The IPC-supervisor telemanipulation setting

$$\begin{pmatrix} \dot{\Delta x_c} \\ \dot{p_c} \\ \dot{\Delta x} \end{pmatrix} = \begin{pmatrix} 0 & 1 & 0 \\ -1 & (-b_v - b_l) & 0 \\ 0 & -1 & 0 \end{pmatrix} \begin{pmatrix} k_c \Delta x_c \\ p_c/m_c \\ k\Delta x \end{pmatrix} + \begin{pmatrix} -1 & 0 \\ b_l & 0 \\ 0 & 1 \end{pmatrix} \begin{pmatrix} v \\ v_c \end{pmatrix} \tag{6.29}$$

$$\begin{pmatrix} F \\ -F_c \end{pmatrix} = \begin{pmatrix} -1 & b_l & 0 \\ 0 & 0 & 1 \end{pmatrix} \begin{pmatrix} k_c \Delta x_c \\ p_c/m_c \\ k\Delta x_c \end{pmatrix} - \begin{pmatrix} b_l & 0 \\ 0 & 0 \end{pmatrix} \begin{pmatrix} v \\ v_c \end{pmatrix}$$

where $\Delta x$, $\Delta x_c$ are the displacement of the springs $k$ and $k_c$ respectively, $p$ and $p_c$ the masses of the robot and controller, $b_v$ is the *viscous* friction of the controller [1] and $b_l$ is the damper related to the *line* that we have to choose in order to obtain the matching condition of Principle 6.4.1, and which corresponds to $B(x)$ in Eq. (6.21).

It can be seen that Eq. (6.29) is of the same form as Eq. (6.21), with in this case the Hamiltonian function given as:

$$H_c(\Delta x, p_c, \Delta x_c) = \frac{1}{2}\left(\frac{p_c^2}{m_c} + k\Delta x^2 + k_c {\Delta x_c}^2\right)$$

Moreover, we have the two power ports $(F, v)$ and $(-F_r, \dot{x})$ where the first one is used to connect to the transmission line and the second to connect to the robot with energy function

$$H_r(p) = \frac{1}{2m} p^2$$

Hence, by choosing $b_l = b$, there will not be any reflected wave from master to slave system, and vice-versa. Hereafter experimental results are shown to explain the behavior of a real master-slave system using this physical controller.

## 6.7 Experiments

In this section real experiments using the controller's structures of the previous section are presented. The controller's structure is the one shown in Figure 6.6. The communication line is an *Internet* connection between the Control-Lab. (Delft University of Technology) and the L.A.R (Lab. of Automation and Robotics - University of Bologna).
In the sequel the laboratory set-up is briefly explained together with interesting results concerning scattering theory and passivity involved in the controller design.

### 6.7.1 Laboratory Set-Up

The laboratory set-up is a symmetric, bilateral, master-slave system, as shown in Figure 6.8. Each side is equipped with a 1-dof manipulator actuated by a DC-motor. Several sensors are available with this set-up:

- encoders for master and slave position measurement;
- tachometer for joint velocity measurement (not used);
- strain gauges for force measurement (not used);
- LEM for current measurements to implement torque servos.

It can be shown [73] that to implement the control scheme of Figure 6.6 it is not necessary to use neither velocity nor force measurements and this is simple and cost effective.

Each master and slave IPC has been implemented in LabView. This package allows to modify the "physical" control parameters (*mass, stiffnesses, damping* ) at run time for testing.

The time-delay between master and slave is due to the Internet connection between the laboratories of Delft and Bologna: the two physical set-ups are both in Delft, but all the scattering signals are sent to Bologna and pinged back before reaching the other side.

LabView offers a simple and efficient way to implement *Client-Server* application. There are two possible communication protocols:

- **TCP-IP protocol**
  TCP/IP enables communication over single networks or multiple interconnected networks (e.g. Internet).
  TCP (Transmission Control Protocol) is a connection-based protocol, which means that sites must establish a connection before transferring data. From a logical point of view, this protocol defines a point-to-point connection, with reliability features.
- **UDP protocol**
  UDP (User Datagram Protocol) is a connectionless protocol. This means that a connection doesn't need to be established with a destination before

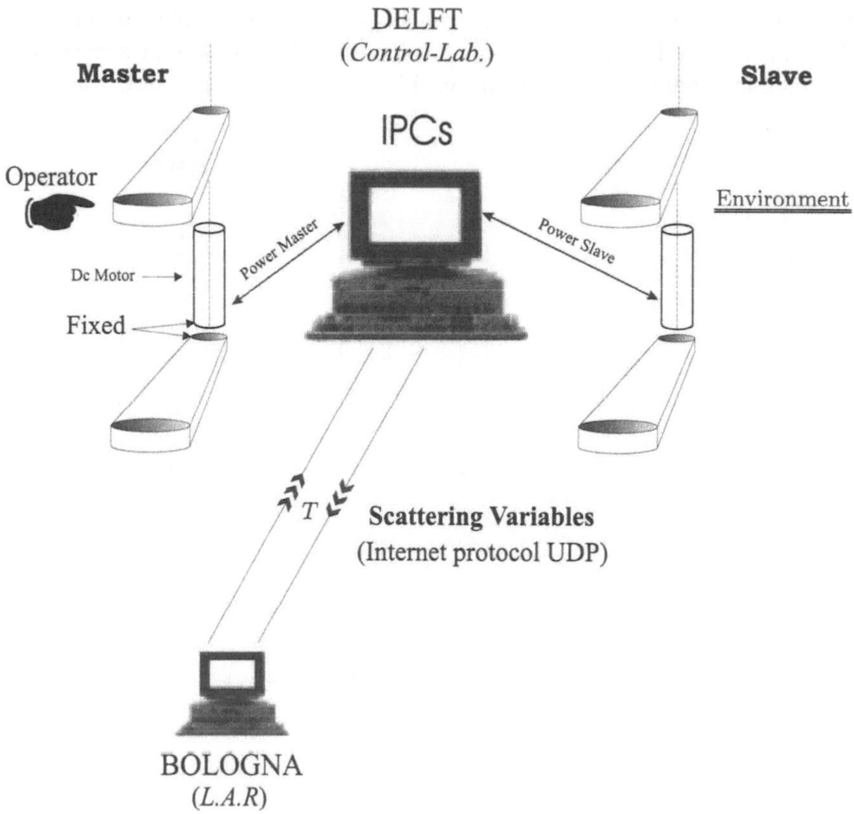


**Fig. 6.8.** Laboratory set-up.

sending or receiving data. Instead, the destination for the data is specified with each datagram sent.
This system offers an easy way to handle data transmission, without any report of errors.

Even if TCP/IP is a reliable way to send/receive data, UDP is more suitable to test the robustness of the algorithm.

### 6.7.2 Case studies

As explained above, we have a time-delay due to the Internet connection between the two system (see Figure 6.8).
The frames sent from each side to Bologna represent the $s^+$ waves. A process running on a machine in Bologna just executes a simple "ping-back" of the received frames so that we can obtain in Delft the delayed $s^-$ waves from the master and slave sides. The estimated time-delay is (about) 200 $msec$, but it is strongly time varying.

**Free motion.** Hereafter some experiments concerning the free motion of the Slave are shown. In this case the operator exerts a force on the master system in order to modify the slave position. Figure 6.9 shows the power variables

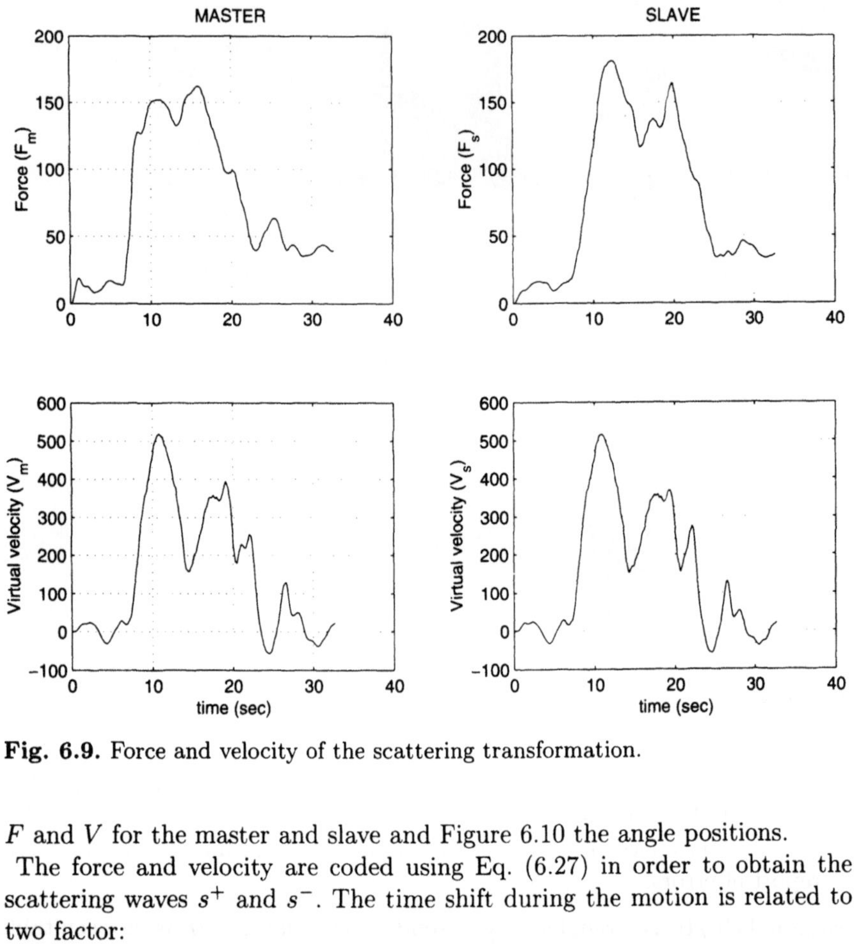


**Fig. 6.9.** Force and velocity of the scattering transformation.

$F$ and $V$ for the master and slave and Figure 6.10 the angle positions.

The force and velocity are coded using Eq. (6.27) in order to obtain the scattering waves $s^+$ and $s^-$. The time shift during the motion is related to two factor:

- Friction. Each side is characterized by a significant mechanical friction on the motor axis which is at this stage not compensated. This causes a stick-slip effect.
- IPC. The "mechanical" IPC has an intrinsic phase delay due the mass-spring-damper dynamic, but this is needed to obtain an intrinsically passive control of interaction.

Figure 6.11 shows the most significant variables of the experiment: $s^+$ and $s^-$.

Note that the system starts to exchange a significant amount of energy through the line as soon as the master starts its movement. After reaching

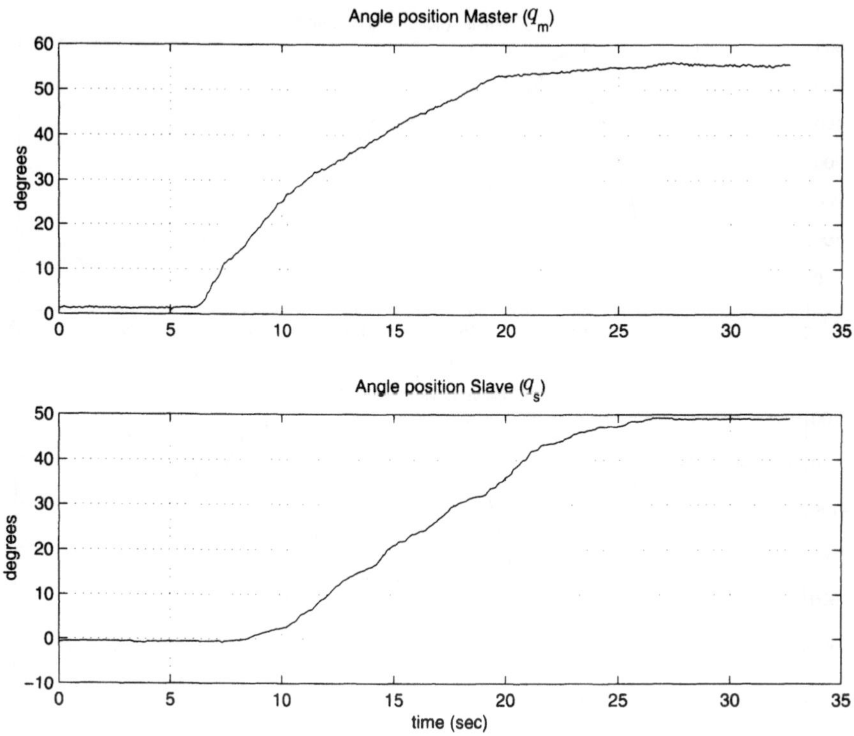


**Fig. 6.10.** Angle positions of the master and the slave.

the final position, the energy is dissipated by the control and no reflection occurs because of the impedance matching.

Figure 6.12 is very interesting, because it shows the power on the master and slave during task execution.
We can see several peaks due to the stick slip behavior, and the mentioned dissipation of energy (asymptotic stability).

**Interaction with the environment.** To verify the behavior of the system during interaction with a passive environment, we have forced the slave to bump against an obstacle after few seconds of free motion. The first plot of Figure 6.13 shows the positions of the master and slave systems during this task. The slave hits an obstacle after 15 degrees of free motion, while the master is been moved from the operator, because of the system's compliance. It's interesting to note what happens to the forces.
During free motion, the force is related to the dynamics of the controller and the robot while during contact the force on the slave increases and it is perfectly reflected to master site.
The overall task is characterized by a stable behavior. Figure 6.14 reports

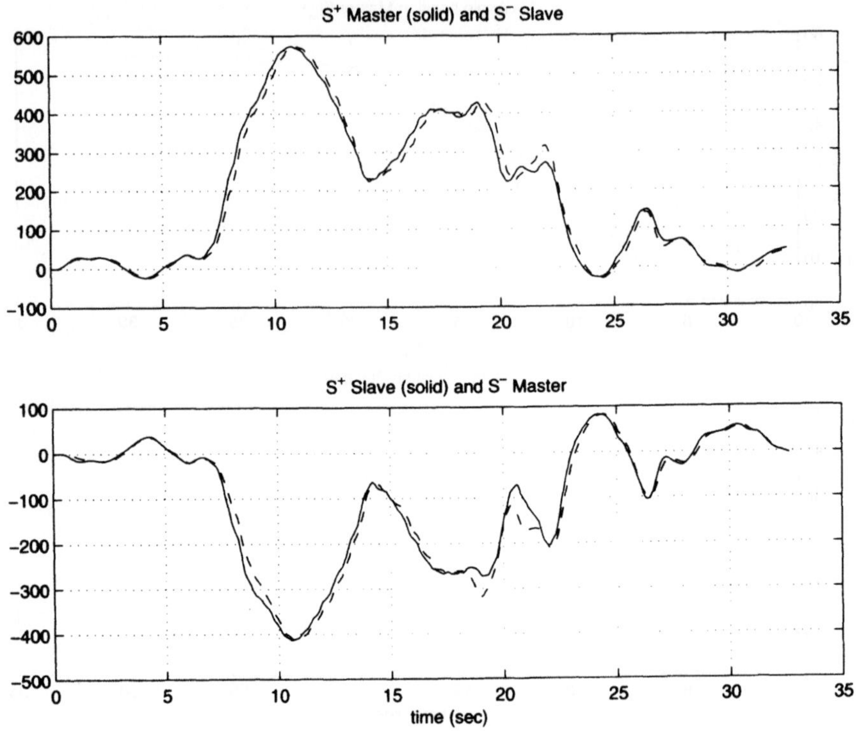


**Fig. 6.11.** Scattering variables (wave variables) flowing through the net.

three plots concerning the power waves $s^+$, $s^-$ and the power transmitted through the network.

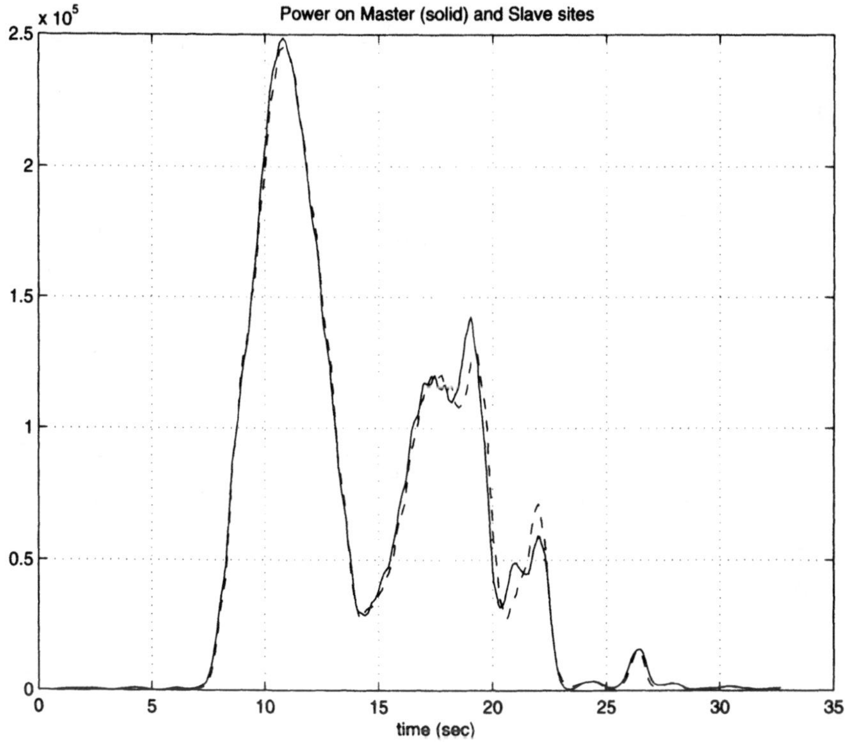


**Fig. 6.12.** Power exchanged in the system.

**Matching condition.** As seen in Sect. 6.4, the matching condition for our one-dimensional case is $b_l = b$, for each master and slave controllers.
Now we'll show what happens if a mismatch occurs, i.e. there is a little shift of either master or slave impedance. In Figure 6.15 a mismatch is forced in the master side.
Note that even if this mismatch is only of 8%, it is enough to cause the system to start swinging.
Indeed we have that the reflected waves bounce back and forth at each side.

## 6.8 Conclusions

In this chapter a general setting for telemanipulation of Port Control Hamiltonian systems has been presented. A new system theoretic condition has been introduced which can be used to test if proper matching is taking place. A possible measure of matching has been also introduced.

It has been shown that the standard form of explicit port controlled Hamiltonian systems is not general enough to obtain matching and it must

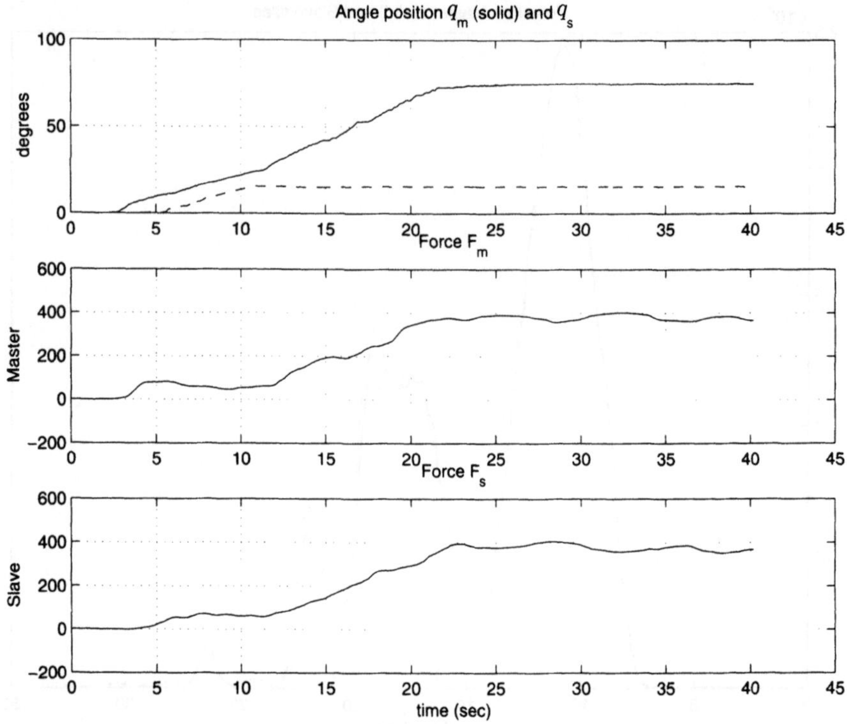


**Fig. 6.13.** Motion of master and slave systems during an interaction task.

be extended by a feed-throw term. This can be shown more generally using directly a network structure as shown in Sect. 6.4.1.

The presented theory is important for the implementation of geometrical telemanipulation where the vector space used $se(3)$ does not have an internal product.

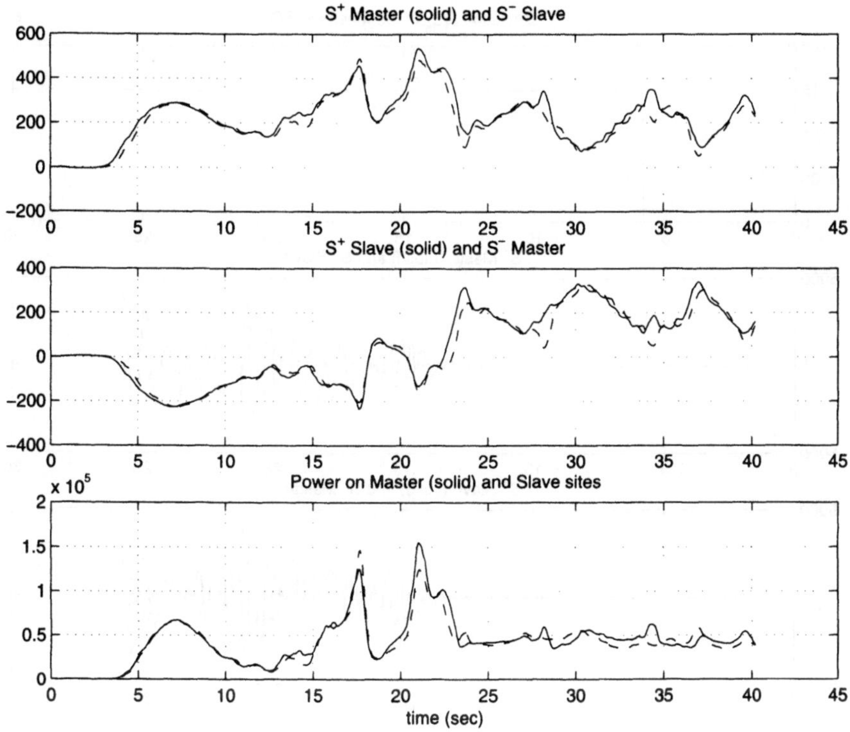


**Fig. 6.14.** Scattering variables and total power exchanged through the communication line (Internet).

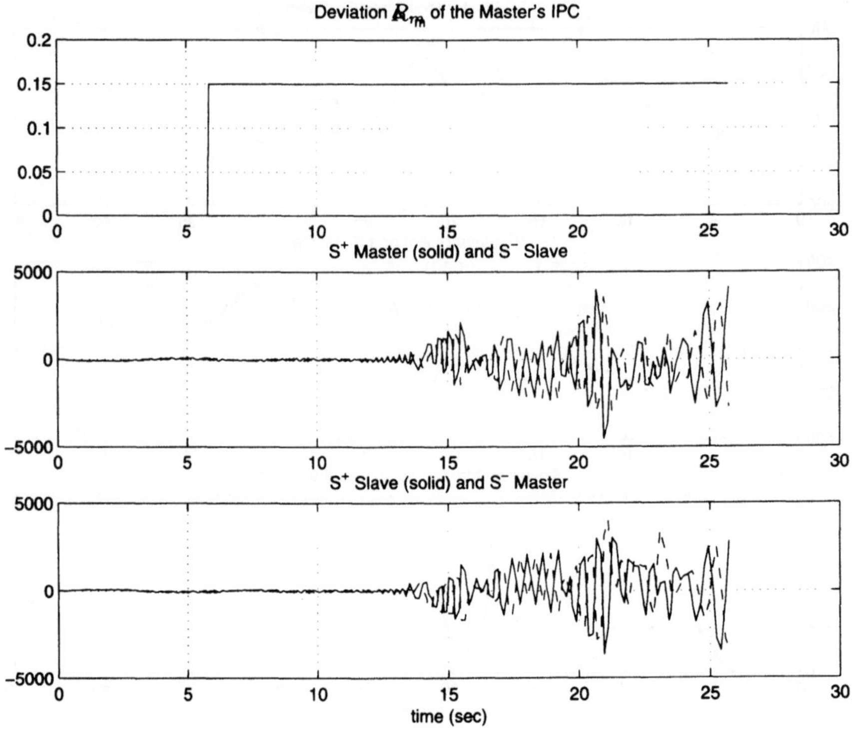


**Fig. 6.15.** Power reflection (oscillating behavior) due to the impedance mismatching at the master side.

# A. Mathematical Background

This appendix should supply sufficient background information to understand the material treated in this thesis. Notions of graph theory, vector spaces, groups, topology, Lie groups and algebras, differential geometry and tensors are introduced here. A lot of concepts will be defined so that any reader has at least a reference for most terms used in this work. Some arguments, like differential geometry, will be given more attention.

The importance of mathematics in engineering studies is unfortunately often underestimated. Complex mechanical mechanisms cannot be analyzed without proper analytical tools. Supplying methods for the analysis of complex systems is the goal of applied mathematics. In this appendix, the concepts used in this work will be presented to allow readers not familiar with them to follow the material. For each of these sections, detailed references will be given where interested readers can find more extensive information on each subject. It is important to stress that the goal of this appendix is NOT to give precise formal definitions and theorems, but to illustrate all the necessary notions intuitively.

## A.1 Basic concepts

This section deals with basic concepts on mappings between sets. Basic notions of set theory are taken for granted; these can be found in any basic book of analysis.

**Definition A.1.1 (Mappings).** *Given two sets $A$ and $B$, we call a map $f$ from $A$ to $B$ and we denote it as:*

$$f : A \to B \; ; \; a \mapsto f(a)$$

*an operation which associates to each element $a \in A$ a unique element $f(a) \in B$. The set $A$ is called the domain of the map and $B$ its co-domain.*

The set of mappings from $A$ to $B$ is usually indicated with $B^A$. We can therefore equivalently write $f \in B^A$.

For this reason, the set of subsets of a set $A$ is often indicated as $2^A$. If for example we say $C \in 2^A$, $C$ is a subset of $A$ and $C \in 2^A$ can be interpreted as a mapping from $A$ to a set of "2" elements. One element in "2" means "belongs to $C$" and the other element means "does NOT belong to $C$".

**Definition A.1.2 (Image).** *The image of a set $C \subset A$ through $f$ is a subset $D \subset B$ for which an element $a \in C$ exists such that $f(a) \in D$. The image of $C$ through $f$ is indicated with $f(C)$. Formally this set can be defined as:*

$$f(C) := \{b \in B;\ s.t.\ \exists a \in C,\ f(a) \in D\}.$$

The image of the full domain $A$ is called the *range* of $f$.

**Definition A.1.3 (Pre-image).** *Given a map $f$ of the previous form, we call*

$$f^{-1}(D) := \{a \in A;\ s.t.\ f(a) \in D\}$$

*the pre-image of $D$ through $f$.*

**Definition A.1.4 (Injective, Surjective, Bijective).** *A map $f : A \to B$ is said to be surjective iff $f(A) = B$, injective if the pre-image of a single element $b \in B$ is a single element in $A$ and bijective iff it is both injective and surjective.*

Surjective means that there is at least one element in the domain which maps through $f$ to any given element in the co-domain. Instead of saying that the map is surjective, it is also often common to say that the map is "onto". Injective means that there is ONLY ONE element which maps through $f$ to any given element in the co-domain. It is also often said that an injective map is "one to one". Bijective means, therefore, that there is a one-to-one relation between the domain and co-domain. It is often said that the map is "one to one and onto". For a bijective $f$, its pre-image corresponds to the usual notion of the inverse of a function.

**Definition A.1.5 (Diffeomorphism).** *A map $\phi : \mathbb{R}^n \to \mathbb{R}^n$ is called a diffeomorphism iff it is bijective and $\phi$ and $\phi^{-1}$ are differentiable.*

**Definition A.1.6 (Mapping composition).** *Given two mappings $f : A \to B$ and $g : B \to C$, there is a unique mapping called the composition of $f$ and $g$ and indicated with $(g \circ f) : A \to C$ defined as:*

$$g \circ f : A \to C\ ;\ a \mapsto g(f(a)).$$

**Definition A.1.7 (Isomorphic sets).** *Two sets $A$ and $B$ are said to be isomorphic if there is a bijective map $f \in B^A$.*

If $A$ and $B$ have a finite number of elements they are isomorphic iff they have the same number of elements. If $A$ and $B$ have an infinite number of elements this is more complex. Such a mapping is called an *isomorphism* between sets. A bijective mapping from a set to itself is called an *endomorphism*.

**Definition A.1.8 (Equivalence relation).** *Given a set $A$, an equivalence relation "$\sim$" on $A$ is a binary relation between elements of $A$, which we indicate with $a \sim b$ such that $\forall a, b, c \in A$:*

1. *Reflexivity: $a \sim a$*
2. *Symmetry: $a \sim b \Leftrightarrow b \sim a$*
3. *Transitivity: $a \sim b$ and $b \sim c \Rightarrow a \sim c$*

An *equivalence class* is one of the largest subset of equivalent elements of $A$. The equivalence class containing $a$, denoted $[a]$, is defined by $[a] := \{b \in A\ ;\ a \sim b\}$. This definition makes clear that if $a \sim b$, $[a] = [b]$.

**Definition A.1.9 (Quotient set).** *We call the set of equivalence classes of $A$ and we indicate it with $A/\sim$, the quotient set of $A$ respect to the equivalence relation $\sim$.*

## A.2 Graph theory

Graph theory studies the properties of graphs. In Chapter 2, graph theory is used to a great extent. The most important concepts are reported here. For a detailed treatment the reader is addressed to [25].

Before defining a graph, we recall that, given two sets $V, W$, with the notation $V \times W$, we indicate the Cartesian product of $V$ and $W$, which corresponds to the set of pairs $(v, w)$, with $v \in V$ and $w \in W$. In the case of a graph, we consider $V \times V$, which is the set of pairs of elements of $V$.

**Definition A.2.1 (Oriented Graph).** *An oriented graph $G$ is a pair $(V, E)$ where $V$ is a set called vertices set and $E \subset V \times V$ is a set called edges set. This set is formed of ordered pairs of elements of $V$. An element $(v_i, v_j) \in E$, with $v_i, v_j \in V$, is called an edge joining $v_i$ and $v_j$.*

We can represent a graph with a set of points called vertices and a set of edges connecting those points: the edge $(v_i, v_j)$ is represented by an arrow going from the vertex $v_i$ to the vertex $v_j$. Graph theory studies the properties of such a graph.

**Definition A.2.2 (Adjacent elements).** *If $(v_i, v_j) \in E$ we say that $v_i$ and $v_j$ are adjacent vertices. If $e_n = (v_i, v_j)$, $e_m = (v_k, v_l)$ and $\{v_i, v_j\} \cap \{v_k, v_l\} \neq \emptyset$, we say that the edges $e_n$ and $e_m$ are adjacent. For an edge $e = (v_i, v_j)$ we say that $e$ is adjacent to $v_i$, $v_j$ and vice versa.*

Adjacent means therefore that somehow the vertices are close to each other.

**Definition A.2.3 (Incidence).** *For an element $e = (v_i, v_j) \in E$, we say that $e$ is incident to $v_i$ and to $v_j$.*

An edge is therefore incident to the two vertices it is connecting in the drawing of the graph.

**Definition A.2.4 (Vertex degree).** *The number of incident edges to a vertex $v$ is called the degree of the vertex $v$.*

**Definition A.2.5 (Edge sequence).** *We call an edge sequence*[1] *of length $n$ for a graph $G = (V, E)$, an ordered list $L = [v_0, e_1, v_1, e_2, \ldots, e_n, v_n]$ for which every $e_i = (v_{i-1}, v_i) \in E$ or $e_i = (v_i, v_{i-1}) \in E$. We say that $L$ is a sequence connecting $v_0$ to $v_n$.*

**Definition A.2.6 (Connected graph).** *A graph $G = (V, E)$ such that, for any two vertices $v_1, v_2 \in V$, there is an edge sequence connecting $v_1$ and $v_2$, is called completely connected.*

Having a connected graph means therefore that, starting from any vertex, we can follow a series of edges in their forward or backward direction to arrive at any other vertex of the graph.

**Definition A.2.7 (Path).** *An edge sequence is called a path iff all the vertices along the sequence are touched only once.*

A path is therefore a kind of "minimal" edge sequence.

**Definition A.2.8 (Circuit).** *A sequence in which only the first and last vertices coincide is called a circuit.*

The last concept is used a great deal in circuit theory and corresponds to what is called a circuit mesh.

**Definition A.2.9 (Tree).** *A completely connected circuit-free graph is called a tree.*

Two factors therefore characterize a tree: 1) the possibility to go from any vertex to any other and 2) the nonexistence of any circuit in the tree.

**Definition A.2.10 (Spanning tree).** *Given a graph $G = (V, E)$, a graph $T(V^*, E^*)$ is called a spanning tree for $G$ iff $V = V^*$, $E^* \subset E$ and $T$ is completely connected and circuit free.*

A spanning tree for a graph $G$ is therefore a graph $T$ that resulted from deleting some edges of $G$. These deletions should not split the graph, to keep connectedness, and there should not be any circuits left in the simplified graph.

**Definition A.2.11 (Forest).** *A circuit-free graph, possibly not connected, is called a forest.*

The components of a forest are a set of disconnected trees.

**Definition A.2.12 (Spanning forest).** *Given a graph $G = (V, E)$, a graph $F(V^*, E^*)$ is called a spanning forest for $G$ iff $V = V^*$, $E^* \subset E$ and $F$ is circuit free.*

[1] Note that in this definition, the orientation of the edges is not important.

## A.3 Concepts from topology

A topological space is intuitively a set of elements together with some information which expresses the "vicinity" of points in this set. This vicinity is expressed by means of a set of sets called *open*, which has to satisfy certain properties. Interested readers can find a review of basic topology concepts in [34].

**Definition A.3.1 (Topological Space).** *A Topological Space is a pair* $(S, \mathcal{O})$ *where* $S$ *is a set and* $\mathcal{O} \subset 2^S$ *is a collection of subsets of* $S$ *called open sets such that the following properties are satisfied:*

1. $\emptyset \in \mathcal{O}$ *and* $S \in \mathcal{O}$
2. $O_1, O_2 \in \mathcal{O} \Rightarrow O_1 \bigcap O_2 \in \mathcal{O}$
3. $\bigcup_{O_i \in \mathcal{O}} O_i \in \mathcal{O}$

*where* $\emptyset$ *is the empty set.*

An example of a topological space is $S = \mathbb{R}^n$ where the set $\mathcal{O}$ can be the set of open balls around any point. Properly said, this topology is induced from the fact that $\mathbb{R}^n$ is a metric space, and we can use this to define the set of open sets.

**Definition A.3.2 (Continuous mapping).** *A mapping between two topological spaces is called continuous iff the pre-image of any open set of the co-domain topological space through* $f$ *is an open set in the domain topological space.*

**Definition A.3.3 (Homeomorphism).** *A bijective mapping* $f$ *between two topological spaces for which* $f$ *and* $f^{-1}$ *are continuous is called a homeomorphism. Two topological spaces for which there exist a homeomorphism are said to be homeomorphic.*

We see later that the concept of homeomorphism is very important for the definition of manifolds. We see that a characteristic of a manifold is that it is locally homeomorphic to $\mathbb{R}^n$, which implies that open sets of $\mathbb{R}^n$ correspond to open sets of the manifold. In this way, we transport the (nice) structure of $\mathbb{R}^n$ to more general entities called manifolds.

## A.4 Groups

Groups are very important for 3D mechanics. A group is a set closed under a composition operation defined for it. Furthermore, a group has a "special element" called the *identity* of the group. This element is such that its composition with any other element of the group results in the same element. For each element, there exist another element, called its *inverse*, such that the composition of the two elements is the identity. Before defining a group, we define an algebraic structure:

**Definition A.4.1 (Algebraic structure).** *An algebraic structure is a pair $(A, T)$ for which $A$ is a set and $T$ is an internal binary operator, which means $T$ is a map of the following form:*

$$T : A \times A \rightarrow A.$$

An algebraic structure can have various properties. The first one is the property of having a special element called the *identity.*

**Definition A.4.2 (Identity element).** *The identity element for an algebraic structure $(A, T)$ is an element $u \in A$ for which:*

$$T(a, u) = T(u, a) = a \ \ \forall a \in A.$$

It can be shown that if this element exists for an algebraic structure, this element is then unique.

**Definition A.4.3 (Associative property).** *An algebraic structure is said to satisfy the associative property iff:*

$$T(T(a, b), c) = T(a, T(b, c)) \ \ \forall a, b, c \in A.$$

**Definition A.4.4 (Commutative property).** *An algebraic structure is said to satisfy the commutative property iff:*

$$T(a, b) = T(b, a) \ \ \forall a, b \in A.$$

**Definition A.4.5 (Inverse element).** *Given an algebraic structure $(A, T)$ with identity element $u$ , we say that $a \in A$ has inverse iff $\exists b \in A$, which is denoted by $a^{-1}$ s.t. $T(a, a^{-1}) = T(a^{-1}, a) = u$.*

It can be shown that $a^{-1}$ is unique for an algebraic structure satisfying the associativity property and having an identity element for any element $a$ which has an inverse $a^{-1}$.

**Definition A.4.6 (Group).** *An algebraic structure $(A, T)$ is called a group if it satisfies the associative property, it has an identity element, and if for each element in $A$ there is an inverse. A group for which the commutative property is also satisfied is called a commutative or Abelian group.*

An example of a group is the set of nonsingular matrices with the operation of matrix multiplication. This is a non-Abelian group since matrix multiplication is not commutative.

**Definition A.4.7 (Subgroup).** *Given a group $(A, T)$, a subgroup of $(A, T)$ is a group $(B, T)$ for which $B \subset A$.*

Note that a subgroup should be a group by itself, which means that for any two elements $a, b \in B$, $T(a, b)$ should belong to $B$. Very often, the operation $T(,)$ is indicated with the operation of composition, for example when we consider the set of bijective mappings from one set to itself. This set with the operation of composition of mappings is a group. We denote in this case $T(a, b)$ as $a \circ b$. For example, the set of permutations of the elements of a finite set is a group of this kind called *group of permutations.*

We can then define when two groups are homeomorphic, which is different than the homomorphism of topological spaces.

**Definition A.4.8 (Groups homeomorphism).** *Two groups $(A, T_A)$ and $(B, T_B)$ are said to be homeomorphic iff there exist a mapping $f : A \rightarrow B$, called homeomorphism between groups, such that:*

$$T_B(f(a), f(b)) = T_A(a, b) \ \ \forall a, b \in A.$$

**Definition A.4.9 (Groups isomorphisms).** *Two groups are said to be isomorphic if they are homeomorphic and the homomorphism is bijective.*

Two groups which are isomorphic are exactly equivalent from an algebraic point of view: there is a one-to-one relation between the elements of the groups and, furthermore, we can use operations on any of the two groups. The result of the two operations on corresponding elements in the groups results in corresponding elements.

## A.5 Vector spaces

The concept of a vector space is very important and used very often in practice. Vector spaces can be defined on any field, such as real, or complex numbers. Here we just use real vector spaces.

**Definition A.5.1 (Vector Space).** *A real vector space $\mathcal{V}$ is a set whose elements are called vectors, with two operations + (addition) and $\cdot$ (scalar multiplication) such that:*

1. *$(\mathcal{V}, +)$ is an Abelian group.*
2. *The scalar multiplication is an operation of the following form:*

$$\cdot : \mathbb{R} \times \mathcal{V} \rightarrow \mathcal{V}$$

*with the following properties $\forall \alpha, \beta \in \mathbb{R}$ and $\forall x, y \in \mathcal{V}$.*

*a)* $\alpha \cdot (\beta \cdot x) = (\alpha \cdot \beta) \cdot x$
*b)* $1 \cdot x = x$
*c)* $(\alpha + \beta) \cdot x = (\alpha \cdot x) + (\beta \cdot x)$
*d)* $\alpha \cdot (x + y) = (\alpha \cdot x) + (\alpha \cdot y)$

An example of a vector space is the set of matrices of a certain dimension. The previous definition lists the property of a vector space. Each vector space has another vector space associated to it, the so-called 'dual vector space', which is the vector space of linear operators from $\mathcal{V}$ to $\mathbb{R}$. The dual space of $\mathcal{V}$ is denoted by $\mathcal{V}^*$. Intuitively, we know that if we consider the vector space of column vectors of $n$ numbers, its dual space must be the vector space of row vectors of $n$ numbers. The result of the application of a row vector to a column vector gives an element in $\mathbb{R}$. The two spaces, even if homeomorphic, are substantially different and this is very important for tensor calculus. The dual vector space of $\mathcal{V}$ is also called the space of 1-forms on $\mathcal{V}$. Before defining the dual space formally, we define a linear map from one vector space to another.

**Definition A.5.2 (Linear map).** *Given two vector spaces $\mathcal{V}$ and $\mathcal{W}$, we call a linear map from $\mathcal{V}$ to $\mathcal{W}$ a map $f$ of the form:*

$$f : \mathcal{V} \to \mathcal{W}$$

*such that:*

$$f(\sum_i \alpha_i v_i) = \sum_i \alpha_i f(v_i) \quad \forall v_i \in \mathcal{V},\ \alpha_i \in \mathbb{R}.$$

The set of all linear maps from $\mathcal{V}$ to $\mathcal{W}$ is indicated with $L(\mathcal{V}, \mathcal{W})$. Clearly $L(\mathcal{V}, \mathcal{W}) \subset \mathcal{W}^{\mathcal{V}}$. This definition also holds when $\mathcal{W} = \mathbb{R}$ since $\mathbb{R}$ also is a vector space. This is useful for what follows.

**Definition A.5.3 (Dual space).** *Given a vector space $\mathcal{V}$, its dual space $\mathcal{V}^*$ is the space of linear maps from $\mathcal{V}$ to $\mathbb{R}$:*

$$\mathcal{V}^* := L(\mathcal{V}, \mathbb{R}).$$

**Definition A.5.4 (Dual product).** *The duality product of $x \in \mathcal{V}$ and $x^* \in \mathcal{V}^*$, which is denoted by $\langle x, x^* \rangle \in \mathbb{R}$, is defined as follows:*

$$\langle , \rangle : \mathcal{V} \times \mathcal{V}^* \to \mathbb{R}\ ;\ (x, x^*) \mapsto x^*(x).$$

Note that the duality product is intrinsically defined for ANY vector space: it is NOT an extra structure which we associate to it !

**Definition A.5.5 (Adjoint map).** *Given any linear map $f : \mathcal{V} \to \mathcal{W}$, we can define a unique map of the form $f^* : \mathcal{W}^* \to \mathcal{V}^*$, called the adjoint map of $f$ and indicated with $f^*$, such that:*

$$\langle w^*, f(v) \rangle = \langle f^*(w^*), v \rangle \qquad \forall v \in \mathcal{V}, w^* \in \mathcal{W}^*. \tag{A.1}$$

**Definition A.5.6 (Linear combination).** *Given a vector space $\mathcal{V}$ and $n$ elements $v_1, \ldots, v_n \in \mathcal{V}$, we call a linear combination of $v_1, \ldots, v_n$ an entity of the following form belonging to $\mathcal{V}$:*

$$\alpha_1 \cdot v_1 + \ldots + \alpha_n \cdot v_n \text{ with } \alpha_1, \ldots, \alpha_n \in \mathbb{R}.$$

**Definition A.5.7 (Linear independence).** *Given n vectors* $v_1, \ldots, v_n \in \mathcal{V}$*, we say that they are linear independent iff:*

$$(\alpha_1 \cdot v_1 + \ldots + \alpha_n \cdot v_n = 0) \Rightarrow (\alpha_1 = \alpha_2 = \ldots = \alpha_n = 0).$$

This means that the vectors are linear independent if the only possible multiplying scalars $\alpha_i$, which result in a null linear combination, are all zero.

**Definition A.5.8 (Subspace).** *Given a vector space* $\mathcal{V}$*,* $\mathcal{A} \subset \mathcal{V}$ *is said to be a subspace of* $\mathcal{V}$ *when any linear combination of elements of* $\mathcal{A}$ *results in an element of* $\mathcal{A}$*:*

$$\sum_i \alpha_i v_i \in \mathcal{A} \quad \forall v_i \in \mathcal{A}, \alpha_i \in \mathbb{R}.$$

**Definition A.5.9 (Spanned subspace).** *Given n vectors* $v_1, \ldots, v_n \in \mathcal{V}$*, the set of all elements resulting from a linear combination of the vectors for any real scalars* $\alpha_i$ *is indicated as* $\mathrm{span}\{v_1, \ldots, v_n\}$.

The $\mathrm{span}\{v_1, \ldots, v_n\}$ is a vector space by itself, which is called a subspace of $\mathcal{V}$ generated by $v_1, \ldots, v_n$. A minimum number of linear independent vectors, whose spanned subspace coincides with the space itself, is called a *base* for the vector space. It can be shown that the number of base vectors of a base is independent of the base chosen. This number is called the *dimension* of the vector space. The importance of a base is that it is characteristic for the vector space and that any of its elements can be generated by a linear combination of elements of a base in a unique way. Eventually we define what is called the direct sum:

**Definition A.5.10 (Direct Sum).** *Given a vector space* $\mathcal{V}$*, we say that* $\mathcal{V}$ *is the direct sum of the subspaces* $\mathcal{A}$ *and* $\mathcal{B}$*, and we denote it with:*

$$\mathcal{V} = \mathcal{A} \oplus \mathcal{B}$$

*iff* $\mathcal{A} \cap \mathcal{B} = \emptyset$ *and* $\forall v \in \mathcal{V}, \exists a \in \mathcal{A}, b \in \mathcal{B}$ *s.t.* $v = a + b$.

It turns out that the vectors $a$ and $b$ of the previous definition are uniquely defined.

### A.5.1 Tensors

Tensors can be considered generalizations of vectors and matrices. Actually they are entities which, after certain coordinates have been chosen, can be represented by multidimensional matrices. They are characterized by the fact that certain operations that can be done on them, called *tensor operations*, are invariant for coordinate changes, and are therefore intrinsic.

Before defining tensors, we need to define the concept of a multilinear map.

**Definition A.5.11 (Multilinear Map).** *Given $n+1$ vector spaces $\mathcal{V}_1,\ldots,\mathcal{V}_n$, $\mathcal{W}$, we say that a map of the form:*

$$L : \mathcal{V}_1 \times \ldots \times \mathcal{V}_n \rightarrow \mathcal{W}$$

*is multilinear iff $\forall i$ the maps*

$$L_i(v_1,\ldots,v_{i-1},v_{i+1},\ldots,v_n) : \mathcal{V}_i \rightarrow \mathcal{W} \; ; \; v_i \mapsto L(v_1,\ldots,v_n)$$

*are linear for each $v_1,\ldots,v_{i-1},v_{i+1},\ldots,v_n$.*

We indicate the set of multilinear operators of the previous form with:

$$L^n(\mathcal{V}_1,\ldots,\mathcal{V}_n;\mathcal{W}).$$

It is possible to see that $L^n(\mathcal{V}_1,\ldots,\mathcal{V}_n;\mathcal{W})$ also has the structure of a vector space. We can now define the set of tensors on a vector space $\mathcal{V}$.

**Definition A.5.12 (Tensors).** *Given a vector space $\mathcal{V}$, we define the set of tensors of order $p+q$ contravariant of order $p$ and covariant of order $q$ or of type $\binom{p}{q}$ with the following vector space:*

$$\mathcal{T}^p_q(\mathcal{V}) := L^{p+q}(\underbrace{\mathcal{V},\ldots,\mathcal{V}}_{q-times},\underbrace{\mathcal{V}^*,\ldots,\mathcal{V}^*}_{p-times};\mathbb{R}).$$

It is possible to see that $\mathcal{T}^1_0(\mathcal{V}) = \mathcal{V}$ and $\mathcal{T}^0_1(\mathcal{V}) = \mathcal{V}^*$. The distinction between covariant and contravariant is fundamental for the invariancy of tensor operations with change of coordinates. For example choose a base $B = \{b_1,\ldots,b_n\}$ for an $n$ dimensional vector space $\mathcal{V}$. We can then consider a unique dual base $\{b^1,\ldots,b^n\}$ for $\mathcal{V}^*$, which is called the *dual base* of $B$, such that:

$$b_i b^j = \delta^j_i$$

where $\delta^j_i$ is called the Kronecker symbol and is equal to 1 if $i = j$ and otherwise 0. Now consider a new base $\tilde{B}$ for $\mathcal{V}$ and its dual base. Suppose that a certain $v \in \mathcal{V}$ has a numerical representation $v_1 \in \mathbb{R}^n$ using base $B$ and $v_2 \in \mathbb{R}^n$ using base $\tilde{B}$. Consider the change of base matrix $A$ such that $v_2 = Av_1$. If we now consider a vector $v^* \in \mathcal{V}^*$ and we express it numerically using the dual base of $B$, we obtain a vector $v^*_1 \in \mathbb{R}^n$ and if we express it in the base dual to $\tilde{B}$ we have another vector $v^*_2 \in \mathbb{R}^n$. It is possible to see that the relation between $v^*_1$ and $v^*_2$ is given by:

$$v^*_1 = A^T v^*_2.$$

This shows that vectors and co-vectors transform in a completely different way. Their transformation is equivalent only for orthonormal coordinate changes. This clearly reflects on the contravariant and covariant parts for the change of coordinates of general tensors. For details on the subject, the reader is addressed to [20].

**Definition A.5.13 ($n-$forms).** *An $n-$form on $\mathcal{V}$ is an antisymmetric $n$ covariant tensor of the following form:*

$$\Omega^n : \underbrace{\mathcal{V} \times \mathcal{V} \times \ldots \mathcal{V}}_{n \text{ times}} \to \mathbb{R} \,;\; (x_1, \ldots, x_n) \mapsto \Omega^n(x_1, \ldots, x_n).$$

*Antisymmetric means that*

$$\Omega^n(x_1, \ldots, x_i, \ldots, x_j, \ldots, x_n) = (-1)^{1-\delta_{ij}} \Omega^n(x_1, \ldots, x_j, \ldots, x_i, \ldots, x_n)$$

*where $\delta_{ij}$ is the Kronecker symbol.*

In general, we denote a tensor with a letter like $T$ without indices, or we indicate it with $T^{i_1,\ldots,i_p}_{j_1,\ldots,j_q}$, or with the multi-index notation[2] $T^{(i)}_{(j)}$ to make explicit that it is of type $(p, q)$. Each of the indexes can have a value from 1 to the dimension of the vector space on which the tensor space is based. For example, a tensor $T_{i,j}$, after coordinates are chosen for the vector space on which it is defined, can be expressed by a matrix (2 indices) whose $i, j$ component represents the $i, j$ component of the tensor in the chosen coordinates.

A typical tensor operation is for example the *annihilation* of a lower index with an upper index. If we have, for example, a vector $v^i$ (tensor of type $(1, 0)$) and a linear map $a^l_k$ (tensor of type $(1, 1)$), we can express the application of the linear map to the vector in implicit tensor notation, as the annihilation of the vector index with the lower index of the linear map defined as:

$$a^j_i v^i := \sum_{i=1}^{n} a^j_i v^i$$

where $n$ is the dimension of the base vector space to which $v^i$ belongs. The linear map $a^j_i$ is numerically represented by a matrix and $v^i$ by a vector and the annihilation operation is nothing else than the multiplication of the matrix and the vector.

It is NOT possible to annihilate two lower indices or two upper ones: this is NOT a tensor operation! It makes no sence to annihilate two vectors or two co-vectors, but it is meaningful to annihilate the unique index of a vector with the unique index of a co-vector: this operation corresponds to the dual product of Def. A.5.4.

## A.6 Lie algebras

Lie algebras are of great importance in the study of 3D motions. A clear and detailed treatment on Lie groups and algebras can be found in [17].

Before defining a Lie algebra, we recall that an internal binary operator $[,] : \mathcal{V} \times \mathcal{V} \to \mathcal{V}$ is called skew symmetric iff:

$$[x, y] = -[y, x] \;\; \forall x, y \in \mathcal{V}.$$

[2] In the multi index notation, $(i) = i_1, \ldots, i_n$.

**Definition A.6.1 (Lie algebra).** *A Lie algebra is a vector space $\mathcal{V}$ taken together with a bilinear skew-symmetric operation* $[,] : \mathcal{V} \times \mathcal{V} \to \mathcal{V}$, *which satisfies the Jacobi identity:*

$$[[x,y],z] + [[y,z],x] + [[z,x],y] = 0.$$

*This operation is often denoted by square brackets and called the commutator.*

If we consider the vector space of square matrices, we can turn it into a Lie algebra by defining the following operation as commutator:

$$[A,B] = AB - BA = -[B,A] \tag{A.2}$$

for any square matrices $A$ and $B$.

Since we are dealing with a vector space, any vector of the space can be expressed as a linear combination of elements of any base. Suppose that we have a finite $n$ dimensional Lie algebra $\mathcal{V}$. Let us indicate a possible base with $v_1, \ldots, v_n$. Since the commutator is a binary internal operation, any elements resulting from an application of the commutator to two elements of the vector space can be expressed as the linear combination of elements of the base. We can therefore consider the result of applying $[,]$ to all the elements of the base and have:

$$[v_i, v_j] = \sum_{l=1}^{n} c_{ij}^l v_l.$$

The scalars $c_{ij}^l$ are called *structural constants* for the Lie algebra and are characterizing elements of it. We can therefore choose a base for any Lie algebra, and find the expression of the structural constants for that base.

## A.7 Differential geometry

Differential Geometry is a branch of Mathematics which studies the geometry of the generalization of concepts such as curves, surfaces and correlated topics. The use of these tools has proven to be of great help for intuition and for structuring concepts. For an excellent and complete treatment of geometric concepts, the reader is addressed to [20, 92]. A concise introduction to the subject can be also found in some chapters of [35, 26, 34]. Other often-cited references are [93, 94, 18]. An excellent treatment on symmetries and Lie groups can be found in [16]. Unfortunately, the importance of differential geometry is often underestimated in basic engineering courses.

### A.7.1 Manifolds

A manifold is intuitively defined as a set which is locally diffeomorphic to $\mathbb{R}^n$ around each of its points. The most classical example of a manifold is

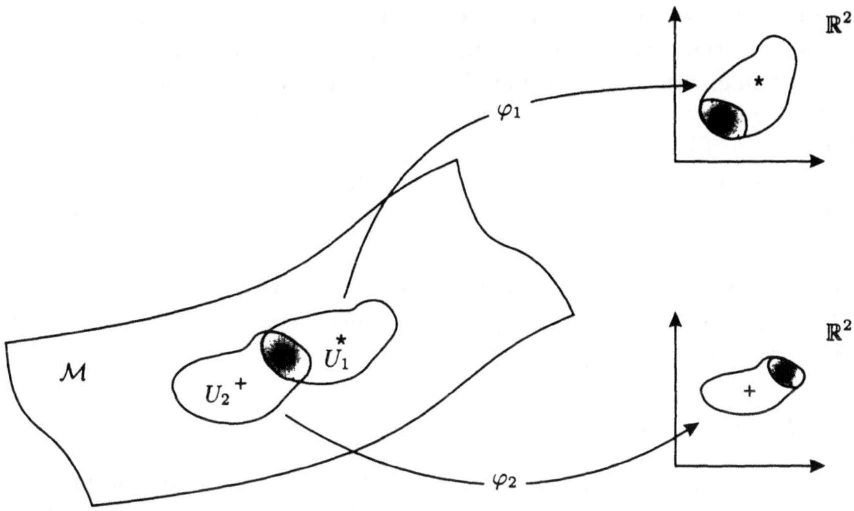


**Fig. A.1.** A two dimensional manifold and its charts.

a smooth surface in space (see Figure A.1): locally we can define a one-to-one mapping between a neighborhood of one of the surface points and a neighborhood of $\mathbb{R}^2$.

The first concept we need to define for a manifold is the one of a chart. This concept comes from cartography, where people try to create an atlas by means of a collection of charts describing different parts of the earth. Some charts in the atlas must overlap to be useful. This overlap creates a kind of continuity in the description of the earth.

**Definition A.7.1 (Chart).** *Given a set $\mathcal{M}$, a local chart on $\mathcal{M}$ is a bijection $\varphi : U \subset \mathcal{M} \rightarrow P \subset \mathbb{R}^n$. We denote this chart with $(U, \varphi)$ (see Figure A.1).*

We can now define an atlas as a collection of charts which overlap in a smooth way in the sense which is hereafter explained.

**Definition A.7.2 (Atlas).** *An atlas on a set $\mathcal{M}$ is a family of charts $(U_i, \varphi_i)$ $i \in I$ where $I$ may be any indexing set, and the three conditions below must be satisfied:*

1. *Continuity: $\varphi_i$ is a homeomorphism (see Def. A.3.3) $\forall i$.*
2. *Covering: $\mathcal{M} = \bigcup_{i \in I} U_i$*
3. *Compatibility: $\forall i, j \in I$ consider the charts $(U_i, \varphi_i)$,$(U_j, \varphi_j)$. If we have that $U_c := U_i \cap U_j \neq \emptyset$, then the function $(\varphi_i o \varphi_j^{-1}) : \varphi_j(U_c) \subset \mathbb{R}^n \rightarrow \varphi_i(U_c) \subset \mathbb{R}^n$ should be smooth.*

The continuity condition ensures that a manifold is a continuous entity, but does not ensure anything concerning differentiability. The covering property ensures that for any part of the manifold there is at least a chart which can be used to study the manifold around that part. The compatibility condition ensures differentiability and smoothness for differential calculus on the manifold. For intuition, just think of a manifold as a curved smooth surface, like the surface of Figure A.1.

In the previous definitions, the charts map to $\mathbb{R}^n$. The integer $n$ is called the dimension of the manifold in a neighborhood of the point of the considered chart.

**Definition A.7.3 (Dimension).** *For a manifold $\mathcal{M}$, we call the dimension of the manifold around a point $p \in U_i$ the dimension of the real space which is the co-domain of the chart $(U_i, \varphi_i)$.*

### A.7.2 Tangent spaces

We can consider the functions which are defined on a manifold. We indicate with $\mathcal{C}^\infty(\mathcal{M})$ the set of infinitely differentiable functions defined on $\mathcal{M}$. In the same way, we indicate with $\mathcal{C}^\infty(p)$ the set of smooth functions defined on an open neighborhood of $p \in \mathcal{M}$. Since we have smoothness properties for a manifold, in a certain point $p \in \mathcal{M}$ of the manifold we can consider the vector space composed of all the tangent vectors to the manifold in the point $p$. This is a vector space because the manifold is locally bijective to $\mathbb{R}^n$. It is possible to think of it by considering all the possible smooth curves passing through $p$. Those curves have a velocity vector in $p$. The set of those velocity vectors is a vector space called the tangent space of $\mathcal{M}$ at $p$. We can give another interpretation of the tangent space which is more useful for analytical purposes.

**Definition A.7.4 (Tangent space).** *We define the tangent space $T_p\mathcal{M}$ to $\mathcal{M}$ at $p$ as the linear space of mappings of the following form:*

$$X_p : \mathcal{C}^\infty(p) \to \mathbb{R}$$

*satisfying the following:*

1. *Linearity: $X_p(\alpha f + \beta g) = \alpha(X_p f) + \beta(X_p g)$*
2. *Leibniz rule: $X_p(fg) = (X_p f)g(p) + f(p)(X_p g)$*

*for each $f, g \in \mathcal{C}^\infty(p)$ and $\alpha, \beta \in \mathbb{R}$.*

If we consider around a point $p$ a chart $(U, \varphi)$ with coordinates $(x_1, \ldots, x_n)$, we indicate a base of the vector space at $p$ with $\{\partial/\partial x_i\}$ so that we can write:

$$X_p = X_1 \frac{\partial}{\partial x_1} + \ldots + X_n \frac{\partial}{\partial x_n}$$

where $(X_1, \ldots, X_n) \in \mathbb{R}^n$ is a local representation of $X_p$ with the coordinates $(x_1, \ldots, x_n)$. The set of all the tangent spaces in any position of the manifold is called the tangent bundle of the manifold:

**Definition A.7.5 (Tangent bundle).** *The tangent bundle of a manifold* $\mathcal{M}$ *is defined as:*

$$T\mathcal{M} := \bigcup_{p \in \mathcal{M}} T_p\mathcal{M}.$$

The dimension of a tangent space at a point $p$ of a constant dimensional manifold $\mathcal{M}$ of dimension $n$, is itself $n$. This can be understood since locally, around $p$, we can study the manifold as being $\mathbb{R}^n$ itself. On the other hand, an element of the vector bundle has dimension $2n$ since we must specify using $n$ coordinates the point $p$ which we consider on the manifold, and then $n$ other coordinates to identify the vector belonging to $T_p\mathcal{M}$.

Since in each point $p$ we have a vector space $T_p\mathcal{M}$, we can consider its dual (see Def. A.5.3). This dual space is called the co-tangent space at $p$ and it is indicated with $T_p^*\mathcal{M}$. The co-tangent space at $p$ is therefore the set of linear operators from $T_p\mathcal{M}$ to $\mathbb{R}$. The set of all the co-tangent spaces can be defined at each point in the same way as we defined the tangent bundle and it is called co-tangent bundle and indicated with $T^*\mathcal{M}$. The vector space $T_p\mathcal{M}$ is also called the fiber of the tangent bundle $T\mathcal{M}$ at $p$.

**Definition A.7.6 (Fiber Projection).** *Given an element* $(p, v) \in T\mathcal{M}$, *we call fiber projection the map which applied to* $(p, v)$ *gives* $v \in T_p\mathcal{M}$. *We denote it with:*

$$\Pi(x, v) = v.$$

**Definition A.7.7 (Canonical Projection).** *We call the following mapping a canonical projection for the tangent manifold on* $\mathcal{M}$*:*

$$\pi : T\mathcal{M} \to \mathcal{M} \ ; \ (x, v) \mapsto x.$$

It is clear that $(\pi(w), \Pi(w)) = w \quad \forall w \in T\mathcal{M}$.

**Definition A.7.8 (Vector Fields).** *We call a smooth mapping of the following form a vector field:*

$$X : \mathcal{M} \to T\mathcal{M}$$

*for which* $\pi \text{ o } X = \text{id}$.

Let us reflect on the previous definition. To each point $p$ of the manifold a vector field associates an element of the tangent bundle which has to be in $p$. In other words, to each point $p$ it associates a "velocity vector" in $p$. The usual integration along paths can therefore be done: given a vector field, we can look for the set of curves parameterized by a scalar $t$ such that the derivative respect to $t$ corresponds to the given vector field.

**Definition A.7.9 (Integration of Vector Fields).** *Given a manifold $\mathcal{M}$ and a vector field $X$ defined on $\mathcal{M}$, we say that a curve $\phi(p, \cdot) : T \to \mathcal{M}$ ; $t \mapsto \phi(p,t)$ is an integral curve of $X$, passing from $p$, where $T$ is an interval of $\mathbb{R}$ containing 0 and $t_0$ iff:*

$$\left.\frac{d\phi(p,t)}{dt}\right|_{t=t_0} = X(\phi(p,t_0)) \quad \textit{and} \quad \phi(p,0) = p.$$

The duals of vector fields are called *co-vector fields*

**Definition A.7.10 (Co-vector Fields).** *We call a smooth mapping of the following form a co-vector field:*

$$X^* : \mathcal{M} \to T^*\mathcal{M}$$

*for which $\pi^*$ o $X^* = \mathrm{id}$, where $\pi^*$ is analogous to $\pi$ but its domain is $T^*\mathcal{X}$.*

Co-vector fields are dual to vector fields, and it is therefore possible to consider the dual product of a co-vector field on a vector field, which would result in a function defined on the manifold. To see that better, let's consider the differential of a function defined on $\mathcal{M}$.

**Definition A.7.11 (Differential).** *Given a smooth function $f \in C^\infty(\mathcal{M})$, we define a co-vector field called differential of $f$ and indicated as $df$ as the unique co-vector field, such that:*

$$df(X)(p) := \langle df(p), X(p)\rangle = \left.\frac{df(\phi(p,t))}{dt}\right|_0$$

*for each vector field $X$ with integral $\phi$.*

The function $df(X)(p)$ is what is also called the *directional derivative* or *Lie derivative*, which is indicated as $L_X f := df(X)$.

The Lie bracket of two vector fields is a third vector field, which is function of the commutativity of the starting vector fields. Consider a point $p \in \mathcal{M}$ of a manifold where two vector fields $X_1$ and $X_2$ are defined. The vector $X_1(p) \in T_p\mathcal{M}$ gives a direction in which we could move from $p$. Suppose we move in that direction infinitesimally on the manifold and indicate with $p_1 := \phi_1(p, dt)$ the arrival point, where $\phi_1(p, \cdot)$ is the integral curve of $X_1$ passing through $p$. In $p_1$ we can now consider the vector $X_2(p_1)$ and we can move analogously for another infinitesimal motion to reach a position $p_2 := \phi_2(p_1, dt)$. We can then do the same going in their opposite directions: first in a direction $-X_1(p_2)$ to reach a point $p_3 := \phi(p_2, -dt)$ and eventually in the direction $-X_2(p_3)$ to reach a final point $p_4 := \phi(p_3, -dt)$. The question is now whether $p_4 = p$. This is the case iff the Lie bracket of $X_1$ and $X_2$ in $p$ is zero. If this is not zero and equal to a third vector field $X_3$, this would mean that we would have reached the point $p_4$ from $p$ directly by moving in a direction $X_3(p)$. We could therefore define the Lie bracket of $X_1$ and $X_2$ as:

$$[X_1, X_2](p) = \frac{d}{dt}\phi_2(\phi_1(\phi_2(\phi_1(p,t),t),-t),-t).$$

Nevertheless, we will later give the definition of the Lie bracket which is usually found in standard books. A classical example of such a concept is a car. Consider a car in a certain position and consider the motion of the steering wheel and the motion forward or backward. We can consider steering a little to the right, move a little forward then steer a little to the left and move a little backward. The position we will reach will NOT be the position from which we started, which means that the Lie bracket of the vector fields representing the steering motion and the vector field representing the motion of the car are NOT commutative.

**Definition A.7.12 (Lie brackets).** *Given two vector fields $X_1$ and $X_2$, we can define a third vector field which is called the Lie bracket of $X_1$ and $X_2$ and indicated with $[X_1, X_2]$, the unique vector field, such that:*

$$[X, Y]f = X(Yf) - Y(Xf).$$

We can generalize the concepts of vector and co-vector fields to concepts of tensor fields. We have seen in Sect. A.5.1 how tensors are defined for a vector space $\mathcal{V}$. If we consider the tangent spaces of a manifold as a vector space for the tensors, we arrive to the concept of a tensor field.

**Definition A.7.13 (Tensor field).** *An operation which smoothly assigns to each point $p$ of a manifold $\mathcal{M}$ a tensor $\mathcal{T}_l^q(T_p\mathcal{M})$ (see Sect. A.5.1) is called a $q$ contravariant and $l$ covariant tensor field.*

At each point $p$, we can therefore consider the vector space of tensors on $T_p\mathcal{M}$. We indicate such a vector space with $T_{q,l}\mathcal{M}_p$. The generalization of tangent and co-tangent bundle is the concept of *tensor bundle*

**Definition A.7.14 (Tensor bundle).** *We call the following set the $q$ contravariant and $l$ covariant tensor bundle:*

$$T_{q,l}\mathcal{M} := \bigcup_{p \in \mathcal{M}} T_{q,l}\mathcal{M}_p.$$

### A.7.3 Distributions

A vector (resp. co-vector) field assigns to each point $p \in \mathcal{M}$ of a manifold a vector belonging to $T_p\mathcal{M}$ (resp. to $T_p^*\mathcal{M}$). A distribution assigns a subspace of $T_p\mathcal{M}$ to each point $p$ .

**Definition A.7.15 (Distributions).** *A distribution on $\mathcal{M}$ is a smooth function which assigns to each point $p \in \mathcal{M}$ a subspace $\Delta(p) \subset T_p\mathcal{M}$. $\Delta$ is called a smooth distribution iff its subspaces in each point are spanned by a set of smooth vector fields:*

$$\exists X_i \ ; \ i \in I \ s.t. \ \Delta(p) = \text{span}\{X_i(p) \ ; \ i \in I\}.$$

*The distribution is called constant dimensional iff for each $p \in \mathcal{M}$, $\Delta(p)$ has the same dimension.*

A co-distribution is defined analogously but as a subspace of the co-tangent spaces $T_p^*\mathcal{M}$.

**Definition A.7.16 (Involutivity).** *Given a smooth distribution $\Delta$ defined as $\Delta = \text{span}\{X_i \ ; \ i \in I\}$, we call this distribution involutive iff $\forall p$ and $\forall i, j \in I$:*

$$[X_i, X_j](p) \in \Delta(p).$$

**Definition A.7.17 (Integral of a Distribution).** *A sub-manifold $\mathcal{N} \subset \mathcal{M}$ is called an integral manifold of a distribution $\Delta$ on $\mathcal{M}$ iff $T_p\mathcal{N} = \Delta(p) \quad \forall p \in \mathcal{N}$.*

If a distribution is integrable, it means that we can patch all the subspaces it defines in each point and consider the so-called involution, which corresponds to the integral manifold. It is not always possible to patch these subspaces. This depends on the involutivity of the distribution. The following fundamental result gives a method to check when it is possible to integrate a distribution.

**Theorem A.7.1 (Frobenius).** *A constant dimensional distribution $\Delta$ is integrable iff it is involutive.*

We can try to intuitively understand the cause of the just given result. If a distribution is not involutive, it means that by moving back and forth along the directions described by the distribution, we can move in directions which are not spanned by the vector fields describing the distribution. This means somehow that the various integral curves of the vector fields cannot be considered as a grid for a sub-manifold because by moving back and forth we would get off the ideal sub-manifold describing the involution of all the tangent spaces described by the distribution.

We can then consider a bigger involutive distribution which contains the starting not involutive distribution.

**Definition A.7.18 (Involutive closure).** *Given a distribution $\Delta$, the smallest involutive distribution containing $\Delta$ is called the involutive closure of $\Delta$.*

The involutive closure of any distribution is integrable for construction. The integral of the involutive closure describes all the points that can be reached if we can move locally back and forth in all the directions of the starting distribution.

### A.7.4 Mappings between manifolds

We can now consider mappings between manifolds.

**Definition A.7.19 (Tangent mappings).** *Given a differentiable mapping $F : \mathcal{M} \to \mathcal{N}$ between two manifolds, we can consider a mapping which is called the tangent map of $F$ at $p$ as a linear map of the form:*

$$F_{*p} : T_p\mathcal{M} \to T_{F(p)}\mathcal{N}$$

*such that*

$$(F_{*p}X)(f) = X(f \text{ o } F)$$

*where $X \in T_p\mathcal{M}$. We call the following map tangent map of $F$:*

$$F_* : T\mathcal{M} \to T\mathcal{N}\ ;\ (p, v) \mapsto (F(p), F_{*p}(v)) \tag{A.3}$$

Notice that $F_*$ maps in the same direction as the starting map $F$, and therefore from the tangent space of $\mathcal{M}$ to the tangent space of $\mathcal{N}$. For co-vectors or contravariant tensors in general, the natural map maps in the opposite direction and it is called therefore *pullback*.

**Definition A.7.20 (Pullback).** *Given a differentiable mapping $F : \mathcal{M} \to \mathcal{N}$ between two manifolds, we can consider the following map called pullback at $p$ of a $k$ covariant tensor $T(p) \in T_{F(p)}\mathcal{N}$:*

$$F_p^* : \mathcal{T}_k^0(T_{F(p)}\mathcal{N}) \to \mathcal{T}_k^0(T_p\mathcal{M})\ ;\ T(F(p)) \mapsto (F^*T)(p)$$

*such that*

$$T(F(p))(F_{*p}v_1, \ldots, F_{*p}v_k) = (F^*T)(p)(v_1, \ldots, v_k)$$

*$\forall v_1, \ldots, v_k \in T_p\mathcal{M}$. The pullback of a $k$ covariant tensor field corresponds to the pullback at each point.*

### A.7.5 Orientation

**Definition A.7.21 (Orientability).** *We say that a manifold $\mathcal{M}$ is orientable if it is possible to define a $C^\infty$ $n-$form $\Omega$ on $\mathcal{M}$ which is not zero at any point. In this case we say that $M$ is oriented by $\Omega$.*

It should be clear that any $\lambda(p)\Omega(p)$ where $\lambda() > 0$ is a $\mathcal{C}^\infty$ function on $\mathcal{M}$ would also be a suitable $n-$form for defining the orientation.

**Definition A.7.22 (Orientation preserving mapping).** *Suppose that we have two manifolds $\mathcal{M}_1$ and $\mathcal{M}_2$ of dimension $n$ and oriented by two $n-$forms $\Omega_1$ and $\Omega_2$ respectively. We say that a diffeomorphism $F : \mathcal{M}_1 \to \mathcal{M}_2$ is orientation preserving, iff $\exists\lambda \in C^\infty(\mathcal{M}_1)$ such that $\lambda(p) > 0 \quad \forall p \in \mathcal{M}_1$ and:*

$$(F^*\Omega_2)(p) = \lambda(p)\Omega_1(p).$$

### A.7.6 Riemannian manifolds

A Riemannian manifold is a manifold on which we can measure distance between points.

**Definition A.7.23 (Riemannian manifold).** *A Riemannian manifold* $(\mathcal{M}, g)$ *is a manifold* $\mathcal{M}$ *together with a symmetric* ($g_{ij} = g_{ji}$)*, positive definite, 2 covariant smooth tensor field* $g_{ij}$ *defined on it, which is called metric.*

A 2 covariant tensor at a point $p$ of a Riemannian manifold is, as already defined, a bilinear mapping which we indicate by using the so-called tensor notation, with:

$$g_{ij}(p) : T_p\mathcal{M} \times T_p\mathcal{M} \to \mathbb{R} \; ; \; (v^i, v^j) \mapsto g_{ij}v^i v^j .$$

The length of a vector $v_i \in T_p\mathcal{M}$ for the given metric is then defined as $\sqrt{g_{ii}v^i v^i}$.

### A.7.7 Connections

If we have a manifold and some tensor field on it, without any extra structure, there is no way to "compare" tensors defined on different points on the manifold. In order to do so, we need an extra structure on the manifold, which is called *differential-geometric connection* or simply *connection*. As the word implies, this structure gives the possibility to "connect" or better compare tensors attached to different points on the manifold and makes possible a lot of operations which otherwise would not be defined. This is not the place to treat connections in details; interested readers are addressed to the excellent, even if a bit old, treatment reported in [20].

**Covariant Differentiation.** The structure of a connection makes possible to define the *covariant differentiation*. This operation can be seen as the generalization of the concept of gradient of a real function to the gradient of a general tensor field of the type $(p, q)$. We see later that the concept of "gradient" indeed permits us to "compare" and "connect" tensors attached to different points.

If we consider a manifold $\mathcal{M}$ and a function $f \in C^\infty(\mathcal{M})$, the gradient representing the differential of $f$ is a co-vector field (tensorfield of type (0,1)). This means that this operation is intrinsically defined and therefore is a tensor operation: the result of the gradient of a function is independent of the coordinates. In this case, a function can be seen as a tensor field of type $(0, 0)$. The gradient of this function increments its lower index of one yielding a co-vector and therefore a tensor field of type $(0, 1)$.

The problem manifests itself when we try to calculate "the gradient" of a tensor field with an order greater then or equal to one, for example the gradient of a vector field. We could think about taking the gradient of each of its components when expressed in coordinates. It turns out that this is NOT a tensor operation (it is not invariant for coordinate changes) and therefore it is not intrinsic. The solution to this problem comes with the definition of the *connection* and a proper definition of the "gradient" operation for a general tensor which is called *covariant differentiation.* Before defining the connection, we define what are called Christoffel symbols.

**Definition A.7.24 (Christoffel symbols).** *Given a manifold $\mathcal{M}$ of dimension $n$ and coordinates $z_i$ we call Christoffel symbols for the manifold in the coordinates $z_i$, a set of $n^3$ functions $\Gamma^k_{pq} : \mathbb{R}^n \to \mathbb{R}$ indexed by $p, q, k = 1, \ldots, n$, such that, if we take other coordinates $z'_i$ of $\mathcal{M}$, in these new coordinates, new functions $\tilde{\Gamma}^{k'}_{p'q'}$ correspond such that:*

$$\tilde{\Gamma}^{k'}_{p'q'} = \frac{\partial z^{k'}}{\partial z^k}\left(\Gamma^k_{pq}\frac{\partial z^p}{\partial z^{p'}}\frac{\partial z^q}{\partial z^{q'}} + \frac{\partial^2 z^k}{\partial z^{p'} z^{q'}}\right). \tag{A.4}$$

A connection can now be defined by means of a set of Christoffel symbols:

**Definition A.7.25 (Connection).** *Given a manifold $\mathcal{M}$, we call a connection on the manifold $\mathcal{M}$ a set of Christoffel symbols $\Gamma^i_{jk}$.*

**Definition A.7.26 (Symmetric Connection).** *A connection $\Gamma^i_{jk}$ is said to be symmetric or torsion free iff $\Gamma^i_{jk} = \Gamma^i_{kj}$.*

**Definition A.7.27 (Affine connection).** *We call a connection affine or Euclidean iff global coordinates exist in which the Christoffel symbols are all identically zero.*

We can now give the definition of a covariant differentiation. The general formula is given for completeness, but it is not important for conceptual reasoning. The important fact is that once we have a connection, we can define a tensor operation called covariant differentiation which, given a tensor field of type $(p, q)$, returns a tensor field of type $(p, q + 1)$.

**Definition A.7.28 (Covariant Differentiation).** *Given a manifold $\mathcal{M}$ together with a connection, we define covariant differentiation as an operation which maps a tensor field of type $(p, q)$ to a tensor field of type $(p, q+1)$, and which is defined by:*

$$T^{(k)}_{(l);r} = \frac{\partial T^{(k)}_{(l)}}{\partial x^r} + \sum_{s=1}^{p} T^{k_1\ldots(k_s\to i)\ldots k_p}_{l_1\ldots l_q}\Gamma^{k_s}_{ir} - \sum_{s=1}^{q} T^{k_1\ldots k_p}_{l_1\ldots(l_s\to i)\ldots l_q}\Gamma^i_{l_s r}$$

*where $(k) = k_1, k_2, \ldots, k_p$, $(l) = l_1, l_2, \ldots, l_q$ and where the notation $k_1 \ldots (k_s \to i) \ldots k_p$ indicates that in the $p-$tuple $k_1 \ldots k_p$ the symbol $k_s$ is to be replaced by $i$. We indicate this operation with the operator $\nabla$:*

$$\nabla_r T^{(k)}_{(l)} = T^{(k)}_{(l);r}.$$

The previous definition would not be meaningful without the following result, which is the consequence of a good choice for the transformation of Christoffel symbols during coordinate changes given in Eq. (A.4):

**Theorem A.7.2.** *The covariant differentiation is a tensor operation: it is invariant for coordinate changes.*

**Parallel Transport.** A consequence of, and an equivalent definition of a connection and the operation of covariant differentiation is the possibility of "transporting" tensors along a curve on the manifold from a point $x_0$ to a point $x_1$. This is actually the reason of the name for this structure: "connection". Before introducing the concept of parallel transport, we generalize the concept of a directional derivative of a function to the concept of a directional derivative for a tensor of any rank.

**Definition A.7.29 (Directional Derivative).** *Given a manifold $\mathcal{M}$, a tensor field of type $(p,q)$ $T^{(p)}_{(q)}$ and a vector field $\xi$, we define the directional derivative of $T^{(p)}_{(q)}$ along $\xi$ the following tensor field which is also of type $(p,q)$:*

$$\nabla_\xi T^{(p)}_{(q)} = \xi^k \nabla_k T^{(p)}_{(q)} := \xi^k T^{(p)}_{(q):k}.$$

The previous definition can be seen to agree with the Lie derivative or a directional derivative of a function along a vector field. It is now possible to define the parallel transport of a tensor along a curve:

**Theorem A.7.3 (Parallel Transport).** *Given a manifold $\mathcal{M}$, a continuous curve $\gamma : [0,1] \to \mathcal{M}$ and a tensor $T$ of any order in $\gamma(0)$, there is a unique tensor at each point on the curve satisfying $(d\gamma/dt)^k \nabla_k T = 0$. This tensor is said to result from the parallel transport of $T$ to the point on the curve where the derivative is calculated.*

In the case of a function, this means that its value on the curve remains constant.

**Theorem A.7.4.** *For an affine connection, the transport of a tensor from a point $a$ to a point $b$ on the manifold $\mathcal{M}$ is independent from the curve that is followed to go from $a$ to $b$*

**Proof.** If the connection is affine, coordinates $x$ exist in which the Christoffel symbols become identically zero. In these coordinates, the covariant differentiation corresponds to the usual gradient of the tensor components and therefore the parallel transport in Euclidean coordinates corresponds to the fact that the tensor components do not vary along the curve. This obviously implies that this operation is independent from the curve since two tensors, $T(a)$ attached to $a$ and $S(b)$ attached to $b$, are parallel iff they have the same components in the Euclidean coordinates ∎

The previous result indeed corresponds with the intuition of the parallel transport of a vector in the Euclidean space from a point to another: of two vectors in the three-dimensional space one is the parallel transport of the other iff in any reference frame they have the same components. Here it is possible to see the difference between Euclidean and curved spaces: a space is curved if the parallel transport depends on the path.

Intuitively, we can think about a small closed path passing through a point $a$. If we transport a tensor from the point $a$ along the curve again to $a$ and we get a different vector than the one we started with, this means that the space is curved in $a$. This is obviously never the case if the connection is affine.

*Remark A.7.1.* Even if the space is Euclidean, we could choose non-Euclidean coordinates for which parallel transport does not longer result in constant components: just think of spheric coordinates.

From what is just said, it should be understandable that given a tensor $T \in T_{q,p}\mathcal{M}_m$ and another point $l \in \mathcal{M}$, there is one and only one $S \in T_{q,p}\mathcal{M}_l$ such that $T$ is parallel to $S$: the parallel transport of $T$ from $m$ to $l$ through any (since the connection is affine) continuous line connecting $m$ to $l$ gives $S$. We say in this case that $T \in T_{q,p}\mathcal{M}_m$ and $S \in T_{q,p}\mathcal{M}_l$ are parallel and we indicate this with $T \parallel S$. It is easy to understand that parallelism is an equivalence relation (see Def. A.1.8) on $T_{q,p}\mathcal{M}$ since it satisfies reflexivity, symmetry and transitivity.

Since we have an equivalent relation, which we indicate as $\parallel$ on $T_{p,q}\mathcal{M}$, we can consider the quotient space.

**Definition A.7.30 (Free Tensors).** *Given a manifold with an affine connection $\Gamma^k_{ij}$, we call the space $T_{p,q}\mathcal{M}/\parallel$ the set of free tensors of type $(p,q)$.*

**Geodesics.** We know what straight lines are in the space in which we live. Can we define the equivalent of straight lines passing through a given point in a general manifold with a connection ? The answer is yes and we call these "generalized straight lines" geodesics.

**Definition A.7.31.** *A curve $\gamma(t) \in \mathcal{M}$ is called a geodesic on the manifold $\mathcal{M}$ and with respect to a connection $\Gamma^k_{ij}$ if the vector field $T^i = (d\gamma/dt)^i$ tangent to $\gamma$ is parallel along the curve itself ,i.e. if the curve parallel transports its own tangent vector:*

$$\nabla_T(T) = 0.$$

It is easy to see that in the case of Euclidean coordinates, the equation of a geodesic corresponds to the one of straight lines.

The existence and uniqueness of a geodesic tangent to a vector $T^i \in T\mathcal{M}$ is a consequence of the existence and uniqueness of standard differential equations.

**Metrics and Connections.** Suppose that we have a Riemannian manifold as in Sect. A.7.6 with a connection defined for it. We can then define what we mean when we say that the connection is compatible with the Riemannian metric.

**Definition A.7.32.** *A connection $\Gamma_{ij}^k$ is said to be compatible with a metric $g_{ij}$ iff the covariant derivative of the metric tensor $g_{ij}$ is identically zero:*

$$\nabla_k g_{ij} \equiv 0 \quad k, i, j = 1, \ldots, n.$$

An important property and consequence of the fact that a connection is compatible with a metric is the following:

**Theorem A.7.5.** *Consider a Riemannian manifold $\mathcal{M}$ with metric $g_{ij}$ for which a connection $\Gamma_{ij}^k$ compatible with $g_{ij}$ is defined. Given a curve $\gamma(t) \in \mathcal{M}$ and two vector fields $T$ and $S$ which are both parallel along the curve, then their scalar product is constant along the curve.*

**Theorem A.7.6 (Levi Civita connection).** *Given a Riemannian manifold $\mathcal{M}$ with a nonsingular metric $g_{ij}$, there exist a unique symmetric connection ($\Gamma_{ij}^k = \Gamma_{ji}^k$) which is compatible with the metric. This unique connection is given in any system of coordinates $x_1, x_2, \ldots, x_n$ by:*

$$\Gamma_{ij}^k = \frac{1}{2} g^{kl} \left( \frac{\partial g_{lj}}{\partial x^i} + \frac{\partial g_{il}}{\partial x^j} - \frac{\partial g_{ij}}{\partial x^l} \right) \tag{A.5}$$

*where $g^{ij} = (g_{ij})^{-1}$; this is known as the Levi-Civita connection for the Riemannian manifold.*

An important corollary of the previous theorem is that if in a certain point the partial derivatives of the metric are all zero, the Christoffel symbols in that point are zero. We can conclude this section with an important remark:

*Remark A.7.2.* Riemannian geometry gives rise to a uniquely defined symmetric connection which corresponds to a unique way in which we can parallel transport tensors along a curve. In this unique way, the scalar product of parallelly transported vectors is constant along this curve. Once we are given a Riemannian manifold, the Christoffel formula allows us to build the unique symmetric connection compatible with the given metric.

**Curvature.** To a connection we can directly associate the concept of curvature of the manifold in a certain point. If the connection is the Levi-Civita one, we can therefore study the curvature of a Riemannian manifold by first considering the Levi-Civita connection and then calculate the curvature.

For this reason, the curvature in a certain point is strictly related to parallel transport. Intuitively, consider an infinitesimal closed curve around a point $p$. If we parallel transport a tensor along this curve back to the starting point, and if the obtained tensor is equal to the starting one, it means that that manifold does not have curvature in $p$; it is said to be flat in $p$.

With Euclidean coordinates, covariant differentiation corresponds to partial derivatives of the components of a tensor. This would imply that in such coordinates, tensors would be parallel iff they have the same components. Such a manifold is called flat or with curvature zero. In such coordinates, the commutativity of partial derivatives would correspond to the commutativity of covariant differentiation. We can use this feature to study curvature: a measure of the curvature is the noncommutativity of covariant differentiation operators.

More formally, with Euclidean coordinates $x^i$ we get:

$$\nabla_i \nabla_j = \frac{\partial}{\partial x^i \partial x^j} = \frac{\partial}{\partial x^i}\frac{\partial}{\partial x^j} = \frac{\partial}{\partial x^j}\frac{\partial}{\partial x^i} = \frac{\partial}{\partial x^j \partial x^i} = \nabla_j \nabla_i$$

where $\nabla_i$ indicates the covariant differentiation in the direction of the coordinate $x^i$. We can therefore say that a manifold is flat in $p$ if for any tensor $T^{(p)}_{(q)}(p)$ we have:

$$(\nabla_i \nabla_j - \nabla_j \nabla_i) T^{(p)}_{(q)}(p) = 0.$$

We can then give the following result:

**Theorem A.7.7 (Curvature and Torsion).** *For any connection $\Gamma^k_{ij}$ on a manifold $\mathcal{M}$ and any vector field $T^i$ on $\mathcal{M}$ we have:*

$$(\nabla_k \nabla_l - \nabla_l \nabla_k) T^i = -R^i_{qkl} T^q + T^p_{kl} \nabla_p T^i$$

*where the 3 covariant, 1 contravariant tensor $R^i_{qkl}$ is called the Riemann curvature tensor and is defined as:*

$$R^i_{qkl} := \frac{\partial \Gamma^i_{qk}}{\partial x^l} - \frac{\partial \Gamma^i_{ql}}{\partial x^k} + \Gamma^i_{pl}\Gamma^p_{qk} - \Gamma^i_{pk}\Gamma^p_{ql}$$

and the 2 covariant, 1 contravariant tensor $T^p_{kl}$ is called the *torsion tensor* and is defined as:

$$T^p_{kl} := \Gamma^p_{kl} - \Gamma^p_{lk}.$$

A connection with a torsion equal to zero is called *torsion free* or *symmetric*. Notice that a manifold with an affine connection as defined in Def. A.7.27 is flat everywhere.

### A.7.8 Euclidean manifolds

Clearly the best example of an Euclidean space is the three-dimensional space in which we live. This space has got an infinite extension[3]. The Euclidean

[3] At least in a simple model in which we do not consider a universe finite expansion as a consequence of the Big-Bang.

space is continuous. Most important of all we can geometrically define concepts like the distance of points, straight lines and parallel transports of objects independently on the path which is used for transport between two points. These concepts are all the concepts we need for a proper and formal definition of an Euclidean manifold.

**Definition A.7.33 (Euclidean Manifold).** *We say that a Riemannian manifold* $(\mathcal{M}, g)$ *is Euclidean if it is flat everywhere considering the Levi-Civita connection (see Theorem A.7.6), and there exist a geodesic passing through* $O, P$ *which is bijective to* $\mathbb{R}$ *for each distinct* $O, P \in \mathcal{M}$.

## A.8 Lie groups

A Lie group $(\mathcal{G}, \, o\,)$ is nothing else than a manifold $\mathcal{G}$ whose points, together with a binary operation " $o$ " defined on them, form a group (see Sect. A.4). A Lie group therefore has a "special" point which is the *identity* of the group. This point is denoted with $e$. An example of a Lie group is the set of nonsingular square matrices with the matrix multiplication as operation $o$. In this case, the element $e$ is the identity matrix.

### A.8.1 Left and right translations

**Definition A.8.1 (Left and Right translations).** *Given a Lie group* $(\mathcal{G}, \, o\,)$, *for every* $g \in \mathcal{G}$, *we call the following map left translation by* $g$:

$$L_g : \mathcal{G} \to \mathcal{G} \; ; \; h \mapsto g \, o \, h,$$

*and right translation by* $g$ *the map*

$$R_g : \mathcal{G} \to \mathcal{G} \; ; \; h \mapsto h \, o \, g.$$

Since $L_g$ and $R_g$ are continuous maps defined on the Lie group, we can consider their tangent maps (see Sect. A.7.4). We can use their tangent maps to map the tangent space at the identity to each tangent space in any point of the Lie group. The tangent space at the identity has the structure of a Lie algebra, as we will see later.

**Definition A.8.2 (Left invariant vector field).** *We say that a vector field* $X$ *on* $\mathcal{G}$ *is left invariant iff*

$$(L_g)_{*h} X(h) = X(g \, o \, h) \; \forall h \in \mathcal{G}.$$

For each $v \in T_e\mathcal{G}$, there is a unique vector field indicated with $X_v$, which is left invariant and equal to $v$ at the identity of the group ($X_v(e) = v$). We can then transform $T_e\mathcal{G}$ into a Lie algebra (see Sect. A.6) by defining the unique vector $v_3$ as the result of the commutator between two vectors $v_1, v_2 \in T_e\mathcal{G}$,such that $X_{v_3} = [X_{v_1}, X_{v_2}]$.

**Definition A.8.3 (Lie algebra).** *For any Lie group $\mathcal{G}$, there exist a unique Lie algebra $\mathfrak{g}$ whose corresponding vector space is $T_e\mathcal{G}$ and for which the algebra commutator results from the previously explained construction is such that:*

$$v_3 = [v_1, v_2] \Leftrightarrow X_{v_3} = [X_{v_1}, X_{v_2}]$$

*where the first square brackets indicate the defined algebra commutator and the second ones are the Lie brackets of vector fields defined in Def. A.7.12.*

One of the most important Lie group operations used for kinematics are the left and right transports at the identity of any vector belonging to the tangent bundle.

**Definition A.8.4 (Left and Right transport to the identity).** *Given any vector $(g, \dot{g}) \in T_g\mathcal{G}$, we call*

$$\Pi\left((L_{g^{-1}})_*(g, \dot{g})\right) \in \mathfrak{g}$$

*the left translation of $(g, \dot{g})$ at the identity and*

$$\Pi\left((R_{g^{-1}})_*(g, \dot{g})\right) \in \mathfrak{g}$$

*the right translation of $(g, \dot{g})$ at the identity, where $\Pi()$ is the fiber projection defined in Def. A.7.6.*

**Definition A.8.5 (Adjoint map).** *The following map is called adjoint map[4] for the Lie group at $h$:*

$$Ad_h : \mathfrak{g} \to \mathfrak{g} \ ; \ v \mapsto \Pi((R_{h^{-1}} L_h)_*(e, v)).$$

### A.8.2 Relations Lie group and algebra

If we indicate with $\phi_v(e, \cdot)$ the integral of $X_v$ passing through $e$, we can give the following definition:

**Definition A.8.6 (Exponential map).** *We define the exponential map in the following way:*

$$\exp : \mathfrak{g} \to \mathcal{G} \ ; \ v \mapsto \phi_v(e, 1).$$

[4] Note that even if the names are the same, this adjoint map and the adjoint map of a linear map as defined in Def. A.5.5 are conceptually different. Which of the two is meant is usually clear from the context.

It can be shown that we actually have $e^{sv} := \exp(sv) = \phi_v(e, s)$ where $s \in \mathbb{R}$ and $v \in \mathfrak{g}$. Furthermore, exp is a local diffeomorphism between a neighborhood of 0 in $\mathfrak{g}$ and a neighborhood of $e$ in $\mathcal{G}$. If we choose then a base $B = \{b_1, \ldots, b_n\}$ of $\mathfrak{g}$, we can express $\forall v \in \mathfrak{g}$:

$$v = \sum_{i=1}^{n} x^i b_i$$

with $(x^1, \ldots, x^n) \in \mathbb{R}^n$. Since exp is a local diffeomorphism, using $\exp(\sum x^i b_i)$, we obtain also a chart around e called *exponential coordinates*. Furthermore, with

$$L_h(e^{\sum x^i b_i})$$

we have coordinates for a neighborhood of $h \in \mathcal{G}$. We can now consider the derivative at the identity of the group $\mathcal{G}$ of the adjoint map.

**Definition A.8.7 (Differential of the adjoint).** *Given any $\xi \in \mathfrak{g}$, we can define the following linear map:*

$$ad_\xi := \left.\frac{d}{dt}\right|_{t=0} Ad_{e^{\xi t}}$$

*which is a linear map within the Lie algebra $\mathfrak{g}$.*

**Theorem A.8.1.** *For any $t_1, t_2 \in \mathfrak{g}$, we have that $ad_{t_1} t_2 = [t_1, t_2]$ where $[,]$ is the Lie bracket of the Lie algebra $\mathfrak{g}$.*

### A.8.3 The Killing Form

Given a Lie Group $\mathcal{G}$ with a corresponding Lie Algebra $\mathfrak{g}$, the Killing form is an operator defined in $\mathfrak{g}$ as follows [20]:

$$\langle v_1, v_2 \rangle := -\operatorname{tr}(\operatorname{ad}_{v_1} \operatorname{ad}_{v_2})$$

where $v_1, v_2 \in \mathfrak{g}$, tr() indicates the tensor operation of trace, and $\operatorname{ad}_x$ is the linear adjoint operator as defined in Def. A.8.7. Notice that $\operatorname{ad}_x$ is a linear operator on the vector space $\mathfrak{g}$ and therefore it is a tensor of type $(1, 1)$. Using the tensor index notation, we get:

$$\langle v_1, v_2 \rangle = -(\operatorname{ad}_{v_1})^i_j (\operatorname{ad}_{v_2})^j_i .$$

### A.8.4 Matrix Lie groups

For a lot of fundamental reasons like Ado's theorem [16], matrix algebras are excellent representatives for any finite dimensional group like the ones we need for rigid body mechanisms.

A matrix Lie group is a group whose elements are square matrices and in which the composition operation of the group corresponds to the matrix product. The most general real matrix group is $GL(n)$ which represents the group of non singular $n \times n$ real matrices. This is clearly a group since the identity matrix represents the identity element of the group, for each matrix, there is an inverse, and matrix multiplication is associative. We will now analyse more in detail features and operations of matrix Lie groups.

**Left and Right maps** If we consider a matrix Lie group $\mathcal{G}$, the operations of left and right translation clearly become:

$$L_G(H) = GH \qquad \text{and} \qquad R_G(H) = HG.$$

We can now consider how velocities are mapped using the previous maps. Suppose that we want to map a velocity vector $\dot{H} \in T_H\mathcal{G}$ to a velocity vector in $T_{GH}\mathcal{G}$ using the left translation and to a vector in $T_{HG}\mathcal{G}$ using right translation. We obtain:

$$(L_G)_*(H,\dot{H}) = (GH, G\dot{H}) \quad \text{and} \quad (R_G)_*(H,\dot{H}) = (HG, \dot{H}G)$$

In particular, if we take a reference velocity at the identity, we obtain:

$$(L_G)_*(I,T) = (G, GT) \quad \text{and} \quad (R_G)_*(I,T) = (G, TG)$$

where $T \in \mathfrak{g}$. With an abuse of notation, we will often indicate:

$$(L_G)_*\, T = GT \qquad \text{and} \qquad (R_G)_*\, T = TG$$

when it is clear that we consider mappings from the identity of the group. On a Lie group, we can define *left invariant* or *right invariant* vector fields. These vector fields are such that the differential of the left invariant and right invariant map leaves them invariant. If we indicate with

$$V : \mathcal{G} \to T\mathcal{G}\ ;\ x \mapsto (x, v)$$

a smooth vector field on the Lie group $\mathcal{G}$, we have seen that this vector field is left invariant if:

$$V(L_g(h)) = (L_g)_* V(h) \quad \forall g, h \in \mathcal{G},$$

and similarly it is right invariant if:

$$V(R_g(h)) = (R_g)_* V(h) \quad \forall g \in \mathcal{G}.$$

For a matrix group, if we take in the previous definitions $h = I$ we obtain respectively:

$$V(G) = GT_L \quad \text{and} \quad V(G) = T_R G$$

where we indicated the representative of the left and right invariant vector fields at the identity with $T_L$ and $T_R$. We can conclude from this that any left or right invariant vector field is characterized completely by its value at the identity of the group. We could now ask ourself: what are the integrals of a left or invariant vector field? From what just said, the integral of a left invariant vector field, can be calculated as the integral of the following matrix differential equation:

$$\dot{G} = GT_L \Rightarrow G(t) = G(0)e^{T_L t} \tag{A.6}$$

where $T_L$ is the value of the vector field at the identity. In a similar way, the integral of a right invariant vector field is:

$$\dot{G} = T_R G \Rightarrow G(t) = e^{T_R t}G(0). \tag{A.7}$$

From this it is possible to conclude that if we take an element $T \in \mathfrak{g}$, its left and right integral curves passing through the identity coincide and they represent the exponential map from the Lie algebra to the Lie group:

$$e : \mathfrak{g} \to \mathcal{G}; T \mapsto e^T.$$

It is easy to show, and important to notice, that integral curves passing through points $H = e^{T_1}$ of right and left invariant vector fields which have as representative in the identity $T_2$, are coincident iff $e^{T_1}e^{T_2} = e^{T_2}e^{T_1}$ which is true iff $[T_1, T_2] = 0$, where the last operation is the commutator of the Lie algebra. But how does the commutator look like for a matrix Lie algebra? Being a Lie group a manifold, we can compute the Lie brackets of vector fields on the manifold. Furthermore, we know that elements of the Lie algebra $\mathfrak{g}$ have a left and right vector field associated to them. We can than calculate the Lie bracket of two left or right invariant vector fields, and if the solution is still left or right invariant, consider the value of the resulting vector field at the identity as the solution of the commutator. We will start with the left invariant case first. Consider we are in a point $G(t) \in \mathcal{G}$ at time $t$. If we have two left invariant vector fields characterized by $T_1, T_2 \in \mathfrak{g}$, the Lie bracket of these two vector fields, can be calculated by moving from $G(t)$ along the vector field correspondent to $T_1$ for $\sqrt{s}$ time, than along the one correspondent to $T_2$, than along $-T_1$ and eventually along $-T_2$. In mathematical terms we have:

$$\begin{aligned} G(t+\sqrt{s}) &= G(t)e^{T_1\sqrt{s}} \to G(t+2\sqrt{s}) = G(t+\sqrt{s})e^{T_2\sqrt{s}} \to \\ G(t+3\sqrt{s}) &= G(t+2\sqrt{s})e^{-T_1\sqrt{s}} \to G(t+4\sqrt{s}) = G(t+3\sqrt{s})e^{-T_2\sqrt{s}} \to \\ G(t+4\sqrt{s}) &= G(t)e^{T_1\sqrt{s}}e^{T_2\sqrt{s}}e^{-T_1\sqrt{s}}e^{-T_2\sqrt{s}} \end{aligned} \tag{A.8}$$

If we look at $\frac{d}{ds}G(t+4\sqrt{s})\big|_{s=0}$, we can approximate the exponentials with the first low order terms and we obtain:

$$G(t+4\sqrt{s}) \simeq G(t)\left(\left(I+T_1\sqrt{s}+\frac{T_1^2}{2}s\right)\left(I+T_2\sqrt{s}+\frac{T_2^2}{2}s\right)\right.$$
$$\left.\left(I-T_1\sqrt{s}+\frac{T_1^2}{2}s\right)\left(I-T_2\sqrt{s}+\frac{T_1^2}{2}s\right)\right)$$
$$\simeq G(t)(I+(T_1T_2-T_2T_1)s+o(s)) \quad \text{(A.9)}$$

which implies

$$\frac{d}{ds}G(t+4\sqrt{s})\bigg|_{s=0} = G(t)(T_1T_2-T_2T_1).$$

From the previous equation, we can conclude that the resulting vector field is still left invariant and it is characterized by the Lie algebra element $T_1T_2 - T_2T_1$. We can therefore define the commutator based on left invariant vector fields as:

$$[T_1, T_2]_L = T_1T_2 - T_2T_1.$$

With similar reasoning, it is possible to show for right invariant vector fields that:

$$\frac{d}{ds}G(t+4\sqrt{s})\bigg|_{s=0} = (T_2T_1-T_1T_2)G(t).$$

and therefore, in this case:

$$[T_1, T_2]_R = T_2T_1 - T_1T_2.$$

We have therefore that:

$$[T_1, T_2]_L = -[T_1, T_2]_R.$$

In the literature, $[,]_L$ is used as the standard commutator and we will adapt this convention.

**Matrix Group Actions.** A group action we can consider for an $n$ dimensional matrix Lie group is the linear operation on $\mathbb{R}^n$. We can therefore define as an action:

$$a(G,P) = GP \qquad G \in \mathcal{G}, P \in \mathbb{R}^n$$

It is easy to see that this group action trivially satisfies all the properties required.

**Adjoint representation .** Using the left and right maps, we can define what is called the conjugation map as $K_g := R_{g^{-1}} L_g$ which for matrix groups results:

$$K_G : \mathcal{G} \to \mathcal{G} \; ; \; H \mapsto GHG^{-1}.$$

But what is the importance of this conjugation map ? To answer this question, we need the matrix group action. Suppose we have a certain element $H \in \mathcal{G}$ such that $Q = HP$ where $Q, P \in \mathbb{R}^n$. What happens if we move all the points of $\mathbb{R}^n$ and therefore also $Q$ and $P$ using an element of $\mathcal{G}$ ? What will the corresponding mapping of $H$ look like ? If we have $Q' = GQ$ and $P' = GP$, it is straight forward to see that:

$$Q' = K_G(H)P'.$$

The conjugation map is therefore related to global motions or equivalently changes of coordinates. We clearly have that $K_G(I) = I$ and therefore the differential of $K_G()$ at the identity is a Lie algebra endomorphism. This linear map is called the adjoint group representation:

$$Ad_G : \mathfrak{g} \to \mathfrak{g} \; ; \; T \mapsto GTG^{-1}.$$

The adjoint representation of the group shows how an infinitesimal motion changes moving the references of a finite amount $G$. Eventually, it is possible to consider the derivative of the previous map at the identity

$$ad_T := \left. \frac{d}{ds} Ad_{e^{Ts}} \right|_{s=0}.$$

This map is called the adjoint representation of the Lie algebra and it is a map of the form:

$$ad_T : \mathfrak{g} \to \mathfrak{g} \qquad T \in \mathfrak{g}$$

If we use the definitions we can see that:

$$\left. \frac{d}{ds} Ad_{e^{T_1 s}} T_2 \right|_{s=0} = \left. \frac{d}{ds} e^{T_1 s} T_2 e^{-T_1 s} \right|_{s=0} = T_1 T_2 - T_2 T_1 = [T_1, T_2]_L$$

which shows that:

$$ad_{T_1} T_2 = [T_1, T_2]_L \tag{A.10}$$

**Examples.** A typical example of a Lie group is the set of homogeneous matrices $SE(3)$ of the form:

$$H := \begin{bmatrix} R & p \\ 0 & 1 \end{bmatrix} \quad p \in \mathbb{R}^3, R \in \mathbb{R}^{3\times 3} \; ; \; R^{-1} = R^T$$

with as group operation the matrix multiplication. For such a Lie group, the Lie algebra is composed of the matrices of the form

$$T_i := \begin{bmatrix} \Omega & v_i \\ 0 & 0 \end{bmatrix} \quad v_i \in \mathbb{R}^3, \Omega_i \in \mathbb{R}^{3\times 3} \ ; \ \Omega_i^T = -\Omega_i.$$

It can be shown that the commutator of the Lie algebra corresponds to:

$$[T_1, T_2] = \begin{bmatrix} [\Omega_1, \Omega_2] & \Omega_1 v_2 - \Omega_2 v_1 \\ 0 & 0 \end{bmatrix}$$

with $[\Omega_1, \Omega_2]$ is the matrix commutator as defined in Eq. (A.2).

# B. Basics of Bond Graphs

This appendix summarizes the basic concepts behind bond graphs [46] which are needed to follow the material treated in this book. Basic ideas and elements are presented. The treatment is in $\mathbb{R}^n$ and not in a coordinate-free framework to make it easier to understand. In Sect. 3.2 some generalizations are presented. Interested readers can consult [95, 96] for a detailed treatment of the subject. Modeling is the process of describing certain features of a system while disregarding others. In *mathematical modeling* we use mathematical relations for our description. Generally, these relations do not have any constraint and are not related to physics. In *physical modeling* on the other hand, we explicitly consider physical concepts and we restrict our description in order to keep some physical laws satisfied. The bond-graph language [46] is built in such a way that the satisfaction of the first principle of thermodynamics, the conservation of energy is enforced. Bond graphs are self-dual graphs [25] in which the vertices correspond to some ideal physical concepts like storage or transformation of energy and the edges describe lossless transfer of energy. Intuitively, the edges, which are called *power bonds*, can be thought of as water pipes which bring water, representing the energy, from one part to the other in the system. Bond graphs are an excellent tool for describing power-consistent networks of physical systems.

In block diagrams, the exchange of information takes place through arrows, which represent variables that a block sets and supplies to another block. If we consider a block $A$ and a block $B$ with an arrow representing a variable $x$ going from $A$ to $B$, it means that $A$ sets the value $x$ and $B$ uses it. This exchange of information is *causal*: $x$ is an *effect* of $A$ and a *cause* for $B$. Often causality is physically artificial and not justified. Consider a linear electrical resistor: there is no reason whatsoever why we should say that such a resistor generates a voltage as a function of the current flowing through it or generates a current as a function of a potential applied to its poles.

In bond graphs, causality is not considered in the modeling phase and it is only necessary for simulation. Very interesting issues about causality in bond graphs have been shown in [33].

The concept of *energy* is certainly one of the most important in physics. The fact that a system has dynamics is the direct consequence that there is an exchange of energy among the various parts of the system. In lumped

**Fig. B.1.** Power Bonds: single-bonds (left) and multi-bonds (right).

physical models, we can describe the system as the network interconnection of basic elements which can store, dissipate or transform energy. These basic ideas hold in any physical domain as explained later.

## B.1 Physical domains

In physics, we can discriminate among domains depending on the kind of energy that a certain part of the system can store. When we are talking about the kinetic energy of a stone thrown in the air, we are reasoning in the translational mechanical domain whereas talking about the potential energy of a capacitor we consider implicitly the electrical domain.

The most important primal domains we can encounter are: mechanical, electromagnetic, hydraulic and thermic. Each of these primar domains but the thermal one, can be further split in two sub-domains:

- mechanical= mechanical potential AND mechanical kinetic
- electromagnetic= electrical AND magnetical
- hydraulic= hydraulic potential AND hydraulic kinetic

The thermic domain is the only one which has no dual sub-domains and this is very much related to the irreversible transformation of energy to the thermal domain.

## B.2 Power conjugate variables

One of the fascinating things about nature is the similarity among domains even if their human perception is very much different at first sight. A typical example is the one of an oscillator: the description of the velocity of a mass attached to a spring is identical to the description of the current flowing through a capacitor and an inductor connected in parallel.

This similarity can be better understood realizing that in each primar domain there are two special variables, called *power conjugate variables* whose product is dimensionally equal to power.

In the mechanical translational domain, the product of a velocity $v$ and a force $F$ is equal to power. In the mechanical rotational domain the product of angular velocity $\omega$ and torque $\tau$ is also equal to power. In the electrical domain the product of current $i$ and voltage $v$ is again dimensionally equal to power. In the thermal domain the product of temperature $T$ and entropy flow

**Table B.1.** Efforts and Flows.

| Domain | Effort | Flow |
|---|---|---|
| Mech. Trans. | force $F$ | velocity $v$ |
| Mech. Rot. | torque $\tau$ | ang. vel. $\omega$ |
| Elec.-Magn. | voltage $v$ | current $i$ |
| Hydraulic | pressure $p$ | flow rate $Q$ |
| Thermic | temp. $T$ | entropy f. $\dot{E}$ |

$E$ is once again dimensionally equal to power. In each domain we call one of these variable *flows* and the others *effort* and we hereafter show the incredible dynamical analogies among the various domains. All the definitions of efforts and flows for some of the used domains are reported in Table B.1. In this appendix the basic concepts will be explained for what are called scalar bondgraphs where efforts and flows are scalars. The concepts can be easily generalised to multi-bonds by considering power as the dual product of an effort and a flow: $e^T f$ (see Figure B.1).

## B.3 The physical model structure and Bond Graphs

In physical modeling, we study a set of atomic elements which represent certain physical phenomenas like the storage of energy, its dissipation, its transformation etc. Each of these elements have as external variables a set of flows and the corresponding dual vectors. This pairing of effort-flow is called an *energetic port* since the dual product of an effort and a flow represents the energy flow through this imaginary port of the element.

As a graphical language for physical modeling, we use bond graphs for the following reasons:

1. It is extremely easy to draw a bond graph of a system.
2. It is mechanical to translate a bond graph to a block diagram or differential equations.
3. Once a few rules are followed, it is impossible to make the common "sign mistakes" of block diagrams.

All the basic elements are connected in a graph whose edges are called *energetic bonds* or simply *bonds*. These bonds represent the flow of energy and can be interpreted as water pipes, where the water represents the energy. The graphical representation of a bond is illustrated in Figure B.1. In the used notation, an effort value is indicated above an horizontal bond and on the left of a vertical one and the dual flow under an horizontal bond and on the right side of a vertical one. It is important to be aware of the following:

- Each bond represents both an effort $e$ and a dual flow $f$ like a voltage and a current or a force and a velocity.

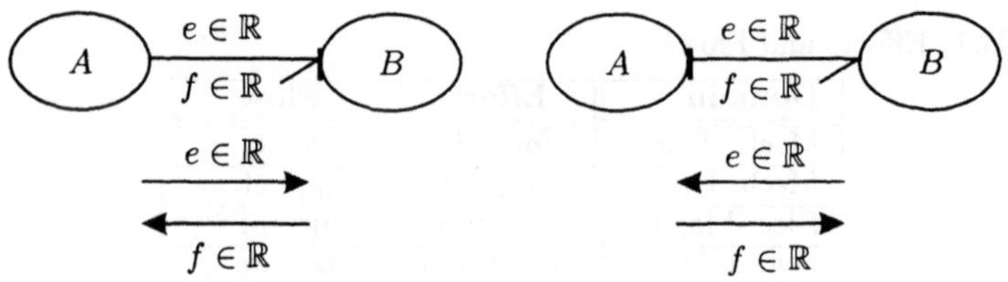


**Fig. B.2.** The Causal Stroke

- The half arrow does not give the direction of either the effort or the flow, but the direction of the positive power $P = e^T f$: if $e^T f > 0$ in Figure B.1 the energy instantly flows from $A$ to $B$.
- The direction of the effort can be if necessary specified by the *causal stroke* as can be seen in Figure B.2 for a single-bond. As a consequence the dual flow goes ALWAYS in the opposite direction for physical reasons. This is because, if an element could supply both the value of the effort and the flow, it could set the power flow $P = e^T f$ independently of the other system. It could therefore extract infinite energy from the other system and this makes no sense.

In Figure B.1 and Figure B.2, the bond connects the power ports of the subsystems $A$ and $B$ and this means that both the effort and flow of those ports MUST be the same: this is an interconnection constraint which specifies how the systems $A$ and $B$ interact.

By means of a set of bonds and junctions which generalize Kirchhoff's laws, we can describe the so called *network structure* which describes how all the basic elements are interconnected and how they interact with one another. This is exactly the same as in electrical networks where we have ideal capacitors, inductors, resistors, current-sources, voltage sources, transformers, gyrators and where we describe their interconnection by means of Kirchhoff's laws.

As in electrical networks all external variables of the basic elements are voltages and currents, in physical modeling we will use the generalization of this concept and use the power conjugate variables efforts and flows as given in Table B.1.

Furthermore, as in electrical networks we can express all the interconnections by means of linear relations relating currents among them and voltages among them, we will show that for lumped physical modeling we can express the general interconnection as linear relations relating efforts among them and flows among them. These relations result purely from a generalization of Kirchhoff's current and voltage laws.

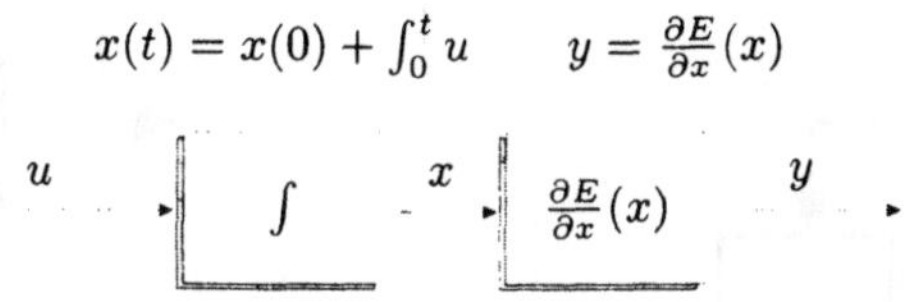


**Fig. B.3.** The general structure of an energetic element

## B.4 Energy Storage and Physical States

One remarkable feature of the storage elements of physical lumped models is that their structure is identical. A storage element is an ideal element which represents the physical storage of energy in a model like ideal capacitors and inductors, masses or springs. Each storage element in what is called causal *integral form*[1] is characterized by:

- An input $u$
- An output $y$
- A physical state $x$
- An energy function $E(x)$ of the state $x$.

as can be seen in Figure B.3.

Note that the energy function is a scalar function of the physical state. The state space equations of the system represented in Figure B.3 are:

$$\dot{x}(t) = u(t) \tag{B.1}$$

$$y(t) = \frac{\partial E}{\partial x}(t). \tag{B.2}$$

In physical modeling, the inputs are *always and only* either efforts or flows like the one reported in Table B.1. Furthermore, if $u$ is an effort, this implies that $y$ is a flow and if $u$ is a flow, this implies that $y$ is an effort. But why is that? An answer is easily given by looking at Figure B.3 and remembering that the product of an effort and a flow is always physical power. By means of these choices, we have achieved that:

> *the power represented by the dual product of $u$ and $y$ is equal to the power supplied to the storage element.*

This is true since by definition we have said that $u$ and $y$ must be two power conjugate variable and since the change in stored energy is equal to:

$$\dot{E} = \frac{dE}{dt} = \frac{dE}{dx}^T \frac{dx}{dt} = \frac{dE}{dx}^T \dot{x} = y^T u = P_{\text{supplied}}.$$

Note that for this reason and for the definition of a power bond,

[1] This name is given by the fact that an integrator appears in Figure B.3.

**Fig. B.4.** The **C** element in its two causal representations.

> *the half arrow power bonds indicating the direction of the positive power of $e^T f = y^T u$ connected to storage elements should be ALWAYS directed towards the storage elements! In Figure B.1, B could be a storage element, but A could be not.*

This is true because $\dot{E} > 0$ ($E$ increases) if energy is supplied to the element and therefore $y^T u > 0$ should indicate energy flowing toward the element with energy storage $E$.

By using the structure of Figure B.3 and the power conjugate variables, we are able to nicely express the change of stored energy $\dot{E}$ as function of the external variable $(e, f)$ of the storage element. Each sub-domain as defined in Sect. B.1 has its own storage element storing the energy type characterizing the sub-domain: the magnetic sub-domain has an ideal element storing magnetic energy and corresponding to an ideal inductor, the electric sub-domain has an ideal element storing electrical energy, corresponding to an ideal capacitor and dual to the previous element as it will be shown later, the mechanical potential sub-domain has an ideal element storing potential energy and corresponding to an ideal spring, the mechanical kinetic sub-domain has an ideal element storing kinetic energy, corresponding to an ideal mass and dual to the previous element and so on.

### B.4.1 Bond graphs representations

Depending whether the input $u$ of the integral form is an effort or the flow we define two dual elements in bond graphs:

- **C** element: it has a *flow* as the input $u$ of the integral causality form of Figure B.3 and as a consequence it has the dual effort as an output $y$.
- **I** element: it has an *effort* as the input $u$ of the integral causality form of Figure B.3 and as a consequence it has the dual flow as an output $y$.

In Figure B.4 the two possible causal representations of a **C** element in bond graphs are shown with around it the corresponding block diagram. On the left of Figure B.4, we see that as defined, a **C** element in its integral form has

**Fig. B.5.** The **I** element in its two causal representations.

a flow $f$ as input, which is also indicated by the causal stroke NOT on the side of the element (effort is an output). Remember that a bond represents a bi-directional signal (effort and flow go in opposite directions). The left side of Figure B.4 is identical to Figure B.3 with a flow $f$ as an input $u$. Furthermore, in the figure $\gamma(q) := \frac{\partial E(q)}{\partial q}$ where we used $q$ instead of $x$ to stress that $q$ is the integral of a flow:

$$q(t) = q(t_0) + \int_{t_0}^{t} f(s)ds.$$

Clearly $q$ is a physical state and it is called in the general case *generalized displacement* to indicate it is the integral of a flow. An energy which is a function of a generalized displacement is called *generalized potential energy*.

On the right side of Figure B.4, we changed the causality and inverted input with output getting what is called the *differential form* due to the presence of the differentiator element. This is shown in the bond graph with the causal stroke on the side of the element which indicates that the effort is now an input instead of an output. In the differential form, we need the inverse of $\gamma$ and we have to differentiate instead of integrate. In the same way as $\gamma$ was the gradient of the energy function, it can be shown that $\gamma^{-1}$ can be calculated as the gradient of a new function which is called *co-energy*. Note that for a **C** element, the co-energy is NOT a function of the state as a proper energy function should be, but a function of an effort:

$$E^*(e) \Rightarrow \gamma^{-1}(e) = \frac{\partial E^*(e)}{\partial e}.$$

We will see examples of energy and co-energy later.

Exactly in an identical way to the **C** element, we have that the dual element **I** has in its integral form an effort as an input. The figure corresponding to Figure B.4 for the dual element **I** is reported in Figure B.5. The identical line of reasoning can be followed inverting effort and flow. On the left of Figure B.5, we see that as defined an **I** element in its integral form has an effort $e$ as input, which is also indicated by the causal stroke on the side of the element. Remember that a bond represents a bi-directional signal (effort and flow go in opposite directions). The left side of Figure B.5 is identical to Figure B.3 with an effort $e$ as an input $u$. Furthermore, in the figure $\gamma(p) := \frac{\partial E(p)}{\partial p}$ where we used $p$ instead of $x$ to stress that $p$ is the integral of an effort:

$$p(t) = p(t_0) + \int_{t_0}^{t} e(s)ds.$$

Clearly $p$ is a physical state and it is called in the general case *generalized momenta* to indicate it is the integral of an effort. An energy which is a function of a generalized momenta is called *generalized kinetic energy.*

On the right side of Figure B.5, we changed the causality and inverted input with output getting what is called the *differential form* due to the presence of the differentiator element. This is shown in the bond graph with the causal stroke NOT on the side of the element which indicates that the effort is now an output instead of an input. In the differential form, we need the inverse of $\gamma$ and we have to differentiate instead of integrate. In the same way as $\gamma$ was the gradient of the energy function, it can be shown that $\gamma^{-1}$ can be calculated as the gradient of a new function which is called *co-energy.* Note that for an **I** element, the co-energy is NOT a function of the state as a proper energy function should be but a function of a flow:

$$E^*(f) \Rightarrow \gamma^{-1}(f) = \frac{\partial E^*(f)}{\partial f}$$

In the standard symobology, the multidimensional **I** elements are indicated with $\mathbb{I}$ and the multidimensional elements **C** elements with $\mathbb{C}$.

To better understand all these concepts which could seem quite abstract at first, we now analyse the scalar elements of the mechanical and electrical domains as an example.

### B.4.2 Mechanical domain

In the mechanical translational domain, from Table B.1 it can be seen that the flow corresponds to a velocity $v$ and an effort to a force $F$. If we start considering the element **C** of Figure B.4 in the translational mechanical domain (the input of Figure B.3 as $u = v$), we have seen that the physical state (a generalized displacement) is the integral of a velocity and therefore a position. The energy (generalized potential energy) of this element must therefore be a function of a position. The element **C** represents therefore in the translational mechanical domain a spring. For an ideal, scalar, linear spring, the stored potential energy is

$$E(x) = \frac{1}{2}kx^2. \qquad \text{(B.3)}$$

This implies that, looking at Figure B.4, we have that the output $y$ is an effort since the input was a flow. Furthermore, since we are in the mechanical domain, an effort corresponds to a force (see Table B.1). Using the energy of Eq. (B.3) we get as an output:

$$y = F = \frac{\partial E}{\partial x} = kx,$$

which is indeed the elastic force generated by a linear spring. It is important to realize that all that has been said holds for ANY properly defined energy function which is the ONLY information characterizing an ideal storage of energy. We could for example take a spring with energy function:

$$E(x) = \frac{1}{2}kx^2 + \frac{1}{4}kx^4, \tag{B.4}$$

which would result in an elastic force of

$$y = F = \frac{\partial E}{\partial x} = kx + kx^3,$$

which is a well defined non-linear ($x^3$ term is present) spring. For the linear spring, the co-energy would be a function of the input in the differential causality of Figure B.4 and therefore a function of the effort $F$:

$$E^*(F) = \frac{1}{2}\frac{F^2}{k},$$

which would give:

$$x = \gamma^{-1}(F) = \frac{\partial E^*(F)}{\partial F} = \frac{F}{k}.$$

If instead of taking the flow as an input in the mechanical domain, we take the effort, we get the **I** element of Figure B.5 which is the storage element of the other sub-domain of the mechanical domain, namely the kinetic mechanical domain. Choosing therefore $u = F$, we know from physics that the integral of a force is equal to the momenta $p$ by Newton's law:

$$F = \frac{dp}{dt} \qquad \text{where} \qquad p = mv. \tag{B.5}$$

Some people would start now wandering and saying that they thought that Newton's law was:

$$F = ma. \tag{B.6}$$

This is NOT the general form of Newton's law because if the mass is time varying, Eq. (B.6) does NOT hold anymore! This can be seen expanding the correct form of Eq. (B.5):

$$F = \frac{dp}{dt} = \frac{dmv}{dt} = m\frac{dv}{dt} + \frac{dm}{dt}v = ma + \underbrace{\dot{m}v}_{\text{in general} \neq 0!}$$

This also shows that

*The "proper physical state" for the element representing the storage of kinetic energy is NOT the velocity $v$ but the momentum $p$.*

This last statement implies that an expression for the kinetic energy must be a function of the momenta $p$ and not of a velocity $v$. Even if $v$ and $p$ are related, as a function the kinetic energy must be a function of $p$ and is:

$$E(p) = \frac{1}{2}\frac{p^2}{m} \tag{B.7}$$

This implies that, looking at Figure B.3, the output $y$ is a flow since the input was an effort as shown on the left of Figure B.5. Furthermore, since we are in the mechanical domain, a flow corresponds to a velocity (see Table B.1). Using the energy of Eq. (B.7) we get as an output:

$$y = v = \gamma(p) = \frac{\partial E}{\partial p} = \frac{p}{m}$$

which is indeed the relation we know between velocity and momenta. The function:

$$E^*(v) = \frac{1}{2}mv^2$$

which is usually called kinetic energy, is properly speaking a *kinetic co-energy* since it is a function of a flow. In the right side of Figure B.5 we get:

$$p = \gamma^{-1}(v) = \frac{\partial E^*(v)}{\partial v} = mv$$

which is indeed what we already know.

### B.4.3 Electrical domain

We can proceed identically as in the mechanical domain and we can realize that contrary to what is often thought, proper physical states in the electrical domain are the charge $q$ and the flux $\phi$ and NOT the current $i$ and the voltage $u$ which are instead respectively the flow and the effort of the electrical domain.

To begin with, consider as an input of Figure B.3 the flow of the electrical domain which corresponds to the current $u = i$ and therefore the corresponding bond graph element is the **C** element of Figure B.4. This implies that the physical state is the integral of $i$. The time integral of a current is a charge $q$ (generalized displacement) which is therefore the physical state of the storage element of the electrostatic domain which has the flow as an input (generalized potential energy storage). The usual ideal linear capacitor energy function is

$$E(q) = \frac{1}{2}\frac{q^2}{C} \tag{B.8}$$

where $C$ is the capacitor constant, and NOT

$$E^*(v) = \frac{1}{2}Cv^2 \tag{B.9}$$

which is instead the *co-energy* and is a function of the effort $v$ (voltage now) and not of a physical state. Using therefore Eq. (B.8) we obtain that the effort output is:

$$y = v = \frac{q}{C},$$

which corresponds to the voltage of an ideal linear capacitor. Once again, we could consider a non linear element with an energy like:

$$E(q) = \frac{1}{2}\frac{q^2}{C} + \frac{1}{4}\frac{q^4}{C} \tag{B.10}$$

and we would get

$$y = v = \frac{q}{C} + \frac{q^3}{C}. \tag{B.11}$$

The energy function is once again the only thing we neet to know to calculate the corresponding input, output and the dynamic equations. Furthermore, using the co-energy we can find an expression of $\gamma^{-1}()$ in Figure B.4:

$$q = \gamma^{-1}(v) = \frac{\partial E^*(v)}{\partial v} = Cv$$

which is the inverse of the previously found relation.

As a last example consider as an input the voltage $v$. The integral of a voltage corresponds to a flux $\phi$ and the corresponding storage element with effort input $v$ is an ideal inductor. The physical state of an inductor is therefore a flux $\phi$ and not the current $i$. We know by now that once we define the energy function we can calculate everything. For a linear ideal inductor we have:

$$E(\phi) = \frac{1}{2}\frac{\phi^2}{L} \tag{B.12}$$

where $L$ is the induction constant, and not

$$E^*(i) = \frac{1}{2}Li^2 \tag{B.13}$$

**Table B.2.** Generalized states

| Domain | Gen.mom. ($\int e$) | Gen.disp. ($\int f$) |
|---|---|---|
| Mechanical Translational | momentum $p$ | displacement $x$ |
| Mechanical Rotational | ang.momentum $m$ | ang. displacement $\theta$ |
| Electromagnetic | flux linkage $\phi$ | charge $q$ |
| Hydraulic | pressure mom $P_p$ | volume $V$ |
| Thermic | NONEXISTENT | entropy $E$ |

which is instead the *co-energy*. Differentiating Eq. (B.12) we obtain the corresponding flow output:

$$i = \frac{\phi}{L}$$

Table B.2 gives a summary of the relations among efforts, flows and states for some sub-domains. In the table $\int e$ indicates the integral of an effort and $\int f$ the integral of a flow.

To conclude the discussion on storage elements it is important to realize that

1. It is important to be aware of what are the real physical states of certain elements like $\phi$ and $q$ and not $i$ and $v$. This becomes very important if we consider time varying capacitors or inductors. The same holds for ANY domain.
2. Once we have an energy function we can directly write the equation of the corresponding element.
3. From the argument of the energy function we can even recognize what physical ideal element it represents i.e. if it is a function of $\phi$ it represents an inductor, if it is a function of $p$ a mass etc.
4. The only ideal physical elements to which a state is associated are elements which represent the storage of energy like the one presented in this section.
5. In bond graph notation the power bond connected to a storage element must be always directed toward the element.

### B.4.4 Duality

As also shown in Figure B.4 and Figure B.5, each physical domain but the thermal one has two storage elements **C** and **I** corresponding respectively to the generalized potential energy storage and generalized kinetic energy storage. Each of these elements is the storage of one of the two sub-domains of a primar domain (see Sect. B.1). These two storage elements are called *dual*. This duality is one of the major concepts in physics since it is the cause of oscillations if dual elements are connected together: a spring with a mass or a capacitor with an inductor.

The thermal domain does NOT have both possibilities and this is the cause of the irreversibility of the transformation of energy to the thermal domain due to a lack of "symmetry" in the domain.

### B.4.5 Extra supporting states

In practical applications, we will sometime use extra states which are needed for specific purposes but do NOT have physical energy associated to it. If for example we want to talk about the position of a mass translating by itself, its physical state is the momenta $p$. The position of the mass can be expressed as the time integral of its velocity which we know to be equal to $p/m$. If we can measure the mass position and not its velocity, the representing state equation would be:

$$\begin{pmatrix} \dot{p} \\ \dot{x} \end{pmatrix} = \begin{pmatrix} 0 \\ p/m \end{pmatrix} + \begin{pmatrix} u \\ 0 \end{pmatrix}$$
$$y = x$$

where the total state is therefore $(p, x)^T$. In the total state of the state space equations, $p$ is a physical state because it is the argument of the energy function of the mass, but $x$ is a supporting state which is needed for analysis and it does not have any physical energy associated to it.

## B.5 Free Energy Dissipation

In the previous section, we have treated storage elements and the states associated to them. To model physical systems, it is also important to treat elements able to describe the irreversible transformation of energy from a domain like the mechanical or electrical one to the thermal domain. It is often talked about "dissipation of energy" which is not correct in the sense that, for the first principle of thermodynamics energy cannot be neither created nor destroyed (dissipated), but only transformed. What it is meant with "dissipation of energy" is actually the dissipation of what is called *free-energy* which is mathematically defined as the Legendre transformation of the energy with respect to the entropy.

If we exclude the thermic domain from our analysis, we can proceed as usual and consider elements like ideal electrical resistors or mechanical dampers which dissipate (free)energy.

An ideal dissipator of free energy is characterized by a purely statical (no-states) relation between an effort and a flow like the following:

$$e = Z(f) \qquad \text{(Impedance form)} \tag{B.14}$$

or

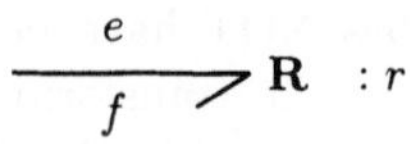


**Fig. B.6.** The irreversible energy transformer or free dissipation element.

$$f = Y(e) \qquad \text{(Admittance form)} \tag{B.15}$$

for which, the following must hold

$$Z(f)f \leq 0 \qquad \text{or} \qquad eY(e) \leq 0 \tag{B.16}$$

The product of the power conjugate port variables $e$ and $f$ gives the energy flowing toward the element. The conditions of Eq. (B.16) ensure that energy can flow just toward the element but cannot be generated by the element because otherwise we would have for example that if $Z(f)f > 0$ would result in:

$$P_{\text{out}} = -P_{\text{in}} = -Z(f)f > 0$$

In Figure B.6 the bond graph representation of this "dissipation" element is reported. Once again, for consistency with the power flow direction,

> *the bond connected to an* ***R*** *element should be ALWAYS directed toward the element!*

### B.5.1 Electrical domain

In the electrical domain, the functions $z(\cdot)$ and $y(\cdot)$ of Eq. (B.14) become for a linear, scalar element

$$u = Ri \qquad \text{and} \qquad i = \frac{1}{R}u$$

which correspond to Ohm's law. For a linear resistive element it does not make therefore any difference if the input is the voltage $u$ (effort) or the current $i$ (flow). The element is said *causally invertible.* In Figure B.6, the ": $r$" would specify the value of the constant $R$ of a linear element ($r = R$).

### B.5.2 Mechanical domain

In the mechanical domain, the functions $z(\cdot)$ and $y(\cdot)$ of Eq. (B.14) become for a linear element

$$F = bv \qquad \text{and} \qquad v = \frac{1}{b}F$$

where $b$ is called the *viscous damping coefficient* and it is a linear model for friction. For a linear damping element makes therefore no difference if the input is the force $F$ (effort) or the velocity $v$ (flow). The element is said again to be *causally invertible.* In Figure B.6, the ": $r$" would specify the value of the constant $b$ of a linear element ($r = b$).

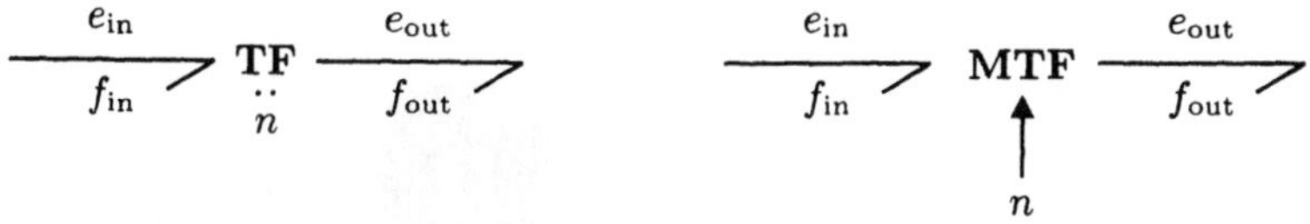


**Fig. B.7.** The ideal transformer element.

## B.6 Ideal Transformations and Gyrations

So far, the elements that we have analyzed are characterized only by one scalar power port: one scalar effort and the corresponding dual scalar flow which are in bond graph represented by only one bond connected to the element. As in electrical networks, there are also elements with two power ports, corresponding to the two power bonds connected to the element.

The elements we introduce now are ideal, power continuous, two port elements which means that in each instant of time, the power flowing into the element from one of the two ports (the input bond) is identical to the one flowing out from the other port (the output bond). This implies that the element cannot store energy inside.

The bond graph representation of an ideal transformer is shown on the left of Figure B.7 where the input and output bonds are respectively the one with positive power flow in and positive power flow out.

The external variables are $(e_{in}, f_{in})$ corresponding to the power flowing in from the input port and $(e_{out}, f_{out})$ corresponding to the power flowing out from the other port. For power continuity we need therefore that

$$P_{in} = e_{in}^T f_{in} = e_{out}^T f_{out} = P_{out}. \tag{B.17}$$

Furthermore, these ideal elements are characterized by the fact that they describe a linear relation between one of the external variable on one port to one of the external variables on the other port. Depending whether this relation is between the flow of one port and the flow of the other port or between the effort of one port and the flow of the other port, we talk respectively of *ideal transformers* or ideal *gyrators*.

### B.6.1 Ideal Transformers

A bond graph representation of a transformer is shown on the left of Figure B.7. An ideal transformer is characterized by a linear relation between flows and, as we will see, by a dependent linear relation between the efforts of the two ports. The characterizing equation is therefore of the form:

$$f_{out} = n f_{in}$$

where $n$ is the linear constant characterizing the transformer and it is also shown on the left of Figure B.7. Due to the power constraint of Eq. (B.17),

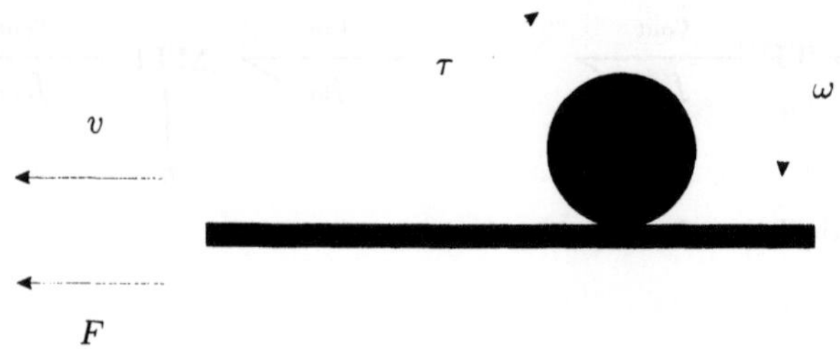


**Fig. B.8.** An ideal transformation

it is now possible to see that the only possible relation between the efforts which satisfies Eq. (B.17) is

$$e_{\mathrm{in}} = n e_{\mathrm{out}},$$

or equivalently

$$e_{\mathrm{out}} = \frac{1}{n} e_{\mathrm{in}}.$$

We can therefore conclude that if the two ports belong to the same physical domain and $n < 1$, the effort at the input port is smaller than the effort at the output port, but the flow at the input port is greater than the flow at the output port.

This kind of element is used to model gear-boxes. Think about going with a race bike which has different gears. We can consider $e_{\mathrm{in}}$ as the torque applied on the axis of the pedal by us and therefore $f_{\mathrm{in}}$ is the angular velocity around the pedals. This port belongs to the mechanical rotational domain. In the same way, we can consider $e_{\mathrm{out}}$ as the corresponding torque applied on the axis of the back wheel of the bike and therefore $f_{\mathrm{out}}$ is the angular velocity of the back wheel of the bike. This latter port is again in the mechanical rotational domain. We know by experience that if we want to go up a hill, we need more torque on the wheel and therefore we need a smaller $n$, but as a consequence, to keep the same velocity of the bike ($f_{\mathrm{out}} = \mathrm{const}$) with a smaller $n$ we will have to cycle much faster since:

$$\underbrace{f_{\mathrm{in}}}_{\text{increases}} = \underbrace{\frac{1}{n}}_{\text{increases}} \underbrace{f_{\mathrm{out}}}_{\text{remains constant}}$$

Another ideal transformation is shown in Figure B.8 which could represent the relation between couple and angular velocity of the back wheel of a bike and linear tension and linear velocity of the chain. In this case we define it with the "proportional constant" $l$:

$$v = l\omega,$$

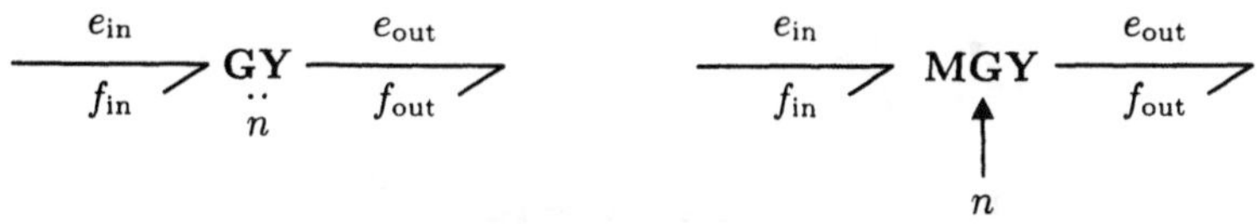


**Fig. B.9.** The ideal gyrator element.

which for power continuity implies

$$\tau = lF.$$

Note that ALWAYS for transformers. the defining constant relates the efforts in one way and also the flows in the other way:

$$\begin{array}{ccc} & \text{velocities} & \\ \begin{array}{c}\text{Rotational}\\ \text{domain}\end{array} & \begin{pmatrix}\omega\\ \tau\end{pmatrix} \begin{array}{c}\longrightarrow\\ \longleftarrow\end{array} \begin{pmatrix}v\\ F\end{pmatrix} & \begin{array}{c}\text{Translational}\\ \text{domain}\end{array} \\ & \text{forces} & \end{array}$$

In case the gain factor $n$ is variable and dependent on another variable of the system, the transformer is called *modulated.* The bond graph symbol is reported on the right of Figure B.7. The symbol has an extra signal arrow indicating where the value $n$ is coming from.

### B.6.2 Ideal Gyrators

Gyrators differ from transformers only because the linear constant relates the effort of the output port with the flow of the input port instead of the effort. With reference to Figure B.9, for a gyrator we have therefore that:

$$e_{out} = nf_{in}$$

where $n$ is the linear constant characterizing the gyrator. Due to the power constraint of Eq. (B.17), it is now possible to see that the only relation between the input effort and the output flow which satisfies Eq. (B.17) is

$$e_{in} = nf_{out},$$

or equivalently

$$f_{out} = \frac{1}{n}e_{in}.$$

The perfect example for a gyrator is the gyrative effect of a DC motor where electrical power flows in and mechanical power flows out. The motor constant $K$ is the sufficient information which relates the input current and the output torque:

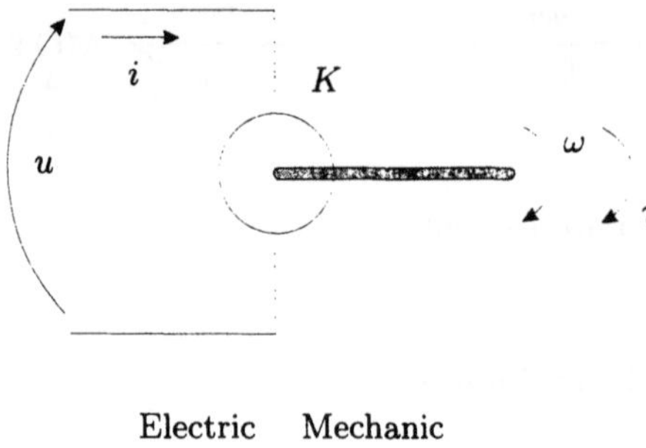


**Fig. B.10.** The ideal gyrator

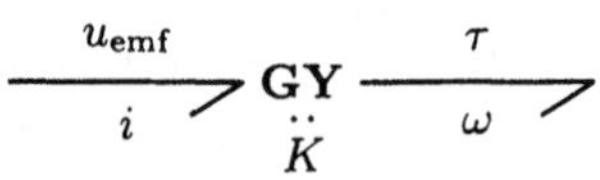


**Fig. B.11.** The ideal gyration of a DC motor.

$$\tau = Ki.$$

Due to power continuity, the only possible relation between the remaining effort and flow is

$$u = K\omega,$$

where $u$ is called the *e.m.f* of the motor.

Often in books the two $K$ of the previous equations are different: this makes no sense because otherwise we would not have conservation of energy! For this element the gyrator is shown in Figure B.11. As with a transformer, the defining constant $K$ relates variables in one direction and the remaining in the other direction:

$$\begin{matrix}\text{Electrical} \\ \text{domain}\end{matrix} \begin{pmatrix} i \\ u \end{pmatrix} \begin{matrix} \longrightarrow \\ \longleftarrow \end{matrix} \begin{pmatrix} \tau \\ \omega \end{pmatrix} \begin{matrix}\text{Rotational} \\ \text{domain}\end{matrix}$$

In case the gain factor $n$ is variable and dependent on another variable of the system, the gyrator is called *modulated.* The bond graph symbol is reported on the right of Figure B.9. The symbol has an extra signal arrow indicating where the value $n$ is coming from.

### B.6.3 Multi-bonds

Multi-bonds are very similar, but the characteristic constant becomes a matrix. Furthermore, if this matrix is time variant, we have what is called a modulated transformer or gyrator.

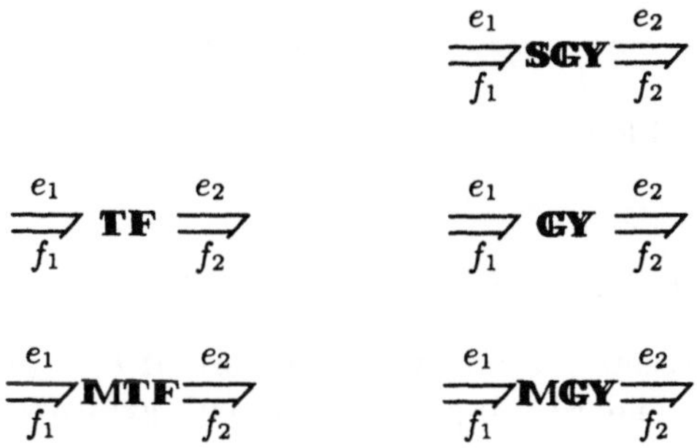


**Fig. B.12.** Transformers and Gyrators

**Transformers: TF, MTF.** If we indicate with $(e_1, f_1)$ the effort-flow pair on one port and with $(e_2, f_2)$ the effort-flow pair on the other port, multidimensional transformers are characterized by the following relation:

$$f_2 = N f_1 \tag{B.18}$$

where $N$ is a matrix of adequate dimensions. For the hypothesis on power continuity, we have:

$$f_2 = N f_1 \overset{e_1^T f_1 = e_2^T f_2}{\Rightarrow} e_1 = N^T e_2 \tag{B.19}$$

which gives the relation of the efforts. An element with such relations can be used as an ideal lossless transformer. If $N$ is constant, the symbol used for multi-bonds is **TF**, as reported in Figure B.12. If $N$ is varying being a function of some other values, it is called a modulated transformer, it is indicated as **MTF** in Figure B.12.

**Gyrators: GY, MGY, SGY.** Transformers are expressed by a relation between flows and by a dual relation between efforts. If we have a relation between a flow of a port and an effort of the other port, we have a *gyrative* effect, which is represented by a gyrator element. The relation characterizing a gyrator therefore has the form:

$$e_2 = N f_1, \tag{B.20}$$

and to have power continuity:

$$e_1 = N^T f_2. \tag{B.21}$$

If $N$ is constant, the bond graph element is reported as **GY** in Figure B.12, but if $N$ is varying, it is indicated with **MGY**.

Gyrators are often represented with only one multidimensional port corresponding in this case to the effort flow pair $((e_1^T, e_2^T)^T, (f_1^T, -f_2^T)^T)$. Such an element is then characterized by the relation:

**Fig. B.13.** Flow and Effort sources

$$e = Sf \quad \text{with} \quad S = -S^T, \tag{B.22}$$

where $(e, f)$ represents the unique power port. If $e = (e_1^T, e_2^T)^T$ and $f = (f_1^T, -f_2^T)^T$, the standard two-port gyrator with gyrative matrix $N$, can be represented with a one-port gyrator with characterizing matrix:

$$S = \begin{bmatrix} 0 & -N^T \\ N & 0 \end{bmatrix}. \tag{B.23}$$

A special gyrator is obtained if $N$ is equal to the identity matrix. This element is called *symplectic gyrator* and is indicated with **SGY** in Figure B.12. This fundamental element only interchanges the efforts and flows at the two ports from an algebraic point of view. It can be used to dualize a **I** element to get a **C** element. The thermal domain, due to its asymmetry, does not have an intradomain[2]**SGY**. Interested readers can consult [45].

## B.7 Ideal Sources

All the elements which we have been treating so far do not generate energy. With this building blocks we would not be able to create any model with which we can interact and to which we could supply energy. To solve this problem we consider two extra elements which are called *ideal flow source* and *ideal effort source.* These two elements are represented respectively on the left and on the right of Figure B.13. Note that these are the only elements from which the power bond direction goes out and this is because the positive flow of energy is the power $P_{\text{source}} = e^T f$.

These two elements correspond respectively to the ideal voltage and current source in the electrical domain. These elements are characterized by the fact that they can supply a certain effort of flow independently of the value of their dual flow and effort.

## B.8 Kirchhoff's laws, junctions and the network structure

In the previous section we have briefly treated all the atomic bricks which are needed to model physical lumped systems. How we place the bricks with

2 *Intra* comes from the Latin *intrā* which means inward. Intradomain means within the same domain.

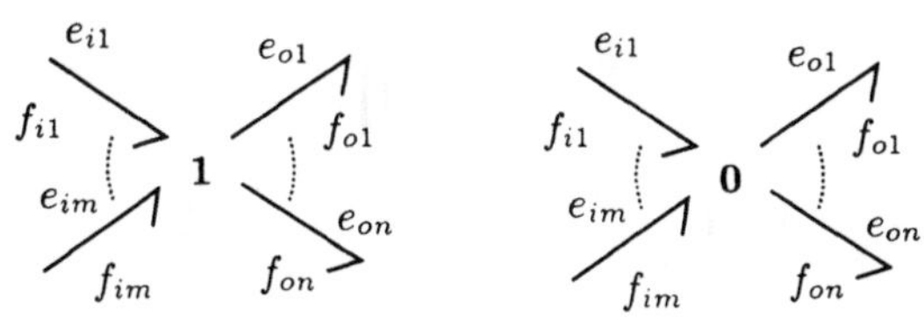


**Fig. B.14.** The junction elements.

respect to each other, will give us a dynamic model. So far we have not said anything on how the basic elements are interconnected and this is absolutely necessary because it strongly characterizes how energy will flow between the various elements and will result in the dynamic equations of the system.

The specification of the interconnection can be done using a generalization of Kirchhoff's laws. The network structure expresses the constraints between the efforts and the flows of all the elements. This represents how elements are connected together. The two basic bond graph elements which represent the generalization of Kirchhoff's laws are the so called **1** junctions, also called *flow junctions* and **0** junctions also called *effort junctions.* The bond graph representation of these elements is illustrated in Figure B.14. Junctions can have any number of bonds attached to them which can be used to connect to other junctions or to the basic elements we have studied to create a total network interconnection. As the transformers, the junctions are power continuous which means that the total power flowing in must be equal to the total power flowing out.

### B.8.1 1-junctions

The **1**-junction is also called flow junction because this junction is characterized by the fact that all bonds connected to it are constrained to have the same flow values. This also implies that causally speaking, only one of the bonds connected to the junction will set the flow value of this junction and all other bonds will use it. The equations characterizing a flow junctions are:

$$f_{i1} = \ldots = f_{im} = f_{o1} = \ldots = f_{on} \qquad \text{(flow equation)},$$

$$\sum_{k=1}^{m} e_{ik} = \sum_{k=1}^{n} e_{ok} \qquad \text{(effort equation)}.$$

It can be easily seen that with the previous equations, the power continuity of the element is ensured. This junction corresponds to the mesh Kirchhoff's law in electrical networks: all the elements in series in a mesh have the same flow since they are traversed by the same current (flow equation). Furthermore, the algebraic sum of their potential must be equal to zero (effort equation).

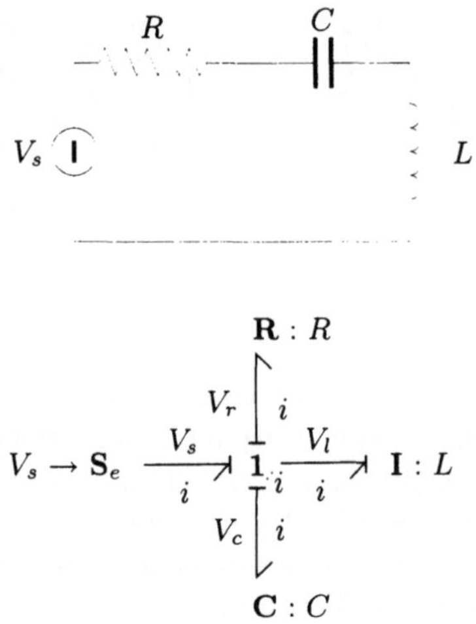


**Fig. B.15.** An electrical example of a **1**-junction.

In Figure B.15 an example is shown of a simple bond graph model of an electrical network. The elements of the electrical network are traversed by the same current and therefore the ideal bond graph elements used to represent them are attached to a flow junction. On each bond are represented the efforts and the flows. Note that, by the fact that all the bonds should point to the **R**, **C** and **I** elements and that the bond of the source should point out of the element, all the signs will automatically result correct! The causality is also reported with the causal strokes. The storage elements are chosen in their integral causality to have proper states. The **I** in its integral causality is the element which "sets" the current of the junction (mesh) and the other elements have this current as input and voltages as outputs instead which are algebraicly summed to supply a value to the inductor. As known in electrical networks, we could not have a current source instead of a voltage source in series with an inductor because this two elements would both try to "set" the current of the junction at the same time, but just one element connected to the junction is allowed to do it. The effort equation of the junction is from Figure B.15:

$$V_s = V_r + V_c + V_l.$$

The equations of the **I** element if we suppose to take it linear are

$$\dot{\phi} = V_l \qquad \text{and} \qquad i = \frac{\phi}{L}.$$

The equations of the **C** element if we suppose to take it linear are

$$\dot{q} = i \qquad \text{and} \qquad V_c = \frac{q}{C}.$$

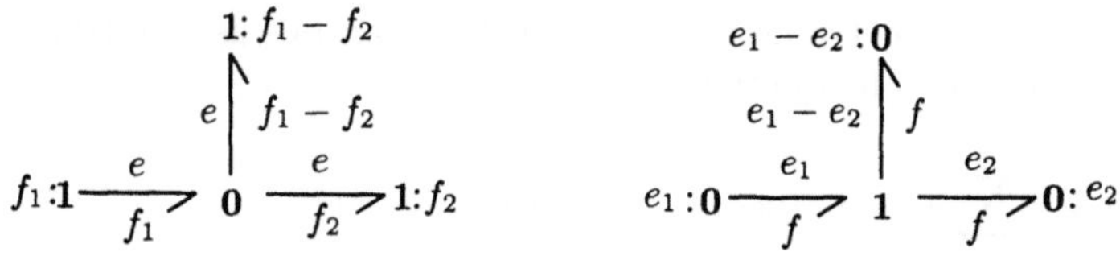


**Fig. B.16.** Flows and efforts differences.

The equation of the **R** element if we suppose to take it linear is

$$V_r = Ri.$$

All these equations uniquely describe the complete dynamics of the system.

### B.8.2 0-junctions

The **0**-junction is also called effort junction because this junction is characterized by the fact that all bonds connected to it are constrained by the junction to have the same effort value at all times. This also implies that, causally speaking only one of the bonds connected to the junction will set the effort value of this function and all other bonds will use it. The equations characterizing an effort junctions are:

$$e_{i1} = \ldots = e_{im} = e_{o1} = \ldots = e_{on} \qquad \text{(effort equation)},$$

$$\sum_{k=1}^{m} f_{ik} = \sum_{k=1}^{n} f_{ok} \qquad \text{(flow equation)}.$$

A similar example as with the **1**-junction could be shown by putting the elements in parallel, and using for causal reasons a voltage source instead of a current source.

The equations of this junction correspond to the current Kirchhoff law which says that in a note the algebraic sum of the currents should be equal to zero.

### B.8.3 Effort difference

Often, as we will see, we need the difference of two efforts of a bond graph to specify the power consistent interconnection with other elements. This is the base for drawing bond graphs from electrical schemes. On the right side of Figure B.16 the way to do it is reported. As shown in the figure, the effort value of the top effort junction is equal to the difference of the effort junction from which the bonds comes out minus the effort value of the junction to which the other bond goes to. Clearly all the flows are the same due to the constraint of the **1**-junction. This relation can be easily verified by just writing the **1**-junction equation:

$$\sum_{k=1}^{m} e_{ik} = \sum_{k=1}^{n} e_{ok} \Rightarrow e_1 = e_2 + e_3 \Rightarrow e_3 = e_1 - e_2$$

### B.8.4 Flow difference

Dually to the previous case, we can get the difference of two flows of a bond graph using the structure shown on the left of Figure B.16. This is very important to model mechanical systems as it will be shown later. As shown in the figure, the flow value of the top flow junction is equal to the difference of the flow junction from which the bonds comes out minus the flow value of the junction to which the other bond goes to. Clearly all the efforts are the same due to the constraint of the **0**-junction. This relation can be easily verified by just writing the **0**-junction equation:

$$\sum_{k=1}^{m} f_{ik} = \sum_{k=1}^{n} f_{ok} \Rightarrow f_1 = f_2 + f_3 \Rightarrow f_3 = f_1 - f_2$$

## B.9 Bond graph modeling of electrical networks

With what learned so far, the process of physical modeling of an electrical network is a straight forward process. In certain cases it is easier to directly draw the final bond graph once you have acquired some inside in the network, but the following way is mechanical.

**Algorithm B.9.1 (Bond graph of electrical networks)**

1. *For each node of the network draw a **0**-junction which corresponds to the potential of that node.*
2. *For each bipole connected between two nodes, use the scheme shown on the right side of Figure B.16 to take the difference of the potentials where a bipole is attached and connect the ideal element to the **0**-junction representing the difference.*
3. *Choose a reference where the potential is zero, and attach an effort source equal to zero to the corresponding **0**-junction.*
4. *Simplify the graph by:*
   - *Eliminating all the junctions where only two bonds are attached to it which have the same continuing direction: one goes in and one goes out.*
   - *Fuse **1** and **0**-junctions which are connected through a single-bond*
   - *Eliminate all the junctions after the 0 reference source which do not add any additional constraint.*

As it will be seen in an example later on, this method is straight forward and ensures that no sign mistakes are made.

## B.10 Bond graph modeling of mechanical systems

With what learned so far, the process of physical modeling of a one dimensional mechanical system becomes a straight forward process. In certain cases it is easier to directly draw the final bond graph once you have acquired some inside in the network, but the following way is mechanical.

**Algorithm B.10.1 (Bond graph of mechanical networks)**

1. *For each moving mass draw a* ***1****-junction which corresponds to the velocity of that mass.*
2. *Add an additional* ***1****-junction representing the inertial reference with a source of flow equal to zero attached to it.*
3. *For each inertia attach a corresponding* ***I*** *element to the one junction corresponding to its velocity.*
4. *For each damper or spring attached between two points, use the scheme shown on the left side of Figure B.16 to take the difference of the velocities of the two points and attach the corresponding element to the* ***1****-junction representing the velocity difference.*
5. *Simplify the graph by:*
   - *Eliminating all the junctions where only two bonds are attached to it and have the same continuing direction: one goes in and one goes out.*
   - *Fuse* ***1*** *and* ***0****-junctions which are connected through a single-bond.*
   - *Eliminate all the junctions after the reference source which do not add any additional constraints.*

## B.11 Examples

We will now treat first a one dimensional example which illustrates all the notions handled so far, and then shorty a multidimensional example.

### B.11.1 DC motor example

In the schematic model of a DC motor reported in Figure B.11.1, we can distinguish 6 interconnected lumps:

- 2 storage elements with corresponding physical states $(\phi, p)$: ideal inductor $L$ and rotational inertia $I$. We have therefore 2 states and the model has therefore physical order 2

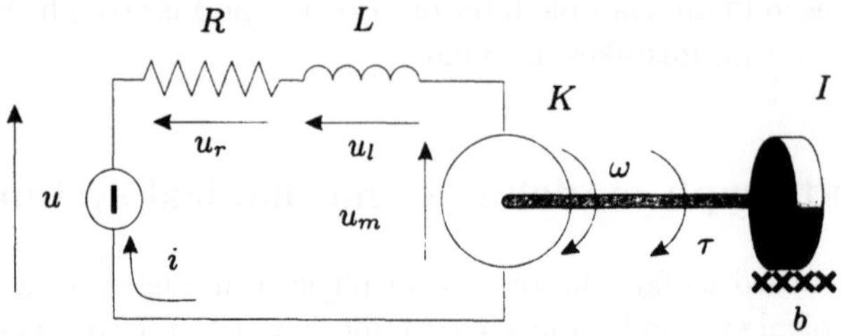


Electric Mechanic

- 2 dissipative elements: the resistor $R$ and the friction $b$.
- 1 gyration effect $K$
- An ideal voltage source $u$

**Elements equations.** We can start by writing all the equations characterizing the various ideal elements.

**Storage elements and physical states.** With reference to Sect. B.4 we obtain:

$$\text{Inertia}\begin{cases}\dot{p} = \tau_I\\ \omega = \frac{\partial E_I}{\partial p} = \frac{\partial}{\partial p}\left(\frac{1}{2I}p^2\right) = \frac{p}{I}\end{cases}$$

$$\text{Inductor}\begin{cases}\dot{\phi} = u_l\\ i = \frac{\partial E_L}{\partial \phi} = \frac{\partial}{\partial \phi}\left(\frac{1}{2L}\phi^2\right) = \frac{\phi}{L}\end{cases}$$

**Dissipating element.** Since the elements are linear, the following equations hold:

$$u_r = Ri, \qquad \tau_b = b\omega,$$

where $\tau_b$ is the dissipating torque.

**Gyration equations.** The gyration equations are:

$$\tau = Ki, \qquad u_m = K\omega.$$

In a lot of books two different constants are used. They MUST be equal for physical reasons as we have seen!

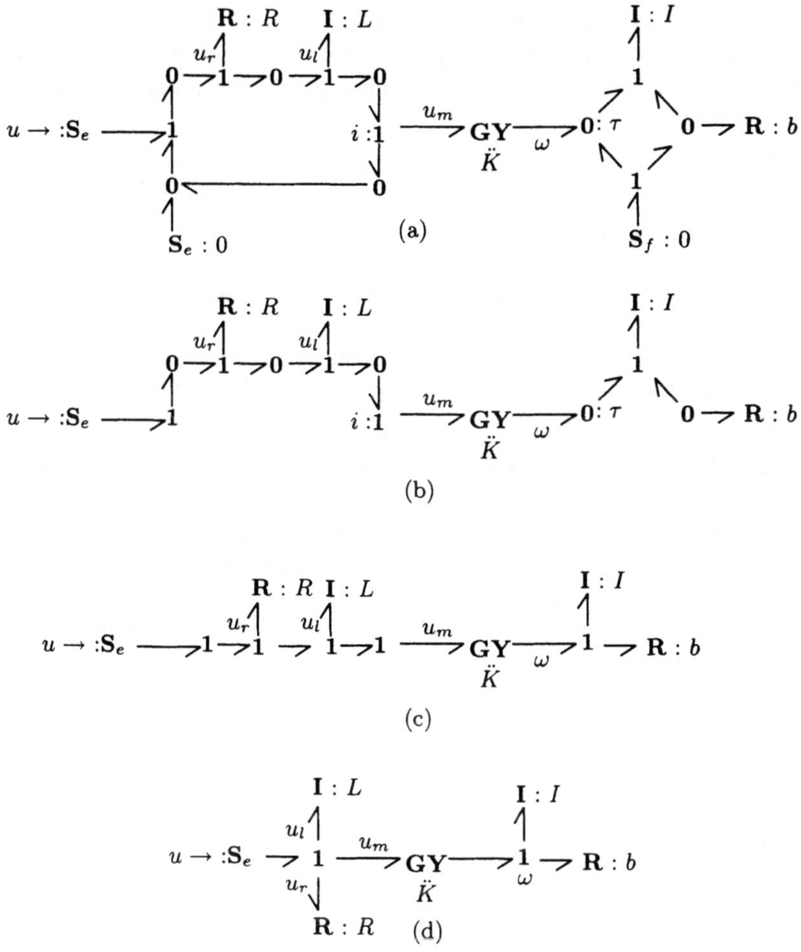


**Fig. B.17.** The bond graph steps of the DC motor model.

**Network interconnection.** We should now look at what kind of mathematical constraints describe the interconnection of the elements described above. This can be done by drawing the bond graph interconnection using Alg. B.9.1, Alg. B.10.1 and the gyrator element which connects the two domains. For this reason the gyrator is called an *intra-domain* element.

The bond graph resulting is shown in Figure B.17(a). In the figure it is possible to see that all the **0**-junctions of the electrical domain are drawn to indicate the connection points of the bipoles. For the mechanical part the two **1**-junctions represent the angular rotation of the wheel (upper junction) and the reference inertial frame (lower junction) where a source of flow equal to zero is attached as explained in Alg. B.10.1. Note that since the gyrator expresses a relation from the flow $i$ to the effort $\tau$, the two junctions directly attached to it are respectively a **1** and a **0** junctions. Note that the torque $\tau$ is applied between the wheel (upper junction in the mechanical part) and the ground (lower junction in the mechanical part). We can now simplify the bond graph by:

1. Eliminating the two zero sources and the junctions attached to it which do not add any information to the remaining junctions as can be easily seen analyzing the junctions equations. The result is reported in Figure B.17(b).
2. Eliminating all the junctions with only two bonds attached to it and obtain Figure B.17(c).
3. Mix all the possible directly communicating junctions of the same type to get the final simplified form shown in Figure B.17(d).

The final bond graph could have also been drawn directly with a bit of insight: the electrical part is a series connection of the source, the resistor, the inductor and the electrical side of the gyrator and this is expressed by connecting these bonds to a **1**-junction; in the mechanical part only the velocity $\omega$ is present. The motor applies a torque to this wheel, but part of it ($\tau_b$) is "stolen" by the dissipating element.

From the final bond graph, we can directly write the equations of the interconnection:

$$u = u_r + u_l + u_m \qquad \text{(electrical } \mathbf{1}\text{-junction)}$$

$$\tau_I + \tau_b = \tau \qquad \text{(mechanical } \mathbf{1}\text{-junction)}$$

**The final equations.** Collecting all the equations we can see that:

$$\dot{p} = \tau_I = \tau - \tau_b = Ki - b\omega = \frac{K}{L}\phi - \frac{b}{I}p,$$

$$\dot{\phi} = u_l = -u_m - u_r + u = -\frac{K}{I}p - \frac{R}{L}\phi + u,$$

which in matrix form are:

$$\dot{\begin{pmatrix} p \\ \phi \end{pmatrix}} = \begin{pmatrix} -\frac{b}{I} & \frac{K}{L} \\ -\frac{K}{I} & -\frac{R}{L} \end{pmatrix} \begin{pmatrix} p \\ \phi \end{pmatrix} + \begin{pmatrix} 0 \\ 1 \end{pmatrix} u \tag{B.24}$$

$$y := \omega = \left(\tfrac{1}{I}\ 0\right) \begin{pmatrix} p \\ \phi \end{pmatrix}.$$

These equations are of the form

$$\dot{x} = Fx + Gu$$
$$y = Hx$$

where $x = (p \quad \phi)^T$ which is a linear time invariant state space form.

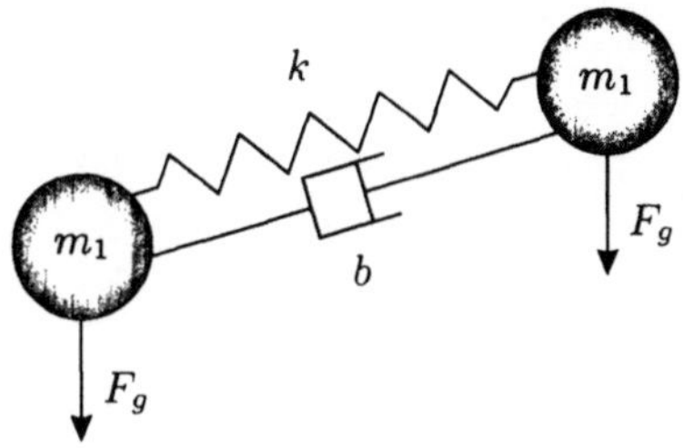


**Fig. B.18.** A mechanical system.

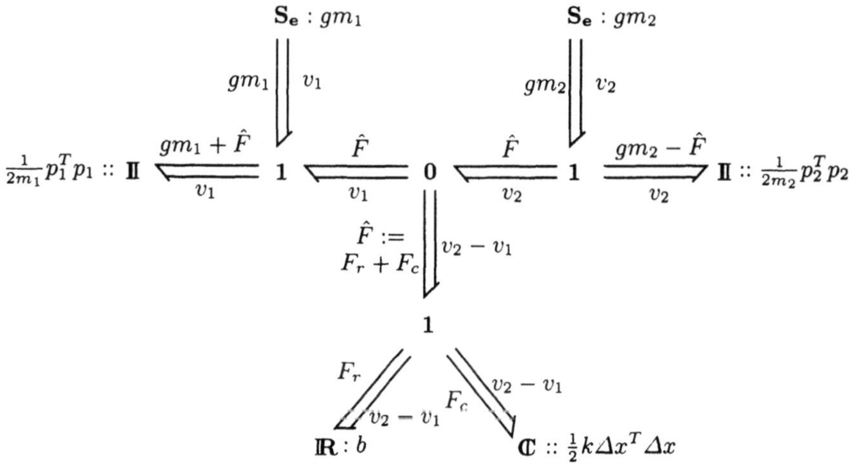


**Fig. B.19.** The bond graph of the 2D mechanical system

### B.11.2 Multidimensional example

As an example of a multidimensional use of bond graphs, consider the mechanical system shown in Figure B.18. The system is a two-dimensional mechanical system composed of two point masses connected to each other by means of an elastic translational spring and a damper. In the figure, the forces $F_g$ represent gravitational forces. The bond graph of the mechanical system is reported in Figure B.19. As usual, on each bond are reported the efforts and flows: the values on the right side and below the half arrows represent flows and on the other side they represent efforts. Notice that all the bonds attached to a 1-junction have the same flows and all the ones attached to a 0-junction the same effort. For energetic elements the notation ":: $E(q)$" indicates the energy function of the element in which $q$ is the energy variable which is equal to the momenta $p_1$ and $p_2$ for the $\mathbb{I}$ elements and to the position difference $\Delta x$ for the elastic elements. The ideal source supplies a constant force equal to the gravitational one for each mass. Finally, the notation ": $b$" for the dissipative element indicates that we have $F_r = b(v_2 - v_1)$.

# Bibliography


1. Stramigioli, S.: From differentiable manifolds to interactive robot control. PhD thesis, Delft University of Technology, Delft, The Netherlands (1998) ISBN 90-9011974-4, `http://lcewww.et.tudelft.nl/~stramigi`.
2. Hogan, N.: Impedance control: An approach to manipulation: Part I-Theory. ASME J. of Dynamic Systems, Measurement and Control **107** (1985) 1–7
3. Hogan, N.: Impedance control: An approach to manipulation: Part II-Implementation. ASME J. of Dynamic Systems, Measurement and Control **107** (1985) 8–16
4. Hogan, N.: Impedance control: An approach to manipulation: Part III-Applications. ASME J. of Dynamic Systems, Measurement and Control **107** (1985) 17–24
5. Stramigioli, S.: Intrinsically passive control and the ipc-supervisor architecture. MIROS Journal, Special Issue on "Force Control of Advanced Robotic Systems **2** (2000)
6. Stramigioli, S., Fasse, E., Willems, J.: A rigorous framework for interactive robot control. Accepted for publication on the International Journal of Control (1997)
7. Stramigioli, S., Maschke, B., Bidard, C.: Network Modeling and Control of Mechanical Systems. In: Modern Control Theory. World Scientific (1999)
8. Stramigioli, S., Maschke, B., Bidard, C.: A Hamiltonian formulation of the dynamics of spatial mechanisms using lie groups and screw theory. In Lipkin, H., ed.: Proceedings of the Symposium Commemorating the Legacy, Works, and Life of Sir Robert Stawell Ball, Cambridge (2000)
9. Stramigioli, S., van der Schaft, A., Maschke, B., Andreotti, S., Melchiorri, C.: Geometric scattering in tele-manipulation of port controlled hamiltonian systems. In: Proceedings of the 39th IEEE Conference on Decision and Control, Sydney (AUS), IEEE (2000)
10. Murray, R.M., Li, Z., Sastry, S.: A Mathematical Introduction to Robotic Manipulation. CRC Press (1994) ISBN 0-8493-7981-4.
11. Karger, A., Novak, J.: Space Kinematics and Lie Groups. Gordon and Breach Science Publishers, New York (1978) ISBN 2-88124-023-2.
12. Ball, R.: A Treatise on the Theory of Screws. Cambridge at the University Press (1900)
13. Lipkin, H.: Geometry and Mappings of screws with applications to the hybrid control of robotic manipulators. PhD thesis, University of Florida (1985)
14. Lipkin, H., Duffy, J.: The elliptic polarity of screws. ASME Journal of Mechanisms, Transmissions, and Automation in Design **107** (1985) 377–387
15. Lipkin, H., Duffy, J.: Hybrid twist and wrench control for a robotic manipulator. ASME J. of Mechanical Design **110** (1988) 138–144

16. Olver, P.J.: Applications of Lie Groups to Differential Equations. Ii edn. Volume 107 of Graduate texts in mathematics. Springer Verlag, New-York (1993) ISBN 0-387-94007-3.
17. Gilmore, R.: Lie Groups, Lie Algebras, and Some of Their Applications. John Wiley & Sons, New York (1974) ISBN 0-471-30179-5.
18. Libermann, P., Marle, C.M.: Symplectic Geometry and Analytical Mechanics. D. Reidel Publishing Company, Dordrecht, Holland (1987) ISBN 90-277-2438-5.
19. Maschke, B.M.: Elements on the Modelling of Mechanical Systems. In: Modelling and Control of Mechanisms and Robots. World Scientific Publishing Ltd., Bertinoro (1996) 1–38
20. Dubrovin, B., Fomenko, A., Novikov, S.: Modern Geometry - Methods and Applications, Part I: The Geometry of Surfaces, Transformation Groups, and Fields. Ii edn. Volume 93 of Graduate texts in mathematics. Springer Verlag, New-York (1992) ISBN 0-387-97663-9.
21. Lončarić , J.: Geometrical Analysis of Compliant Mechanisms in Robotics. PhD thesis, Harvard University, Cambridge (MA) (1985)
22. Selig, J.M., Rooney, J.: Reuleaux pairs and surfaces that cannot be gripped. International Journal of Robotics Research **8** (1989) 79–87
23. Bidard, C.: Dual bases of screw-vectors for inverse kinestatic problems in robotic. In: 4th Int. Workshop on Advances in Robot Kinematics, Ljubljana (1994)
24. Bidard, C.: Screw-vector bond graphs for multibody systems modelling. In Granda, J., Cellier, F., eds.: Proc. of the Int. Conf. on Bond Graph Modeling ICBGM93, San Diego (1993) 195–200
25. Recski, A.: Matroid Theory and its Applications in Electrical Network Theory and in Statics. Volume 6 of Algorithms and Combinatorics. Springer Verlag, Hungary (1989) ISBN 0-387-15285-7.
26. Nijmeijer, H., van der Schaft, A.: Nonlinear Dynamical Control Systems. I edn. Springer Verlag, New York inc. (1990) ISBN 0-387-97234-X.
27. Maschke, B.M., Bidard, C., van der Schaft, A.: Screw-vector bond graphs for the kinestatic and dynamic modeling of multibody systems. In: Proc. of the 1994 ASME Int. Mechanical Engineering Congress and Exposition. Volume DSC 55-2., Chicago, USA (1994) 637–644
28. Brockett, R.W.: Mathematical Theory of Networks and Systems. In: Robotic manipulators and the product of exponentials formula. Springer (1984) 120–129
29. Luenberger, D.G.: Optimization by vector space Methods. John Wiley & Sons, New York (1968)
30. Jongkind, W.: Dextrous Gripping. PhD thesis, Delft University of Technology, Delft, The Netherlands (1993) ISBN 90-9005806-0.
31. Melchiorri, C., Sandri, M., Stramigioli, S., Wenhao, L.: Some system control aspects for a robotic hand/arm manipulator. In: Proceedings of the ECC93, Groningen (1993) 552–557
32. Salisbury, J.K., Craig, J.J.: Articulated hands: Force control and kinematic issues. International Journal of Robotics Research **1** (1982)
33. van Dijk, J.: On the role of bond graph causality in modelling mechatronic systems. PhD thesis, University of Twente, Enschede, The Netherlands (1994)
34. Abraham, R., Marsden, J.E.: Foundations of Mecahnics. Ii edn. Addison Wesley (1994) ISBN 0-8053-0102-X.
35. Arnold, V.: Mathematical Methods of Classical Mechanics. Ii edn. Springer Verlag (1989) ISBN 0-387-96890-3.
36. Park, F., Bobrow, J., Ploen, S.: A lie group formulation of robot dynamics. International Journal of Robotics Research **14** (1995) 609–617

37. van der Schaft, A., Maschke, B.M.: The Hamiltonian formulation of energy conserving physical systems with external ports. Archiv für Elektronik und Ubertragungstechnik **49** (1995) 362–371
38. van der Schaft, A., Maschke, B.M.: Mathematical modeling of constrained Hamiltonian systems. In: Proc. 3rd Nonlinear Control Systems Design Symposium, NOLCOS95, Tahoe City, California, USA (1995)
39. Maschke, B.M., van der Schaft, A.: Interconnected Mechanical systems. Part2: The dynamics of spatial mechanical networks. In: Modelling and Control of Mechanical Systems. Imperial College Press, London (1997) 17–30 ISBN 1-86094-058-7.
40. Willems, J.C.: Paradigms and puzzles in the theory of dynamical systems. IEEE Trans. on Automatic Control **36** (1991) 259–294
41. Willems, J.C., J.W.Polderman: Introduction to Mathematical Systems Theory: A Behavioral Approach. Springer Verlag, New York (1998)
42. Fliess, M., Lévine, J., Martin, P., Rouchon, P.: Flatness and defect of non-linear systems: introductory theory and examples. International Journal of Control **61** (1995) 1327–1361
43. Kron, G.: Tensors for Circuits. 2nd edn. Dover books on advanced mathematics. Dover, New York (1959)
44. de Vries, T.J.: Conceptual design of controlled electro-mechanical systems, a modeling perspective. PhD thesis, University of Twente, Enschede, The Netherlands (1994)
45. Breedveld, P.: Physical Systems Theory in Terms of Bond Graphs. PhD thesis, Technische Hogeschool Twente, Enschede, The Netherlands (1984) ISBN 90-90005999-4.
46. Paynter, H.M.: Analysis and Design of Engineering Systems. M.I.T. Press, Cambridge, Massachusetts (1960) Course 2.751.
47. Lewis, A.D.: Aspects of Geometric Mechanics and Control of Mechanical Systems. PhD thesis, California Institute of Technology, Pasadena, U.S.A. (1995)
48. Maschke, B.M., van der Schaft, A., Breedveld, P.C.: An intrinsic Hamiltonian formulation of network dynamics: Non-standard poisson structures and gyrators. Journal of the Franklin institute **329** (1992) 923–966 Printed in Great Britain.
49. Courant, T.: Dirac manifolds. Trans. American Math. Soc. 319 (1990) 631–661
50. Maschke, B.M., van der Schaft, A.: Port controlled Hamiltonian systems: modeling origins and system theoretic properties. In: Proc. 3rd Nonlinear Control Systems Design Symposium, NOLCOS'92, Bordeaux (1992) 282–288
51. Murray, R.M.: Nonlinear control of mechanical systems: A lagrangian perspective. In: Proc. 3rd Nonlinear Control Systems Design Symposium, NOLCOS'95, Lake Tahoe, CA, U.S.A. (1995)
52. Golo, G., Breedveld, P., Maschke, B., van der Schaft, A.: Geometric formulation of generalized bond graph models: Generalised junction structure. Submitted to the Journal of the Frankling Institute (2000)
53. Lozano, R., Brogliato, B., Egeland, O., Maschke, B.: Dissipative Systems, Analysis and Control. Communication and Control Engineering. Springer, London (2000)
54. van der Schaft, A., Maschke, B.M.: On the Hamiltonian formulation of nonholonomic mechanical systems. In: Reports on Mathematical Physics. Volume 34. Elsevier Science (1994) 225–233
55. Maschke, B.M., van der Schaft, A., Breedveld, P.C.: An intrinsic Hamiltonian formulation of the dynamics of lc-circuits. Trans. IEEE on Circuits and Systems I:Fundamental Theory and Applications **42** (1995) 73–82

56. Maschke, B.M., van der Schaft, A.: Interconnection of systems: the network paradigm. In: Proc. of the $35^{th}$Conf. on Decision and Control, Kobe, Japan (1996) –
57. A.M.Bloch, P.E.Crouch: Representation of dirac structures on vector spaces and nonlinear lcv-circuits. In G. Ferraya, R.Gardner, H., H.Sussman, eds.: Proc. of Symposia in Pure mathematics, Differential Geometry and Control Theory. Volume 64. (1999) 103–117
58. Dalsmo, M., van der Schaft, A.: On representations and integrability of mathematical structures in energy-conserving physical systems. SIAM Journal of Control and Optimization **37** (1999) 54–91
59. Dorfman, I.: Dirac structures and integrability of nonlinear evolution equations. John Wiley (1993)
60. Blankenstein, G.: Implicit Hamiltonian Systems: Symmetry and Interconnection. PhD thesis, University of Twente, Enshede (2000)
61. van der Schaft, A.: $L_2$-Gain and Passivity Techniques in Nonlinear Control. Communications and Control Engineering. Springer Verlag (2000) ISBN 1-85233-073-2.
62. Won, J., Hogan, N.: Nodicity and nonlinear interacting dynamic systems. In K.Danai, ed.: Proceedings of the ASME dynamical Systems and Control Division. Volume 58 of DSC., ASME (1996) 615–621
63. Lončarić , J.: Normal forms of stiffness and compliance matrices. IEEE Trans. on Robotics and Automation **3** (1987) 567–572
64. Fasse, E.D.: On the spatial compliance of robotic manipulators. ASME J.of Dynamic Systems, Measurement and Control **119** (1997) 839–844
65. Huang, S.: The Analysis and Synthesis of Spatial Compliance. PhD thesis, Marquette University, Milwaukee, Wisconsin (1998)
66. R.Roberts: Minimal realisation of spatial stiffness matrix with simple springs connected in parallel. IEEE Trans. on Robotics and Automation **15** (1999) 953–958
67. Fasse, E.D., Breedveld, P.C.: Modelling of elastically coupled bodies: Part i: General theory and geometric potential function method. ASME J. of Dynamic Systems, Measurement and Control **120** (1998) 496–500
68. Fasse, E.D., Breedveld, P.C.: Modelling of elastically coupled bodies: Part ii: Exponential- and generalized-coordinate methods. ASME J. of Dynamic Systems, Measurement and Control **120** (1998) 501–506
69. Fasse, E.D., Broenink, J.F.: A spatial impedance controller for robotic manipulation. IEEE Trans. on Robotics and Automation **13** (1997) 546–556
70. Fasse, E.D.: Some applications of screw theory to lumped-parameter modeling of visco-elastically coupled rigid bodies. In: Ball Symposium, Cambridge (2000)
71. van der Schaft, A.: Interconnection and Geometry. In: The Mathematics of Systems and Control. University of Groningen (1999) 203–218
72. E.R.Kandel, J.H.Schwartz, T.M.Jessell: Principles of Neural Science. 4th edn. McGraw-Hill, New York (2000)
73. Stramigioli, S., Maschke, B.M., van der Schaft, A.: Passive output feedback and port interconnection. In: Proc. Nonlinear Control Systems Design Symposium1998, Enschede, The Netherlands (1998)
74. Anderson, R.J.: Passive computed torque algorithms for robots. In: Proceedings of the 28th Conference on Decision and Control, Tampa, Florida (1989) 1638–1644
75. Berghuis, H.: Model-based Robot Control: from Theory to Practice. PhD thesis, University of Twente, Enschede, The Netherlands (1993) ISBN 90-9006110-X.
76. Slotine, J.J.E., Li, W.: Applied Nonlinear Control. Prentice Hall (1991) ISBN 0-13-040890.

77. Colgate, J.E.: The Control of Dynamically Interacting Systems. PhD thesis, Massachusetts Institute of Technology, Cambridge, U.S.A. (1988)
78. Blankenstein, G., van der Schaft, A.: Closedness of interconnected dirac structures. In: Prooceedings of the NOLCOS 98. Volume 2. (1998) 381–386
79. Ortega, R., Loria, A., Kelly, R., Praly, L.: On passivity-based output feedback global stabilization of euler-lagrange systems. In: Conference on Decision and Control. (1994)
80. Stramigioli, S.: Creating artificial damping by means of damping injection. In K.Danai, ed.: Proceedings of the ASME Dynamic Systems and Control Division. Volume DSC.58., Atlanta, (GE) (1996) 601–606
81. Koditschek, D.E.: Robot kinematics and coordinate transformations. In: Proceedings of the 24th Conference on Decision and Control, Ft. Lauderdale, FL (1985)
82. Bedrossian, N.S., Spong, M.W.: Remarks on robot dynamics: Canonical transformations and riemannian geometry. Report (1989)
83. Stramigioli, S.: A novel impedance grasping strategy based on the virtual object concept. In: Theory and Practice of Control and Systems, Alghero, IEEE, World Scientific (1998) 783–788 ISBN 9810236689.
84. Polland, N.S., Lozano-Pérez, T.: Grasp stability and feasibility for an arm with an articulated hand. IEEE (1990)
85. Stramigioli, S., Melchiorri, C., Andreotti, S.: Passive grasping and manipulation. Submitted to the IEEE Transactions of Robotics and Automation (1999)
86. Anderson, R.J., Spong, M.W.: Bilateral control of teleoperaators with time delay. IEEE Transactions on Automatic Control **34** (1989) 494–501
87. Niemeyer, G., Slotine, J.J.E.: Stable adaptive teleoperation. IEEE Journal of Oceanic Engineering **16** (1991) 152–162
88. Maschke, B.M., van der Schaft, A.: Hamiltonian systems, pseudo-poisson brackets and their scattering representation for physical systems. In: Int. Symp. on Motion and Vibration Control, 17th ASME Biennal Conf. on Mechanical Vibration and Noise, Las Vegas, Nevada (1999) 12–15
89. Maschke, B.M., van der Schaft, A., Bidard, C.: An extension of scattering variables to spatial mechanisms. In: Proceedings of the 3rd IMACS Conference on Mathematical Modelling, MATHMOD. (2000)
90. van der Schaft, A.: $L_2$-Gain and Passivity Techniques in Nonlinear Control. 2nd revised and enlarged edn. Springer Communications and Control Engineering series. Springer-Verlag, London (1999) first edition Lect. Notes in Control and Inf. Sciences, vol. 218, Springer-Verlag, Berlin, 1996.
91. Maschke, B.M., van der Schaft, A.: Port controlled hamiltonian representation of distributed parameter sytems. In: Workshop on modeling and Control of Lagrangian and Hamiltonian Systems, Princeton (2000)
92. Dubrovin, B., Fomenko, A., Novikov, S.: Modern Geometry - Methods and Applications, Part II:The geometry and topology of manifolds. Volume 104 of Graduate texts in mathematics. Springer Verlag, New-York (1985) ISBN 0-387-96162-3.
93. Boothby, W.M.: An Introduction To Differentiable Manifolds and Riemannian Geometry. 516'.36. Academic Press, Inc., New York (1975) ISBN 0-12-116050-5.
94. Spivak, M.: A Comprehensive Introduction to Differential Geometry. Ii edn. Publish or Perish, Inc. (1979) ISBN 0-914098-79-9.
95. Rosenberg, R.C., Karnopp, D.C.: Introduction to Physical System Dynamics. McGraw-Hill (1983)
96. Karnopp, D.C., Margolis, D.L., Rosenberg, R.C.: System Dynamics, A Unified Approach. John Wiley & Sons (1990)

# Index

# Lecture Notes in Control and Information Sciences

**Edited by M. Thoma and M. Morari**

**1997–2000 Published Titles:**

**Vol. 231:** Emel'yanov, S.V.; Burovoi, I.A.; Levada, F.Yu.
Control of Indefinite Nonlinear Dynamic Systems
196 pp. 1998 [3-540-76245-0]

**Vol. 232:** Casals, A.; de Almeida, A.T. (Eds)
Experimental Robotics V: The Fifth International Symposium Barcelona, Catalonia, June 15-18, 1997
190 pp. 1998 [3-540-76218-3]

**Vol. 233:** Chiacchio, P.; Chiaverini, S. (Eds)
Complex Robotic Systems
189 pp. 1998 [3-540-76265-5]

**Vol. 234:** Arena, P.; Fortuna, L.; Muscato, G.; Xibilia, M.G.
Neural Networks in Multidimensional Domains: Fundamentals and New Trends in Modelling and Control
179 pp. 1998 [1-85233-006-6]

**Vol. 235:** Chen, B.M.
$H_\infty$ Control and Its Applications
361 pp. 1998 [1-85233-026-0]

**Vol. 236:** de Almeida, A.T.; Khatib, O. (Eds)
Autonomous Robotic Systems
283 pp. 1998 [1-85233-036-8]

**Vol. 237:** Kreigman, D.J.; Hagar, G.D.; Morse, A.S. (Eds)
The Confluence of Vision and Control
304 pp. 1998 [1-85233-025-2]

**Vol. 238:** Elia, N. ; Dahleh, M.A.
Computational Methods for Controller Design
200 pp. 1998 [1-85233-075-9]

**Vol. 239:** Wang, Q.G.; Lee, T.H.; Tan, K.K.
Finite Spectrum Assignment for Time-Delay Systems
200 pp. 1998 [1-85233-065-1]

**Vol. 240:** Lin, Z.
Low Gain Feedback
376 pp. 1999 [1-85233-081-3]

**Vol. 241:** Yamamoto, Y.; Hara S.
Learning, Control and Hybrid Systems
472 pp. 1999 [1-85233-076-7]

**Vol. 242:** Conte, G.; Moog, C.H.; Perdon A.M.
Nonlinear Control Systems
192 pp. 1999 [1-85233-151-8]

**Vol. 243:** Tzafestas, S.G.; Schmidt, G. (Eds)
Progress in Systems and Robot Analysis and Control Design
624 pp. 1999 [1-85233-123-2]

**Vol. 244:** Nijmeijer, H.; Fossen, T.I. (Eds)
New Directions in Nonlinear Observer Design
552pp: 1999 [1-85233-134-8]

**Vol. 245:** Garulli, A.; Tesi, A.; Vicino, A. (Eds)
Robustness in Identification and Control
448pp: 1999 [1-85233-179-8]

**Vol. 246:** Aeyels, D.; Lamnabhi-Lagarrigue, F.; van der Schaft, A. (Eds)
Stability and Stabilization of Nonlinear Systems
408pp: 1999 [1-85233-638-2]

**Vol. 247:** Young, K.D.; Özgüner, Ü. (Eds)
Variable Structure Systems, Sliding Mode and Nonlinear Control
400pp: 1999 [1-85233-197-6]

**Vol. 248:** Chen, Y.; Wen C.
Iterative Learning Control
216pp: 1999 [1-85233-190-9]

**Vol. 249:** Cooperman, G.; Jessen, E.; Michler, G. (Eds)
Workshop on Wide Area Networks and High Performance Computing
352pp: 1999 [1-85233-642-0]

**Vol. 250:** Corke, P. ; Trevelyan, J. (Eds)
Experimental Robotics VI
552pp: 2000 [1-85233-210-7]

**Vol. 251:** van der Schaft, A. ; Schumacher, J.
An Introduction to Hybrid Dynamical Systems
192pp: 2000 [1-85233-233-6]

**Vol. 252:** Salapaka, M.V.; Dahleh, M.
Multiple Objective Control Synthesis
192pp. 2000 [1-85233-256-5]

**Vol. 253:** Elzer, P.F.; Kluwe, R.H.; Boussoffara, B.
Human Error and System Design and Management
240pp. 2000 [1-85233-234-4]

**Vol. 254:** Hammer, B.
Learning with Recurrent Neural Networks
160pp. 2000 [1-85233-343-X]

**Vol. 255:** Leonessa, A.; Haddad, W.H.; Chellaboina V.
Hierarchical Nonlinear Switching Control Design with Applications to Propulsion Systems
152pp. 2000 [1-85233-335-9]

**Vol. 256:** Zerz, E.
Topics in Multidimensional Linear Systems Theory
176pp. 2000 [1-85233-336-7]

**Vol. 257:** Moallem, M.; Patel, R.V.; Khorasani, K.
Flexible-link Robot Manipulators
176pp. 2001 [1-85233-333-2]

**Vol. 258:** Isidori, A.; Lamnabhi-Lagarrigue, F.; Respondek, W. (Eds)
Nonlinear Control in the Year 2000 Volume 1
616pp. 2001 [1-85233-363-4]

**Vol. 259:** Isidori, A.; Lamnabhi-Lagarrigue, F.; Respondek, W. (Eds)
Nonlinear Control in the Year 2000 Volume 2
640pp. 2001 [1-85233-364-2]

**Vol. 260:** Kugi, A.
Non-linear Control Based on Physical Models
192pp. 2001 [1-85233-329-4]

**Vol. 261:** Talebi, H.A.; Patel, R.V.; Khorasani, K.
Control of Flexible-link Manipulators Using Neural Networks
168pp. 2001 [1-85233-409-6]

**Vol. 262:** Dixon, W.; Dawson, D.A.; Zergeroglu, E.; Behal,A.
Nonlinear Control of Wheeled Mobile Robots
216pp. 2001 [1-85233-414-2]

**Vol. 263:** Galkowski,K.
State-space Realization of Linear 2-D Systems with Extensions to the General nD (n>2) Case
248pp. 2001 [1-85233-410-X]

**Vol. 264:** Baños, A.; Lamnabhi-Lagarrigue, F.; Montoya, F.J.
Advances in the Control of Nonlinear Systems
344pp. 2001 [1-85233-378-2]

**Vol. 265:** Ichikawa, A.; Katayama, H.
Linear Time Varying Systems and Sampled-data Systems
376pp. 2001 [1-85233-439-8]

MIX
Papier aus verantwortungsvollen Quellen
Paper from responsible sources
FSC® C105338

If you have any concerns about our products,
you can contact us on
**ProductSafety@springernature.com**

In case Publisher is established outside the EU,
the EU authorized representative is:
**Springer Nature Customer Service Center GmbH**
**Europaplatz 3, 69115 Heidelberg, Germany**

Printed by Libri Plureos GmbH
in Hamburg, Germany